THE LAW AND PRACTICE ON DISASTER ISSUES

THE LAW AND PRACTICE ON DISASTER ISSUES

Edited by

Oluyemisi A. Bamgbose

Safari Books Ltd.
Ibadan

Published by
Safari Books Ltd
Ile Ori Detu
1, Shell Close
Onireke
Ibadan.
Email: info@safaribooks.com.ng
Website: http://safaribooks.com.ng

First Published 2018

ISBN: 978-978-55448-2-4

DEDICATION

To God Almighty
and
To all victims of Disaster

Table of Contents

List of Figures

List of Tables

Foreword

This book is relevant today in view of many crises all around the world resulting into Disaster. Disaster affect millions of lives each year, destroying properties and the environment, maiming, displacing and killing people. These result in insecurity, scarcity of food and medicine, crime, infringement of rights, loss of lives and many more. Globally, natural and man-made hazards are catastrophic events that affect different countries dependent on the peculiarity of the locality. Nigeria, like many other developing countries, is periodically affected by various forms of natural and man-made hazards. Disaster can be natural, like flood, drought, locust swarm, tornado or man-made disasters like human stampede, toxic waste and most recent occurrences of terrorism and wars notably in Iraq and Syria and in Nigeria, especially in the North-East region, which particularly start as insurgency and develop into a disaster.

Though it is not always possible to completely eliminate a risk that will culminate into a disaster, the resulting damage can be minimised largely by dissemination of information, planning and prompt action. The enormity of the disaster problem globally, calls for a proactive approach that will ensure effective disaster reduction at all levels. The substantial intervention of government is required in all cases of disaster and in cases of very large magnitude, humanitarian assistance is required.

In Nigeria, the use of security agencies is at times required depending on the magnitude of the disaster. The Nigeria Police, the Nigeria Security and Civil Defence, Fire Brigade and Peace Corps are the first contacts. The military most often come in when the disaster has assumed a larger proportion. Military forces are always very effective in counter disaster roles. The flexible organizational structure, capability of sustained operation in all weathers and conditions by day and night and well trained management system of the military forces make them particularly well suited for effective disaster relief operations. The military can offer assistance to the victims in engineering, communications, transport, rescue and medical services. The command structure and discipline in the military make for effective and quick relief in disaster area.

Disasters can be addressed from a diverse range of topics. One of which is the legal angle. From the legal perspective, the issue of disaster affects individuals and their rights, families especially children and women, communities, the environment and the economies of countries to mention a few. The articles in the book *Law and Practice on Disaster Issues* address legal matters in Disasters and have significantly addressed these area and more, therefore making this book very apt and relevant. I sincerely hope that the articles will provide a basis for further enhancing legal discourse and protection in disaster issues.

This book has revealed a lot of facts on "Law and Practice of Disaster." The writers of the chapters are selected experts in the different fields of Law and they have done justice to the book project. It is a great piece of work that I strongly recommend to all and sundry.

OLADAYO POPOOLA
Maj. Gen. (Rtd) OFR, LL.B, BL mni

Preface

Law and Practice on Disaster Issues is the first and major publication in Nigeria to present legal materials from diverse fields of Law in a single volume on disaster issues. The idea to publish this book came as a result of the interest I developed in disaster issues, after more than 3 decades of teaching at the Faculty of Law, University of Ibadan, Nigeria and being approached by the then Head of Department of Geography of the University in 2010 to develop the content of a new course titled "Legal Issues in Disaster Risk Management." Prior to this date, in November, 2009, the University of Ibadan was selected by the Federal Government of Nigeria, as one of the six Universities to be funded by the National Emergency Management Agency to offer Master's Degree and Diploma in Emergency and Disaster Management Courses. In the course of developing the curriculum and teaching, I discovered the dearth of publications on disaster issues from legal perspectives, making it an arduous task for the students to carry out their research. This motivated me in writing the book *Law and Practice in Disaster Issues.* As the foundation lecturer in the course and an Associate Lecturer in the Department of Geography in the University of Ibadan, Nigeria to date, I became passionate about legal issues in disaster and developed very keen interest in putting together a book.

The book comprises of 14 well written chapters, with each chapter having between 8,000 and 14,000 words. The chapters cover a general outline on disaster issues and different areas of Law including human rights, international law, criminal law, disability law, law of torts, family law, childs right, law of trespass, humanitarian law, social security, reproductive/sexual rights and corporate law.

The contributors were carefully selected from my knowledge and experience of their abilities, specialisation and integrity. This selection was a herculean task considering the fact that the introduction of disaster into legal training is a recent development. The end result is an assembly of astute scholars, the best minds, whose interests and expertise are not in doubt.

A major strength of this book is the in-depth work of the contributors who all are lecturers in Universities in Nigeria, United Kingdom and South Africa. The book therefore contains in one volume, a detailed work of experts on diverse areas of law on disaster issues, and it is a very practical and resource material for students at all levels in tertiary institutions, practitioners and policy makers.

To ensure easy reading of this book, several tools have been employed. The arrangement of the contributors in the book follow the sequence of the chapterisation and a comprehensive index is included.

The task of publishing this book has been enormous and involved the contribution of many people.

First, I acknowledge the Post-Graduate School and colleagues in the Department of Geography, both of University of Ibadan, Nigeria. I commend most especially, the contributors for the brilliant write-ups and for meeting the tight deadlines. I appreciate my research and personal assistants and the publishers, Safari Books Limited.

Oluyemisi Bamgbose

ACKNOWLEDGEMENTS

I thank the Almighty God, the giver of wisdom and insight, for the grace to bring this book into reality.

I also appreciate the Department of Geography, University of Ibadan, for the opportunity given to me in 2010 to be the foundation lecturer in the course "Legal Issues in Disaster Risk Management", which led to my interest in the area of Disaster Issues.

My heartfelt gratitude as well goes to the Postgraduate School of the University of Ibadan, Ibadan for the collaboration and support in the publishing of this book and ensuring the dissemination of its content to the world at large.

The efforts of the contributors who are all accomplished scholars and were carefully selected based on their expertise, specialisations and experience are equally recognised.

I am grateful to Professor Adeniyi Olatunbosun, Dean, Faculty of Law, University of Ibadan; Professor Charles Olatubara, Dean, Faculty of Environmental Design and Management, University of Ibadan; Professor Abiodun Bankole, Department of Economics, University of Ibadan; Professor Lekan Sanni, Department of Urban and Regional Planning, University of Ibadan; Professor Grace Adejuwon, Department of Psychology, University of Ibadan; Professor C.O Adekoya, Department of Jurispudence and International Law, Olabisi Onabanjo University, Ago Iwoye; Dr. Olusiyi Ipingbemi, Acting Head of Department, Urban and Regional Planning, University of Ibadan; Dr. Olugbenga Akingbehin, Department of Public Law, University of Lagos, Akoka; Dr. Yingigba Akinyemi, Department of Geography, University of Ibadan who all painstakingly reviewed the brilliant contributions.

Similarly, credit also goes to General Oladayo Popoola rtd, a fine, astute and disciplined gentleman, a man of many parts, former governor of Oyo and Ogun States, a professional soldier and general in the Nigerian Army and a distinguished and accomplished lawyer, for writing the foreword to this book at a short notice.

My personal and research assistants, Chioma Njoku and Sola Aleru, who worked tirelessly during the collation of all contributions, Bukola Ochei and Bose who were available to help, were all wonderful during the period.

Chief Joop Berkhout and the indefatigable team at Safari Books Limited did a great editorial and design work on this book. I am grateful to them.

Finally, my husband, Honourable Justice Olatokunbo Bamgbose is appreciated for his support, advice and encouragement during the course of writing the book. I also thank my children, Dr. Ademola and Oluwadamilola Bamgbose, Olaoluwatofunmi Bamgbose and Morewaoluwa Bamgbose for encouraging me to link "our" conservative discipline of law to a new found area in disaster issues.

Prof. Oluyemisi Bamgbose
January 8th 2018

About the Contributors

1. **Professor Oluyemisi Adefunke Bamgbose** Professor of Criminal Law and Criminology, Member of the Governing Council, University of Ibadan, Ibadan, Director, Women's Law Clinic and Immediate Past Dean at the Faculty of Law, University of Ibadan, Nigeria with a teaching academic career since 1987 and also a member of the Governing Council of Chrisland University, Abeokuta, Ogun State. She studied law and obtained her Master's degree at the University of Lagos Akoka. She specialises in women, children and human right issues and has 80 (Eighty) publications made up of articles, books and technical reports in English language and translation into other languages in both national and international outlets. Professor Bamgbose, a Fellow, Salzburg Seminar Austria (1996), was a Visiting Professor at Marquette University School of Law, Milwaukee, Wisconsin USA 2001; Saint Louis University School of Law, Saint Louis, Missouri, USA in 2008 and the University of Lagos, Lagos in 2014. She is an Associate Lecturer, Department of Geography, University of Ibadan, Nigeria. (Legal Issues in Disaster Risk Management since 2010).

2. **Deborah D. Adeyemo** *holds a* LL.B from the prestigious Obafemi Awolowo University (OAU), Ile-Ife, Osun State Nigeria and an LL.M from the University of the Western Cape, Cape Town South-Africa. She is a barrister and solicitor of the Supreme Court of Nigeria and currently undergoing a PhD in Law at the University of Ibadan. Deborah is also a Fellow of the German Academic Exchange Service (DAAD) Programme at the South African-German Centre for Criminal Justice, Cape Town, South Africa. Her primary research interest is in the field of public law with specific interest in issues relating to criminal justice and crime prevention. Her current research interest centres on transitional justice and international criminal law. She is currently a lecturer in the Public Law department at the University of Ibadan, Ibadan, Nigeria where she teaches Criminal law and Constitutional law.

3. **Dr. Idiat Funmilola Akande** graduated from Usmanu Danfodio Sokoto in 1996, and proceeded to the Nigerian Law School Lagos in 1997. She was called to Nigerian Bar in February 1998. She did her National Youth Service Corps between 1998 and 1999 at Government College Katsina. She proceeded for her LL.M in the year 2000 in Ahmadu Bello University, Zaria and graduated in 2004 with an LL.M Degree. In the same year 2004, she enrolled for her PhD Degree in Ahmadu Bello University and graduated in 2008 with the Doctor of Philosophy of Law (PhD). Dr. Idiat Funmilola Akande is currently a lecturer in Faculty of Law Public Law Department, Ahmadu Bello University,

Zaria, Kaduna State, Nigeria. Her areas of specialisation are International Law, Criminal Law, Constitutional Law and Administrative Law, all of which are in Public Law Department.

4. **Dr. Akeem Olajide Bello** is a Senior Lecturer, Department of Public Law, Faculty of Law, University of Lagos. He teaches Criminal Law, Human Rights and Environmental Law at the undergraduate level and Comparative Criminal Law and Procedure at the Masters Level. Dr. Bello was formerly Senior Special Assistant to the Governor of Lagos State on Public and Constitutional Law (2007-2011) and Law Reform Commissioner, Lagos State Law Reform Commission (2012-2015). He developed for UNODC Nigeria, a Curriculum for Training of Prosecutors, a Training Manual for Prosecutors and trained prosecutors drawn from the Federal and State Ministries of Justice. He is member of the International Society for Reform of Criminal Law, Vancouver, Canada.

5. **Dr. Bukola R. Akinbola** is a Senior Lecturer, Department of Jurispudence and International Law, University of Ibadan, Nigeria. She has acted as Head, Department of Public & International Law from 1st of August 2014 to the 29th of September 2014. Upon the emergence of four departments which replaced the former two dept at the Faculty of law in the University of Ibadan, Dr. Akinbola became the first Acting Head, Dept of Jurisprudence & International Law from 30th September 2014 to 31st July 2017. Dr. Akinbola has worked in various areas of law, including Environmental Law and Human Rights. She is particularly passionate about the law on the rights of persons with disabilities. In this book, she examines the links between the law, disability and disaster management.

6. **Dr. Nwudego Nkemakonam Chinwuba** is an Associate Professor, Department of Private and Property Law, Faculty of Law, University of Lagos, Akoka, Lagos. She holds the Taslimi Elias Prize in Jurisprudence. Thoughts on the structure and theory of law have underlined her academic and research endeavours. She is the author of Jurisprudence and Legal Education: an assessment of their link to Nigeria's Developmental Aspiration. The author empathises with the vulnerable and poor and connects the question of poverty and vulnerability to a poor synthesis of Law; its role, application and operation in any given society. The author's works are geared towards addressing the plight of the poor and vulnerable, and how a conducive environment of communality and the operation of the law as a pivot for order can bring about a fundamental change in any society and particularly so, in Nigeria.

7. **Roseline O. Ehiemua** is currently undertaking a PhD in Law in University of Benin with research interests in Family Law and matrimonial property rights of spouses. She lectures in the Department of Private and Property Law, in the Faculty of Law of Ambrose Alli University, Ekpoma, Edo State. She teaches

Family Law, Equity and Trusts Law, and Engineering Law, and has researched extensively in the use of law clinics to enhance legal training and pedagogy and presented papers in this area within and outside the country. She has also been published in reputable law journals on variety of law subjects.

8. **Dr. Omolade Olunike Olomola** is the immediate past Acting Head of Department of Private and Property Law, University of Ibadan. She is actively involved with the implementation of the Clinical legal education mode of teaching in Nigerian Universities. Her research interests include Family Law; Reproductive Health and Sexual Rights; Child Rights, Gender Justice and International Humanitarian Law. She won the prestigious MacArthur Foundation funding grant award which she utilised during the fall semester of 2009 at the University of Toronto (International Reproductive and Sexual Health Rights Programme), Canada, where she understudied the curriculum on Reproductive and Sexual Rights Law. She has professional certifications in different areas of Legal research pedagogy. She is also involved with Ethics training on HIV/AIDS Counselling and Training (HCT) of peer investigators. She is currently the Acting Dean, Faculty of Law, Elizade University, Ilara Mokin, Ondo State.

9. **Professor Chukwu Amari Omaka** studied in reputable universities in Nigeria and the United States. A Senior Advocate of Nigeria (SAN), Professor of Environmental Law, Fellow of the Institute of Charted Mediators/ Conciliators (FICMC), Dean Faculty Law Ebonyi State University/Director of EBSU Law Clinic. He is a thorough-bred academic, widely travelled scholar/ external examiner of postgraduate programmes within and outside Africa. A well-trained environmentalist and well certificated law clinician, including Certificate in Streetlaw and Oral Advocacy (South Africa) 2006; Clinical Legal Education (Valencia, Spain) 2011; GAJE TOT in Clinical Legal Education, National Law University New Delhi (2013). Awarded United Nations and African Union Fellowships in International Law in Accra Ghana in August and December 2016, respectively. Over the years, Omaka have used his career to defend the poor free of charge through his office Professor Amari Omaka Public Interest Law Clinic, located at 14 Ojeowere Street, Abakaliki.

Agom Caroline Ucha is a lecturer at the Faculty of Law, Ebonyi State University and currently a doctoral degree scholar in law at Abia State University Uturu Nigeria.

10. **Dr. Gloria D. Shajobi-Ibikunle** is a Senior Lecturer at the Department of Public & International Law, Faculty of Law, University of Abuja. She was the immediate past Director, Planning & Academic Matters, Nigerian Law School, Abuja. Dr. Ibikunle is committed to reform of the criminal justice system particularly as it concerns the persons who transverse the hallway of the criminal justice system and has written several articles in this regard. She is a

Fellow of the Chartered Institute of Administration. She is the Chairperson of SOAR Initiative, an NGO committed to stopping sexual based offences against minors.

11. **Dr. Elijah Oluwatoyin Okebukola** is a Senior Lecturer in the Faculty of Law, Nasarawa State University, Keffi, Nigeria. He teaches international law, international humanitarian law, international refugee law and components of international criminal law. He is the Head of Department of Public and International Law as well as Programme Manager of the postgraduate diploma studies in Refugee and International Humanitarian Law. He has consulted for international organisations including the UNODC. He is a Resource Person to the National Institute of Legislative Studies, Nigeria. He is a Barrister and Solicitor of the Supreme Court of Nigeria as well as Solicitor of the Supreme Court of England and Wales (non-practising). He frequently appears as counsel before the superior courts in Nigeria.

12. **Dr. Ademola Oluborode Jegede** is a Senior Lecturer in the Department of Public Law, University of Venda, Thohoyandou, South Africa. He holds degrees from Obafemi Awolowo University Ile-Ife, University of Ibadan and the Centre for Human Rights, Faculty of Law, University of Pretoria. He has been a research visitor to the Centre for International Environmental Law, USA and Human Rights Institute at Abo Akademi, Finland. His research focuses on the interface of climate change with human rights of vulnerable groups and general international human rights law.

Untalimile Crystal Mokoena is a law lecturer in the Department of Public Law at the University of Venda, South Africa. She holds LL.B and LL.M degrees, University of Venda. Her research interests are in the interface of social security issues with general international human rights law.

13. **Folake Tafita** is a law teacher, Senior Lecturer in the Department of Private & Property Law, University of Ibadan. Graduate, University of Toronto, Canada with a Masters in Reproductive and Sexual Health Law; Fellow, Mac Arthur Fellowship in (University of Toronto) Reproductive and Sexual health Law. Prior to her career in law, she qualified as a trained nurse having obtained a Diploma in nursing, from the University College Hospital, Ibadan, Nigeria. She is currently a PhD student, working on reproductive and sexual health rights and maternal mortality in Nigeria. Other research interests of the contributor includes; health law and policy, international humanitarian law, Clinical legal education, women and children's rights. Folake Tafita as a scholar and researcher has collaborated with other fellow scholars with publications in international and reputable journals. She has attended and presented her works singly and jointly at both national and international conferences.

14. **Damilola Odetola** is a doctoral researcher at the University of Kent, where she also teaches law. She studied law at the University of Ibadan and attended the Nigerian Law School. At both institutions, she graduated with first class honours and as the best graduating student. She obtained her Master's degree from the University of Oxford in 2014. Damilola has wide ranging experience in commercial law, corporate insolvency and debt restructuring, having worked in variety of legal positions in Nigeria and London. Her current research aims to bring an African perspective to the corporate insolvency and debt restructuring discourse. Damilola is a member of the Insolvency Lawyers Association (ILA), Business Recovery & Insolvency Practitioners Association of Nigeria and INSOL International. She has served as Associate Editor of Oxford University Commonwealth Journal and is currently an Associate Research Fellow and Editor at the Institute of Oil, Gas, Energy, Environment and Sustainable Development, Nigeria.

1

AN OUTLINE OF NATURAL AND MAN-MADE DISASTERS

Oluyemisi A. Bamgbose

Abstract

On a daily basis, a part of the world experiences a disaster. Disasters are global occurrences with devastating effects. Certain occurrences are natural, as they arise without direct human involvement. These are natural disasters. Though it is not possible to control the events from occurring, efforts can be made to alleviate the effect on human lives and environment. There are also occurrences that are as a result of human activities, mistakes or malicious intent referred to as man-made disasters. In certain situations, there are multiple disasters which are referred to as complex disasters. The severity or effect of any disaster is dependent on the management or response, prior, during or after the occurrence.

This chapter outlines the different classes of disaster, discusses selected events resulting in disaster and mitigating steps in reducing the effects. Some recommendations are made and in conclusion, it is advocated that there should be very effective awareness programmes for different communities prone to specific disasters and that there should be collective efforts in addressing situations of disasters.

Keywords: Natural Disaster, Man-Made Disaster, Complex Disaster.

MEANING AND DEFINITION

The word "disaster" derives from Middle French *désastre* and from Old Italian *disastro,*[1] which in turn comes from the Greek prefix "bad star."[2] The root of the word "disaster" ("bad star" in Greek) comes from an astrological theme in which the ancients used to refer to the destruction or deconstruction of a star as a disaster.[3] Cooplar further states that the term disaster "is derived from the Latin roots *dis-* and *astro*, meaning "away from the stars" or, in other words, an event to be blamed on an unfortunate astrological configuration."[4]

The term "disaster" is synonymous with devastation, catastrophe, calamity and tragedy. Different authors have however made attempts to describe or define disaster.[5] Syed defines disaster as an overwhelming ecological disruption occurring on a scale sufficient to require outside assistance.[6] Oztaysi *et al* quoted Fritz, and defined disaster as

> an event, concentrated in time and space, in which a society or a relatively self-sufficient subdivision of a society, undergoes severe danger and incurs such losses to its members and physical appurtenances that the social structure is disrupted and

1 Definitions Available at http://www.definitions.net/definition/Disaster. (last accessed 27th November 2017).

2 Definition Available athttp://www.definitions.net/definition/Disaster. (last accessed 27th November 2017.

3 supra

4 D.P. Coppola, *Introduction to International Disaster Management,* (2006, Elsevier, Amsterdam).

5 David A. McEntire *Disaster Response and Recovery: Strategies and Tactics for Resilience.* (2015, Wiley and Sons) 37.

6 Syed, M.H. (2009) Encyclopaedia of Disaster Management, Himalaya Publishing House, India.

the fulfilment of all or some of the essential functions of the society is prevented.[7]

The International Federation of Red Cross and Red Crescent Societies (IFRC) describes disaster as the combination of hazard, vulnerability and inability to reduce the potential consequences of risk.[8] McEntire posits that disasters are usually as a result of hazards and may trigger off a disaster[9] and defines a hazard as "a physical, technological, or intentional agent such as an earthquake, industrial explosion, or terrorist bombing.[10]

A disaster is a natural or man-made hazard that has come to fruition, resulting in an event of substantial extent causing significant physical damage or destruction, loss of life, or drastic change to the environment. Chroust says that a disaster is a sudden unplanned event that causes great damage or serious loss to an organisation.[11] Chroust *et al* further defines disaster as a natural or man-made event that negatively affects life, property, livelihood or industry often resulting in permanent changes to human societies, ecosystems and environment.[12] A disaster is therefore any tragic event with great loss stemming from events which include earthquakes, floods, catastrophic accidents, fires and explosions.

Disasters is the consequence of inappropriately managed risks. The risks are the product of hazards and vulnerability. When

7 Öztaysi B., Behret H., Kabak Ö., Sarı I.U., Kahraman C. "Fuzzy Inference Systems for Disaster Response". Chapter 4. In: Vitoriano B., Montero J., Ruan D. (eds) Decision Aid Models for Disaster Management and Emergencies. Atlantis Computational Intelligence Systems, vol 7. (2013,Atlantis Press, Paris) 75

8 Url-1: Available at http://www.ifrc.org/en/what-we-do/disaster-management/about-disaster-management/ (Last accessed 24th November 2017)

9 David A. McEntire *Disaster Response and Recovery: Strategies and Tactics for Resilience.* (2007, Wiley and Sons)

10 *Ibid*

11 Gerhard Chroust ICT Support for Disaster Management Available at https://www.researchgate.net/profile/Gerhard_Chroust/publication/287568853_ICT_support_for_disaster_management/links/56ae1ac308ae28588c61aaf6/ICT-support-for-disaster-management.pdf Accessed 27th November 2017.

12 G Chroust, O Giinther, M Roth, H Sturmand PZiehesberger "First Responder in Regional Disaster: A Case of Social Responsibility". In Social Responsibility: Range of Perspectives per Topic and Countries. (2015, Bentham Science Publishers) 80.

a hazard has effect on vulnerable people it is called a disaster while hazards that strike in areas with low vulnerability are not considered a disaster, as is the case in uninhabited regions.

Classification

Disasters are said to be as a result of hazards.[13] It has been opined that with the intervention of human actions, hazards can be prevented from resulting into disasters. It may be argued, that all disasters are as a result of human failure to introduce appropriate disaster management measures and therefore all disasters are human-made.

McEntire states that hazards are present for many different reasons.[14] He further states that some hazards naturally occur in the environment, whereas others are the result of human activity, mistakes or malicious intent.[15] Hazards resulting in disasters may be divided into two classifications. Disaster can therefore be divided into two, namely natural or man-made disasters. Schoitsch has suggested a third classification. He opines that at times, natural disaster can be interwoven leading to what is generally termed, complex disasters. According to the International Federation of Red Cross and Red Crescent Societies, some disasters can result from several different hazards or, more often, to a complex combination of both natural and man-made causes and different causes of vulnerability.[16] Examples of such complex combination given by the International Federation of Red Cross and Red Crescent Societies are food insecurity, epidemics, conflicts and displaced populations.[17] The third classification is termed complex disaster. The different classifications are hereby discussed.

13 David A. McEntire *Disaster Response and Recovery: Strategies and Tactics for Resilience*. (2007, Wiley and Sons).

14 *Ibid*

15 *Ibid*

16 Complex/manmade hazards: complex emergencies- Available athttp://www.ifrc.org/en/what-we-do/disaster-management/about-disasters/definition-of-hazard/complex-emergencies/Accessed 27th November 2017.

17 *Ibid*

A. Natural Disaster

Natural disasters are naturally occurring physical phenomena caused either by rapid or slow onset events.[18] It is a consequence when a natural hazard affects humans and/or the built environment. Natural disasters arise without direct human involvement. Natural hazards kill tens of thousands of people and destroy billions of naira and millions of dollars of habitat and property each year. Natural disaster can become more severe because of human actions or inactions prior, during or after the disaster itself. The effects of a natural disaster may be escalated if there is no appropriate emergency management or a coordinated response. In such cases, the result is huge financial loss, severe environmental damage and or massive loss of human lives. Though it may not be possible to control nature and to stop the development of natural phenomena, however, efforts could be made to avoid disasters and alleviate the effects on human lives, infrastructure and property.

The rapid growth of the world's population and its increased concentration often in hazardous environment has escalated both the frequency and severity of natural disasters. In developing countries, the tropical climate, unstable land forms, poor or no budgetary allocation for disaster prevention and lack of planning are cumulating factors that make effects of natural disasters more severe in these regions.

Natural disasters have been classified according to the origins. According to the IFRC, the classification are as follows:[19] Geophysical origin which include earthquakes, volcanic eruptions and landslides; climatic origin examples of which are drought, flood, cyclone, locust, forest fire, extreme temperatures, drought and wildfires; hydrological origin which include avalanches and floods; meteorological origin like cyclones and storms/wave

18 Types of disasters – IFRC Available at https://www.bing.com/search?q=Natural%20disasters%20are%20naturally%20occurring%20physical%20phenomena%20caused%20either%20by%20rapid%20or%20slow%20onset%20events&form=SWAUA2 Accessed 26th November 2017.

19 Types of disasters: Definition of hazardUrl-1: http://www.ifrc.org/en/what-we-do/disaster-management/about-disaster-management/ accessed 24-11-2017.

surges, and biological origin which include epidemics and insect/ animal plagues.

As mentioned earlier in this paper, though natural disaster is said to arise without direct human involvement, human or human related factors may escalate and or aggravate the impact.

Factors that may escalate natural disaster

Rapid growth of the world's population coupled with urbanisation is noted as major causes of escalation in natural disasters in the 21st century. The effects these will have on the environment has been a cause of concerns for experts. While this trend has been noted in Europe, Middle East and North Africa, this does not exclude other parts of the world.

Brauch in a study of the Mediterranean, analysed the relationship between urbanization in relation to population growth and climate change and natural disasters as outcomes of environmental stress for the Mediterranean space.[20]

Other escalating factors include increased concentration often in hazardous environment aggravating both the frequency and severity of natural disasters; deforestation and non-engineered constructions which make the disaster-prone areas more vulnerable and in many developing nations, poor or no budgetary allocation for disaster prevention has led to loss of lives and properties.

It is apt to however state that, though it may not be possible to control nature and to stop the development of natural phenomena, but efforts could be made to avoid disasters and alleviate their effects on human lives, infrastructure and property.

20 Hans Günter Brauch Urbanization and Natural Disasters in the Mediterranean: Population Growth and Climate Change in the 21st Century Available in http:// www.bvsde.paho.org/bvsacd/cd46/cap11-urbani.pdf Accessed 25th November 2017.

A Précis of the Varying Forms of Selected Natural Disasters

There are certain major adverse events that may be predictable. There are others that are not predictable and the occurrence cannot be controlled. These are known as natural disasters. There are many events that fall into this category. However, for the purpose of this work, selected examples will be discussed.

I. Avalanche

Meaning

An avalanche is when a large batch of snow on the side of a mountain slides quickly down the mountain. It is a rapid flow of snow down a sloping surface. In an avalanche, lots of material or mixtures of different types of material fall or slide rapidly under the force of gravity.

Types of Avalanche

Avalanches are often classified by what they are made of. There are snow avalanche, ice avalanche, rock avalanche or soil avalanche. A mixture of these is classified as debris avalanche.

Causes of Avalanches

An avalanche is caused when a build-up of snow is released down a slope, and it is one of the major dangers faced in the mountains in winter. Avalanches are typically triggered in a starting zone from a mechanical failure in the snowpack when the forces on the snow exceed its strength but sometimes only with gradually widening. Weather might be the biggest factor that can cause an avalanche. It plays a huge role in the development of snow on the ground.

Snowpack which builds up over the season, and during winter, may be strong or weak. Avalanches occur if the bond is weak or even if you have a strong bond in the upper layers of the snowpack, and the layers beneath it are weak. The presence of humans in these areas trigger off avalanches.

The terrain of a place may aggravate an avalanche. Snow built on a steep slope may cause an avalanche to happen. Avalanches develop on slopes between 25°- 55° and it starts on slopes between the 35°- 45° range.

Effect of Avalanches

Because they are so sudden and the snow can move very quickly and with a lot of strength, an avalanche can sometimes be dangerous for whoever or whatever is in its path.

Most Prone Areas of Avalanches

Mountainous areas throughout the Arctic and temperate regions and those which have slope angles between 25°C and 60°C are at risk of avalanche. Avalanche is more severe in Europe than in North America especially the western part.[21] Avalanche-prone areas are Peru near the Andes Mountain, Argentina, Italy especially the Swiss Alps Mountain, the French Alps, Canada, Afghanistan, Slovakia, Austria, Mount Everest in Nepal, and Iceland.[22]

Examples of Disasters Resulting From Avalanches

On May 31st 1970, an earthquake off the coast of Peru caused a substantial section of the north slope of Mt. Huascaran to collapse, killing over 202,000 people. Situated in the Cordillera Blanca, the world's highest tropical mountain range. The avalanche moved down hill burying the towns of up to 300 feet of rock and debris.[23]

Over a 24-hour period during World War 1, Italian and Austrian forces died in a series of avalanches that occurred at the Tyrolean Alps on December 13th 1916 with about 10,000 casualties caused by a mixture of heavy snowfall and man-made explosives.[24]

21 Available in https://people.uwec.edu/jolhm/eh3/group4/references/nsidc.htm.

22 *Ibid*

23 The 10 Deadliest Avalanche Events in History Available in http://unofficialnetworks.com/2013/04/30/10-deadly-avalanches/

24 *Ibid*

It is reported that on September 4th 1618 the entire town of Plurs, Switzerland was wiped off the face of the earth when a massive avalanche, buried the town. There were about 2,427 casualties.[25]

In Asia, on April 7th 2012, an avalanche hit a Pakistani military base near Siachen Glacier region, trapping and killing soldiers and civilians.[26]

In Wellington, Washington, United States of America on March 1st 1910 there was an avalanche after a severe snow fall for several days resulting in several feet of snow. A few days after as a result of a lightning strike, a slab of snow broke loose from the side of Windy Mountain during a violent thunderstorm. A ten-foot high mass of snow, half a mile long and a quarter of a mile wide, fell toward the town killing many people.[27]

Protective/mitigating Measures

Explosives are used to trigger avalanches on potentially unstable slopes so that the avalanches will occur when people are not endangered.

Devices such as avalanche rakes (large reinforced fencing) are used on slopes to hold snow in place, and diversion structures such as dams or wedges are used at the base of the slope, split or deflect the snow in an avalanche.

Afforestation of the avalanche-prone area is adopted to hold down soil and trees.

Issuing early warning of an impending avalanche is an effective measure adopted after predicting the occurrence of avalanches through scientific analysis.

25 *Ibid*
26 *Ibid*
27 *Ibid*

II. Drought

Meaning

Because of regional differences and perceptions, it is difficult to give a concise definition of drought. However, it is generally said to be a deficiency of precipitation over an extended period of time — usually a season or more — resulting in a water shortage for some activity, group, or environmental sector.[28] Considering drought from the operational perspective, it is a protracted period of deficient precipitation resulting in extensive damage to crops, resulting in loss of yield.[29]

Types of Droughts[30]

Meteorological drought, Agricultural drought, Hydrological drought and Socio-economic drought.

Causes

Drought can be caused by not receiving rain or snow over a period of time. Another cause of drought is the change in the wind patterns that move clouds and moisture through the atmosphere causing a place not to receive its normal amount of rain or snow over a long period of time. Annual dry seasons in the tropics significantly increase the chances of a drought developing and this may subsequently cause bush fires. Periods of heat can aggravate and significantly worsen drought conditions by hastening evaporation of water vapour.

Effects

Like other hazards resulting in disaster, this natural phenomenon may result in reduction in water quality, hunger, war, civil unrest, malnutrition, famine, habitat change, mass migration, reduced

28 University of Nebraska–Lincoln UNL Institute of Agriculture and Natural Resources UNLWater Available in https://water.unl.edu/drought/whatisdrought. Accessed 27th November 2017.

29 *Ibid*

30 *Ibid*

electricity production, the risk of contracting waterborne diseases such as diarrhoea, a major killer of children around the world, due to use of contaminated water. Drought threatens the livelihood of people and causes widespread death and famine.

Most prone areas of drought

In Africa, countries like Zimbabwe, Angola, Ethiopia, Somalia, Swaziland and Algiers have all experienced disasters as a result of drought. In the United States of America, States like Arizona, Alabama, Louisiana, Georgia and New Mexico had similar experiences of disaster as a result of drought. In Asia, India, Madhya Pradesh, along with the western states of Rajasthan and Gujarat have suffered the effect of drought and in Europe, Southern Europe has experienced drought.

Examples of Disasters resulting from Drought

In Africa, Zimbabwe, a country formerly known as the food basket of Africa, suffered perennial shortages, erratic rains due to climate change in many years which led to severe drought in 2016.[31] In July 2016, the then President, Mugabe, declared a state of disaster in regard to severely affected areas and declared an emergency of drought.[32] A UNICEF report in 2011 stated that in Ethiopia, an estimated 8 million of Ethiopia's 60 million people were at risk due to East Africa drought recorded as one of the worst drought in decades.[33] The drought resulted in a food crisis across Kenya, Somalia, Ethiopia, and Djibouti.

According to a UNICEF report, in India, nearly 130 million people living in 12 states were seriously affected by what was termed the worst drought in 100 years.[34] Madhya Pradesh, and

31 Zimbabwe declares 'state of disaster' due to drought 5 February 2016 Available in https://www.theguardian.com/world/2016/feb/05/zimbabwe-declares-state-of-disaster-drought-robert-mugabe Accessed 30th November 2017.

32 *Ibid*

33 Drought Disaster Available in https://www.unicef.org/drought/drought-countries.htm Accessed 30 November 2017.

34 *Ibid*

some of the western states of Rajasthan and Gujarat and Andhra Pradesh in the south were affected by severe drought caused by failure of monsoon rains.[35] There were similar reports of drought in Pakistan, sparking fears of a massive humanitarian crisis[36] and UNICEF also having concern for three countries in Southeast Asia namely Indonesia, the Philippines and Timor-Leste.[37]

In the United States of America, Dylan Stableford reported that Georgia, Arizona, New Mexico, Texas and some other states recorded very severe drought in 2012.[38] It is reported that in 2017, the Southern Europe, especially in the regions of Castile and Leon, the largest cereal growing region in Spain, has been particularly badly affected, with the effect of drought with crop losses estimated at around 60 to 70 percent.[39]

Protective /Mitigating measures

Some of the protective measures include dams, cloud seeding, desalination, drought monitoring, land use, outdoor water-use restriction and rainwater harvesting.

35 *Ibid*

36 Southeast Asia Globe editorial "Southeast Asia facing severe drought" February 12, 2016 Available in http://sea-globe.com/southeast-asia-facing-severe-drought/. Accessed November 30th 2016.

37 Dylan Stableford "U.S. declares drought-stricken states largest natural disaster area ever" July 12 2012 Available in https://www.yahoo.com/news/blogs/lookout/us-natural-disaster-area-drought-150130308.html Accessed on 30th November 2017.

38 Dylan Stableford "U.S. declares drought-stricken states largest natural disaster area ever" July 12 2012 Available in https://www.yahoo.com/news/blogs/lookout/us-natural-disaster-area-drought-150130308.html Accessed on 30th November 2017.

39 Isla Binnie and Paul Day "One of worst droughts in decades devastates South Europe crops"July 14, 2017 Available in https://www.reuters.com/article/us-europe-farming-drought/one-of-worst-droughts-in-decades-devastates-south-europe-crops-idUSKBN19Z1XW Accessed on 30th November 2017.

III. Heat in Excess

Meaning

Excessive heat or heat wave is a prolonged period of excessively hot weather, which may be accompanied by high humidity. In the United States of America, heat wave can be defined as a period of at least two consecutive days of excessive hot weather. However, the definition of heat wave vary. In North-Eastern US, heat wave is typically defined as three consecutive days where the temperature exceeds 90 degree F (32.2 degree C),[40] in Demark, and Netherlands, it exceeds 28 degree C (82.4 degree F) and in Sweden, weather exceeding 250c (77.00f). In Adelaide, heat wave is defined as five consecutive days at or above 35 degree C (95 degree F).[41] Therefore, heat wave or excessive heat definition varies as the weather vary in each and every country.[42]

Causes

A heat wave can occur in certain regions when a low pressure offshore and high pressure inland combine to form a Bergwind.[43] It may also occur in other regions when a high pressure system originating in the Gulf of Mexico becomes stationary just off the Atlantic Seaboard (typically known as a Bermuda High.[44]

40 Sofia Aivalioti "Electricity Sector Adaptation To Heat Waves"Sabin Center for Climate Change Law, Columbia Law School "January 2015Available in http://columbiaclimatelaw.com/files/2016/06/Aivalioti-2015-01-Electricity-Sector-Adaptation-to-Heat-Waves.pdf" Accessed on 7th December 2017.

41 Extreme heat sends SA temperatures above 40 degrees Celsiushttp://www.abc.net.au/news/2015-12-17/summer-heat-spell-grips-much-of-south-australia/7037720.

42 Q Yin and Jinfeng Wang "The association between consecutive days' heat wave and cardiovascular disease mortality in Beijing, China" 2017 Feb 23in BMC Public Health. 2017; 17: 223. Available in https://www.ncbi.nlm.nih.gov/pmc/articles/PMC5322604/ Accessed on 8th December 2017.

43 "Climate Change in Africa" Climate Emergency InstituteAvailable in https://www.climateemergencyinstitute.com/climate_change.html Accessed on 8th December 2017.

44 Available in https://sites.google.com/a/student.glenview34.org/horrifying-heat-waves/Accessed on 8th December 2017.

Effects of excessive heat (heat wave)

The consequences of heat in excess can be subdivided into parts:

Health effects: Excessive heat causes: Heat exhaustion, heat stroke, heat syncope, sun burn, heat rash, thousands of deaths from hyperthermia, and psychological effects;

Sociological/economic effect: Catastrophic crop failures, widespread power outages due to increased use of air conditioning;

Physical effects: Excessive heat also do cause roads and highways to buckle, water lines to burst and power transformer to detonate (causing fire).

Most prone areas of Heat Wave/Excessive Heat

Heat wave occur mostly in Denmark, United States of America Netherlands, Western State of South Africa, Adelaide in Australia.

Examples of Disasters Resulting From Heat Wave

An example is the European Heat Wave in 2003.[45] The 2003 European heat wave, resulted in a health crisis in several countries as well as a drought which led to a crop shortage. In the Ukraine, 75% of wheat crops were lost, and at least 14,802 casualties were recorded in France alone.[46] The extreme temperatures also resulted in forest fires and counter-intuitively, even flooding.[47]

Protective/ Mitigating Measures

Measures that can be taken include early warning signs, information on heat shelters and availability of water at train stations and strategic places.

45 Josef 25 Worst Natural Disasters Ever Recorded September 20 2017 Available in http://list25.com/25-worst-natural-disasters-recorded/Accessed on 8th December 2017.

46 Available in https://drive.uqu.edu.sa/_/maatia/files/Disaster%20Part1.pdfAcce ssed on 8th December 2017.

47 Josef 25 Worst Natural Disasters Ever Recorded September 20 2017 Available http://list25.com/25-worst-natural-disasters-recorded/.

IV. Hurricane, also called Typhoon or Cyclone

Meaning

Hurricane is a type of tropical cyclone. The cyclone is accompanied by thunderstorms and in the Northern hemisphere, a counter clockwise circulation of winds where the earth surface. It is an intent rotating organic weather system that processes maximum sustained winds exceeding 119km/hr (74mph). It form and intensify over tropical organic region. It is not the wind that causes the majority of the death and destruction, but water/ flood, from tidal surge.

Types of hurricane

Hurricane is categorized according the strength of their winds using the Saffir-Simpson.

There are different names for different types of hurricane some of which are: Andrea, Barry, Jerry, Katrina, Nestor; Olga; Pablo; Sandy Harvey, Irma, and Maria.

Effect of hurricane

Some of the effects are:

a. *Storm surge* – It is the fast uprising of sea level that happens when a hurricane approaches coat. Another effect is the rip tides. Rip tides are strong sea current. Heavy rainfall, tornados and high winds come with hurricane.
b. *Economic loss and Heritage loss* – The effect of hurricanes cannot be overestimated. As a result of Hurricane Andrew in 1992, the United Nations put the Everglades in Florida in the USA on the list of World Heritage sites in Danger.[48] In 2017, Hurricane Irma was reported to have again threatened the Everglades in Florida.[49] The Everglades is acclaimed

48 Maddie Stone Why the Everglades Might Never Look the Same After Hurricane (9 October 2017) Available in Irmahttps://gizmodo.com/irma-is-but-the-latest-disaster-to-strike-the-everglade-1803072231. Accessed 30th November 2017.

49 Maddie StoneWhy the Everglades Might Never Look the Same After Hurricane (9 October 2017) Available in Irmahttps://gizmodo.com/irma-is-but-the-latest-disaster-to-strike-the-everglade-1803072231. Accessed 30th November 2017.

to be "an internationally-recognized ecological treasure, home to alligators, panthers, manatees, crocodiles, more than 300 species of birds, and hundreds of rare or endemic plants."[50]

Most Prone Areas of Hurricane

In 2017, the United States of America witnessed one of the worst hurricane seasons with very "hyperactive" Atlantic Hurricane seasons which began since June 1st 2017 till November 2017. Specifically, Louisiana, New Orleans, Caribbean Island, Trinidad and Tobago, Gulf of Mexico are all prone to hurricane.

Examples of Disasters caused by Hurricane

An example of a hurricane is, Hurricane Andrew In August 1992. The storm was responsible for 65 fatalities and US $26 billion in property damages. Hurricane Andrew became the 5th costliest hurricane in US history after Hurricanes Katrina, Wilma, and Ike. On August 25th 2017, Category 4 hurricane Harvey happened in Texas causing around $180 billion in damage, a bit lower than the largest estimates of Hurricane Katrina.[51] Hurricane Harvey flooded one-third of Houston, forcing 39,000 people into shelters and damaging 203,000 homes.[52] In October 2017, the State of Florida in the United States of America was hit by Hurricane Irma which was a natural disaster that reached the North American shores with an estimated damage of up to $100 billion.[53]

In Europe, in autumn 2017, Hurricane Ophelia struck in Ireland and the United Kingdom.[54] Three people were killed,

---

50 *Ibid*

51 Josef 25 Worst Natural Disasters Ever Recorded September 20th 2017 Available in http://list25.com/25-worst-natural-disasters-recorded/ Accessed 30th November 2017.

52 *Ibid*

53 Josef 25 Worst Natural Disasters Ever Recorded September 20th 2017Available in http://list25.com/25-worst-natural-disasters-recorded/.

54 "Hurricane Ophelia: Three people die as storm hits Ireland" BBC News17th October 2017 Available in http://www.bbc.com/news/uk-41627442 Accessed 7th December 2017.

22,000 people were left without electricity and internet was cut off for some households across the UK.[55] In 2014, Hurricane Gonzalo, struck the United Kingdom on October 21st causing serious problem of power in about 18, 000 homes[56] and there were cancelation of several flights.

Protective/Mitigation Measures

Some of the following measures may be adopted; planning for safe and effective evacuation, developing hurricane program training, aids, quick response and recovery and post- storm assessment.

v. Locust/Insect swarm

Meaning

A large insect that lives in hot countries and flies in large groups, destroying all plants and crops of that area. A plague of locusts is a devastating natural disaster. These infestations have been feared and revered throughout history. Unfortunately, they still wreak havoc today.

Types

Locusts are related to grasshoppers and the two insects look similar. The desert locust is a type of locust which is notorious.[57] According to *National Geographic,* "A desert locust swarm can be 460 square miles (1,200 square kilometers) in size and pack between 40 and 80 million locusts into less than half a square mile (one square kilometer)."[58] Other types of locusts which may

55 Ophelia: What a Tropical Storm Is Doing in Europe" BBC News Available in http://www.newsweek.com/hurricane-ophelia-what-tropical-storm-doing-europe-683647 Accessed 7th December 2017.

56 Available in http://www.cbsnews.com/news/battered-bermuda-cleans-up-after-hurricane-gonzalo/.

57 "Locust/grasshopper Operations Guidebook Revision: 2nd January 89" US Agency for International Development Available in http://pdf.usaid.gov/pdf_docs/PNABC136.pdf Accessed November 4th 2017.

58 "Locust" National Geographic Available in http://channel.nationalgeographic.com/animals/bugs/articles/locust/.

cause a disaster include the African Migratory Locust, the Red Locust, the Brown Locust and the Senegalese Grasshopper.[59]

Causes

When environmental conditions produce many green plants and promote breeding, locusts can congregate into thick, mobile, ravenous swarms.

Effect

Each locust can eat its weight in plants each day, so a swarm of such size would eat 423 million pounds (192 million kilograms) of plants every day.[60] They are a severe agricultural threat. Desert locust plagues may threaten the economic livelihood, while locust swarms devastate crops and cause major agricultural damage attendant human misery, famine and starvation.[61] Generally it adversely impact global food security. Locust swarms are typically in motion and can cover vast distances.[62]

Most Prone Areas

They occur in many parts of the world, but today locusts are most destructive in sustenance farming regions of Africa. They are also found in Middle East and Asia.

Examples of Disaster as a result of Locust Swarm

It is recorded that between 1985 and 1987, after a period of severe drought, there was so much rain which created a conducive environment for the locust in the Southern Africa to the Horn of Africa and from there, across the Sahel to Northern Africa.[63] Locust swarm was also recorded in Karco of the Union of South Africa

---

59 *Ibid*

60 "About Locust" National Geographic December 15 2009 Available in https://www.nationalgeographic.com/animals/invertebrates/group/locusts/ Accessed 7th December 2017.

61 *Ibid*

62 *Ibid*

63 *Ibid*

and Southern Botswana, Southern Zambia and Namibia.[64] Other areas that had been affected are the Sahelian nations including the Cape Verde Islands, parts of North Africa and Southwest Asia.[65]

Protective/ Mitigating Measures

There are different measures adopted to mitigate against the disaster of locust swarm. Factors to be considered in adopting any measure include the size of the area affected, accessibility to the terrain, environmental concerns including human habitation. Other measures adopted include the use of aerial or ground equipment or through the application of pesticide for locust control.

VI. TORNADO

Meaning

A tornado is a very narrow, violently rotating column of air that extends from the base of a cumuliform cloud (such as a thunderstorm) to the ground.[66] It is a powerful column of winds around a center of low atmospheric pressure. It looks like a large black funnel hanging down from a storm cloud. The narrow end will move over the earth, wiping back and forth. The wind inside a tornado spiral inward and upward with a lot of speed and power.[67] It creates an internal vacuum that then sucks up anything it passes over. When the tunnel touches a structure, the fierce winds have the ability to tear it apart. This makes the tornado the most dangerous storm known to mankind. A tornado can form quickly, sometimes in a minute or less. It can travel across the ground at high speed, and then just suddenly vanish.[68]

64 *Ibid*

65 *Ibid*

66 "Severe Weather 101" National Severe Storm Laboratory Available in http://www.nssl.noaa.gov/education/svrwx101/tornadoes/Accessed on 9 December 2017

67 The Tornado June 29 1998 Available in www.tsgc.utexas.edu/stars/tornado.html Accessed on 9th December 2017.

68 Available in http://www.academia.edu/23710233/OOPL_Slides_ENVIRONMENTAL_HAZARDS Accessed on 9th December 2017.

Types/Classification

There are basically two types. These are classic tornadoes which is said to form from the mesocyclone of a supercell[69] and the land spouts tornadoes which is of a weaker type.[70]

Causes of Tornado

Tornadoes conditions are formed when different temperature and humidity form to cause thunderclouds.

There has been some discussions and possible linkages between climate change and tornado outbreak, as nearly almost all always happen following a severe weather event.[71]

Effects

The effect of tornadoes could be environmental as it could cause destruction of buildings and trees. It could also be economic as the effect could result in food shortage and rebuilding of major structures. Injury and loss of lives are very common effects of tornadoes.

Most Prone Areas

The most tornado-prone area in the United States is called Tornado Alley.[72] The states that are mostly affected include Texas, Oklahoma, Kansas and Nebraska. Other regions prone to tornadoes include Argentina, Australia, Bangladesh, Canada, Germany, and the United Kingdom.[73]

---

69 Available in http://www.answers.com/Q/How_many_types_of_tornado_are_there Accessed on 9th December 2017.

70 *Ibid*

71 Karl Mathiesen"Extreme weather already on increase due to climate change, study finds" 27 April 2015 Available inAvailable inhttps://www.theguardian.com/environment/2015/apr/27/extreme-weather-already-on-increase-due-to-climate-change-study-findsAccessed on 9th December 2017.

72 "Tornado Alley: The Most Tornado Prone Region in the World" available in http://www.worldatlas.com/articles/what-is-tornado-alley-where-are-tornadoes-most-likely-to-occur-where-is-the-tornado-belt.html Accessed on 4th November 2017

73 "Severe Weather 101"National Severe Storm Laboratory Available inhttp://www.nssl.noaa.gov/education/svrwx101/tornadoes/ accessed on 8th December 2017.

Examples of Disaster as a result of Tornadoes

On May 27th 1997, it was reported that multiple tornadoes resulting in disaster occurred in the Williamson and Travis counties in central Texas causing injuries, deaths, and an estimated 20 million dollars in personal and commercial insured losses.[74] Between April 25th and 28th, 2011, a tornado resulted in a disaster known as the "Super Outbreak."[75] Erdman reports that this occurrence was reported to be one of the worst tornado outbreaks in U.S. history.[76] The report stated that the tornado claimed more than 300 lives, injured 2,775 and was responsible for $10.8 billion in total damage, making it the costliest tornado outbreak in U.S. history. In 2013, Oklahoma experienced back-to-back tornado events; the deadliest was in Moore, Oklahoma, where 25 people died and almost 400 were injured.[77]

Protective/Mitigating Measures

While you can't stop a tornado from hitting, there are mitigating measures against the devastating effects. In areas prone to tornadoes safety measures such as staying prepared by having safety kits, like batteries, torch, water, food, staying tuned to weather announcements and knowing where to shelter, like in the basement, are some of the measures.[78] Flying and falling objects are the major causes of injuries and death during a tornado. It is therefore important that adequate preparation is made when there is an impending tornado.

74 "Tornado Disaster -- Texas, May 1997" Centers for Disease Control and Prevention MMWR November 14, 1997 / 46(45);1069-1073 Available in https://www.cdc.gov/mmwr/preview/mmwrhtml/00049839.htm Accessed on 5th December 2017.

75 Jon Erdman "April 2011 Tornado Super Outbreak: 17 Things That Shocked Us Most" Tornado NewsAvailable in https://weather.com/storms/tornado/news/april-2011-superoutbreak-tornadoes Accessed 5th December 2017.

76 Jon Erdman "April 2011 Tornado Super Outbreak: 17 Things That Shocked Us Most" Tornado NewsAvailable in https://weather.com/storms/tornado/news/april-2011-superoutbreak-tornadoes Accessed 5th December 2017.

77 "Staying Safe in a Tornado" Centers for Disease Control and Prevention Available in https://www.cdc.gov/features/tornadosafety/index.htmlAccessed 5th December 2017.

78 "Staying Safe in a Tornado" Centers for Disease Control and Prevention Available in https://www.cdc.gov/features/tornadosafety/index.html.

VII. Tsunami

Meaning

Tsunami is a Japanese word with the English translation: "harbor wave." It is derived from a Japanese term meaning wave (*"nami"*) in a harbour (*"tsu"*).[79] A tsunami is a series of ocean waves with very long wavelengths (typically hundreds of kilometers). Tsunamis has been referred to as "tidal waves." Tsunamis have no connection with tides; the popular name, tidal wave, is entirely misleading.[80] A tsunami's impact upon a coastline is dependent upon the tidal level at the time a tsunami strikes, tsunamis are unrelated to the tides. (Tides result from the gravitational influences of the moon, sun, and planets).

Types

According to a report by the International Tsunami Information Center of UNESCO, tsunamis can be classified into the following categories.These are local tsunami, meteorological tsunami, micro tsunami, ocean-wide tsunami, regional tsunami, teletsunami or distant tsunami.[81]

Causes

According to Windle, tsunami is caused by large-scale disturbances of the ocean, such as earthquakes, landslide, volcanic eruptions, explosions, meteorites.[82] It is further stated that for tsunamis that are generated by underwater earthquakes, the amplitude (i.e. wave height) of the tsunami is determined by the amount by which the

79 Available in 2017 International Tsunami Information Center | A UNESCO/IOC-NOAA Partnership Available in http://itic.ioc-unesco.org/index.php?option=com_content&view=article&id=1860:1-tsunami-classification&catid=2118&Itemid=2441Accessed 7th December 2017.

80 *Ibid*

81 *Ibid*

82 Lauren Windle "What is a tsunami, how are they formed and what are some of the worst disasters?"The Sun 7 September 2017 Available in https://www.thesun.co.uk/tech/1992625/tsunami-how-formed-earthquake-worst-disasters/Accessed 7th December 2017.

sea-floor is displaced. Similarly, the wavelength and period of the tsunami are determined by the size and shape of the underwater disturbance.[83]

Effects

Tsunamis are among the most devastating natural calamities. The earthquake-generated waves can quickly engulf low-lying land and bring widespread destruction and death. They can deposit sand and debris far inland from where they came ashore.

Tsunamis have great erosion potential, stripping beaches of sand that may have taken years to accumulate and undermining trees and other coastal vegetation. Capable of inundating, or flooding, hundreds of meters inland past the typical high-water level, the fast-moving water associated with the inundating tsunami can crush homes and other coastal structures. The effects of a tsunami are devastating. The effects of the tsunami on the country during this period range from destruction and damage, death, injury, millions of dollars in financial loss, and long-lasting psychological problems for the inhabitants of the region.

Tsunami victims suffer psychology problems in the days and weeks after the destruction. This could even continue for years – often their entire life time. A study by the World Health Organisation on survivors of the tsunami in Sri Lanka on December 24th 2004 confirmed this.[84]

83 *Ibid*

84 Frank Neuner and Elisabeth Schauer, Claudia Catani, Martina Ruf and Thomas Elbert Post-tsunami Stress: A Study of Posttraumatic Stress Disorder in Children Living in Three Severely Affected Regions in Sri Lanka Available in ,https://pdfs.semanticscholar.org/6691/3c8cdbee95010126f4d4f25793dea3cd83d3.pdf Accessed on 7th December 2017.

Most Prone Areas

Although tsunamis are seen in every ocean on Earth, 80 percent of it occur in the Pacific "Ring of Fire" – The Pacific Ocean.[85] Other areas prone to tsunamis are Kolkata, India, Tokyo-Yokohama, Japan,[86] Coast of Chile, Southern part of New Zealand, coast of Mexico and Turkey.[87]

Examples of Disaster as a result of Tsunami

One of the recorded disasters that occurred as a result of tsunami is the Indian Ocean Tsunami of 2004 which occurred on December 26th 2004. According to a UNESCO report, the majority of that Indian Ocean tsunamis were generated by earthquakes, volcanic eruptions, landslides, and unknown causes and the death toll was about one thousand people.[88] In 2011, the Tohoku earthquake and tsunami took place with 15,884 deaths resulting in the tsunami, making it one of the most expensive natural disasters ever recorded, costing the US equivalent of $300 billion.[89]

Protective/mitigating

Two ways to find out if a tsunami may be coming and which is protective are discussed below.

1. **Natural warning**
   Strong ground shaking, a loud ocean roar, or the water receding unusually far exposing the sea floor are all nature's warnings that a tsunami may be coming.

85 https://borgenproject.org/10-worst-tsunamis/.

86 Jenni Ryall "These are the riskiest cities in the world for natural disasters' New. com Available in http://www.news.com.au/world/these-are-the-riskiest-cities-in-the-world-for-natural-disasters/news-story/6b59d0d41548b999869a0f344a0d10cb Accessed 5th December 2017.

87 Lauren Windle "What is a tsunami, how are they formed and what are some of the worst disasters?" The Sun 7 September 2017 Available in https://www.thesun.co.uk/tech/1992625/tsunami-how-formed-earthquake-worst-disasters/ Accessed 7th December 2017.

88 "Indian Ocean Tsunami 2004" http://itic.ioc-unesco.org/index.php?option=com_content&view=category&id=1136&Itemid=1373.

89 https://borgenproject.org/10-worst-tsunamis.

2. Official warning

A tsunami warning may be issued. Tsunami warnings might come via radio, television, telephone, text message, door-to-door contact by emergency responders, NOAA weather radios, or in some cases, by outdoor sirens.

VIII. VOLCANIC ERUPTION

Meaning

The word 'volcano' is derived from the name 'volcano', a volcanic island in the Aeolian island of Italy, whose name in turn originates from Vulcan, the name of a god of fire in roman mythology.[90] A volcano is an opening, or rupture in a planet's surface or crust which allows hot magma, volcanic ash and gases to escape from the magma chamber below the surface.

Types of volcanic eruption

The variety of volcanic eruption makes them very difficult to classify, for this reason they are placed into two general categories:

Explosive Volcanoes

Volcanoes are explosive where magma are produced by the melting of sub-ducted oceanic plates and other materials which is thicker (and more viscous), consequently blocking magma conduit inside the volcano, traps and compresses gases, causing pressure to build and creating conditions for a possible explosive eruption. An example of those that are explosive is Mount Helens.

Effusive Volcanoes

Effusive volcanic eruptions such as Hawaii on the other hand are relatively gentle and produce enormous volumes of lava annually. They produce low viscosity magma (with large releases of gases) that is very fluid and cools to form a dark basaltic rock.

90 Amina Alhassan "Volcanoes' science of violence" June 18 2011Available in https://www.dailytrust.com.ng/news/general/volcanoes-science-of.../6956.html Accessed on 5th December 2017.

Causes

Volcanoes erupt because of density and pressure. In a report, the cause of volcanic eruption that causes disaster was stated thus:

> The lower density of the magma relative to the surrounding rocks causes it to rise (like air bubbles in syrup). It will rise to the surface or to a depth that is determined by the density of the magma and the weight of the rocks above it. As the magma rises, bubbles start to form from the gas dissolved in the magma. The gas bubbles exert tremendous pressure. This pressure helps to bring the magma to the surface and forces it in the air, sometimes to great heights.[91]

Effects of volcanic eruptions

Volcanic eruptions indeed produce hazardous materials which have tremendous effects on the environment, climate and physical health of the victims so exposed and they are associated with deterioration of social and economic conditions. These effects include problems of respiratory systems (clogging lungs), eyes and skin, psychological effects, injuries, transportation and communication problems, building collapses and power outrage etc. However, soils around volcanoes have been found to be good sites for crops production as the soil therein tends to be fertile.

Most Prone Areas of Volcanoes

The "Ring of Fire" that encircles the Pacific Ocean – which stretches up the West Coast of the Americas, around and across to Asia, looping down to the east of Japan, before overwhelming much of Indonesia and the Philippines and whipping around Australasia – boasts the most, with 452 volcanic eruptions.[92]

---

91 "Why are there volcanoes? Why do they erupt?" Oregon state University Available in http://volcano.oregonstate.edu/why-are-there-volcanoes Accessed on 5th December 2017.

92 Huge Morris "Mapped : The countries with the most volcanoes – where does Indonesia rank" 27 November 2017 Available in http://www.telegraph.co.uk/travel/maps-and-graphics/mapped-the-worlds-most-dangerous-volcanoes/

Mount Agung, on the Indonesian island of Bali,[93] United States of America, Russia, Japan, Chile, Ethiopia, and Papua New Guinea.

Examples of Volcanoes resulting in Disaster

One of the volcanic eruptions resulting in a disaster is the Nevado Del Ruiz volcano eruption in Colombia in 1985.[94] It is reported that Nevado Del Ruiz, known locally as "the Sleeping Lion", had not erupted for nearly 150 years.[95] The stratovolcano is said to be the second largest volcano-related disaster of the 20th century.[96] The volcano produced an enormous flow that buried and devastated the town of Armero in Tolima on November 13th 1985.[97] What became known as the "Armero tragedy" caused the death of 25,000 people.[98]

Protective/ Mitigating Measures

To prevent or at least reduce the impact of this hazard, a series of measures must be taken before, during and after a volcanic eruption. The preparation of hazard maps helps to determine whether a volcano is potentially hazardous and to assess the risk. Monitoring of volcanoes by satellite needs to increase in order to detect possible changes. The public must be informed and educated on the result of possible dangers. Emergency and evacuation plans must be worked out.

---

93 Huge Morris "Mapped : The countries with the most volcanoes – where does Indonesia rank" 27th November 2017. Telegraph Available in http://www.telegraph.co.uk/travel/maps-and-graphics/mapped-the-worlds-most-dangerous-volcanoes/hhttp://www.telegraph.co.uk/travel/maps-and-graphics/mapped-the-worlds-most-dangerous-volcanoes/

94 "1985: Volcano kills thousands in Colombia" BBC on this day Available in http://news.bbc.co.uk/onthisday/hi/dates/stories/november/13/newsid_2539000 /2539731.stm

95 *Ibid*

96 "10 years since Dec 26, 2004 Indian Ocean Tsunami" International Tsunami Information Center Available in http://itic.ioc-unesco.org/index.php?option=com_content&view=category&id=1136&Itemid=1373.

97 *Ibid*

98 *Ibid*

IX. Flood

Meaning

Flood can be defined as an overflow of large quantities of water onto a normally dry land. Floods make an enormous impact on the environment and society.

Types/ Classification

There are different types of flood or classification of flooding. Floods may be classified as flash floods, rapid on-set floods and slow on-set floods.[99] Flash floods are extremely dangerous. Flood is also classified as coastal, urban, riverine, fluvial, reservoir and pluvial flooding.[100] Flood has also been classified as riverine, dam failure and storm surge.[101]

Causes of Flooding

Flooding happens in many ways due to overflow of streams, rivers, lakes or oceans or as a result of excessive rain.[102] Flooding may be caused by a number of natural causes or physical factors.

Effects of Flooding

Floods destroy drainage systems in cities, causing raw sewage to spill out into bodies of water, buildings can be significantly damaged and even destroyed, toxic materials such as paint, pesticide and gasoline can be released into the rivers, lakes, bays, and ocean, killing maritime life. Floods may also cause millions

99 Chris Dolan "Hazard-Wise: Classroom Resources for Teachers on Natural Hazards & Disasters"Available in https://books.google.com.ng/books?isbn=0788127314- Accessed on 8th December 2017.

100 James Hubbard "Types of flood and flooding impact" Ambiental Environment assessment September 20th, 2012 http://www.ambiental.co.uk/types-of-flood-and-flooding-impact/.

101 "Types of Flooding" State Emergency Service Available in https://www.ses.vic.gov.au/get-ready/floodsafe/types-of-floods.

102 River flooding and management BBC Available in http://www.bbc.co.uk/schools/gcsebitesize/geography/water_rivers/river_flooding_management_rev1.shtml.

of dollars' worth of damage to a city, both evicting people from their homes and ruining businesses. Floods cause significant amounts of erosion to coasts, leading to more frequent flooding if not repaired. Loss of livestock and farmland are serious effects in countries that are dependent on agriculture.

There are benefits that may accrue from flooding. Floods do make a slight positive impact on the environment. Floods spread sediment containing beneficial nutrients to topsoil that might never arrive there otherwise.

Areas prone to Flooding

Countries prone to flood include Bangladesh, China, India, and Cambodia. Any plain low-lying area adjacent a river, lagoon or lake is also more likely to have floods anytime the water level rises.

Examples of Disaster as a result of Flooding

In 1928, the Saint Francis Dam in Los Angeles was filled with 12.5 billion gallons of water — enough water for one year for a population of 1.2 million people. However, on March 12th 1928, the dam broke, unleashing a 78 foot wall of water. It obliterated houses, ranches, automobiles, animals, and people. In roughly one hour, more than 500 people were killed and Santa Paula was overrun by water.[103]

On Friday, April 18th 1997, the Red River flooded over the dikes into Grand Forks, North Dakota[104] spreading over a large area of Grand Forks and East Grand Forks. About 60,000 people were rendered homeless and downtown Grand Forks was left in flames. The damage was so extensive that it was weeks before people could return and rebuild their homes, and their lives.[105]

---

103 Case Study: St. Francis Dam (California, 1928) Available in http://damfailures.org/case-study/st-francis-dam-california-1928/ Accessed on 7th December 2017.

104 "1997 Floods in the Red River of the North and Missouri River Basins in North Dakota and Western Minnesota" US Geographical Survey Available in https://pubs.usgs.gov/of/1997/0575/report.pdfAccessed on 7th December 2017.

105 *Ibid*

The Mozambique flood that took place between February and March 2000 after very heavy rain claimed over 800 lives.[106] Another flood claiming lives took place in January 2013,[107] and in February 2017, there was another disaster in Mozambique as a result of flood.[108]

Protective/ Mitigating Measures

Flood management techniques include river engineering, afforestation and planning controls to restrict urban development on floodplain.

x. EARTHQUAKE

Meaning

An earthquake (also known as a quake, tremor or temblor) is the result of a sudden release of energy in the Earth's crust that creates seismic waves. Earthquakes are measured using observations from seismometers.

Types/classification of Earthquakes

Earthquake could be classified according to magnitude, size and intensity. According to magnitude, the earthquake can be classified as great, major, strong, moderate, light, minor and insignificant.[109] According to size and intensity, they are classified as fore-shock, main-shock and after-shock.

---

106 Anita Powell! February 2" Deadly Floods Hit Mozambique" Available inhttps://www.voanews.com/a/united-nations-reports-devastating-damages-after-mozambique-flood/1595319.html Accessed on 8th December 2017.

107 *Ibid*

108 Tomas Cumbana "Mozambique: Storm Dineo Kills Seven and Destroys 20,000 Homes" Newsweek 17 February 2017 Available in013http://www.newsweek.com/mozambique-storm-dineo-557810Accessed on 30th November 2017.

109 "How earthquakes are measured" CBS NEWS Available in https://www.cbsnews.com/pictures/how-earthquakes-are-measured/3/ Accessed on 30 November 2017; "earthquake magnitude - Michigan Technological University" Available in www.geo.mtu.edu/UPSeis/magnitude.html.

Causes of Earthquakes

There are different causes of earthquake. It could be as a result of natural cause by movement in the earth crust as a result of geological fault, volcanic eruptions and landslide.[110] Stress built up at point of weakness may cause a rock to deform. Earthquake are said to last for a few minutes and this is not the immediate cause of the disaster which follows. It is the aftermath triggered by it that results into disaster. These include fire, flood, tsunami, building collapse, gas explosion, fire, avalanches and such other secondary causes.

The cause of earthquake can also be induced due to activities on land. Activities such as mining, drilling, blasting could trigger off an earthquake.

Most earthquake-related deaths are caused by the collapse of structures and the construction practices play a tremendous role in the death toll of an earthquake.[111]

Effects of Earthquakes

Earthquakes can result in landslides, tsunamis, fires due to gas rupture, collapsed buildings and loss of lives.

Areas Prone to Earthquakes

Botswana,[112] Mexico, Tajikistan in India and the Indian subcontinent are generally prone areas of earthquake.

---

110 Peace Ezebuiro "Earthquake: Everything you must know about this catastrophic disaster" Available in https://answersafrica.com/earthquake-everything-you-must-know-about-this-catastrophic-natural-disaster.html Accessed 30th November 2017.

111 Allen, Clarence R. and Nordquist, John M. Foreshock, Main Shock, and Larger Aftershocks of the Borrego Mountain Earthquake. In: The Borrego Mountain Earthquake of April 9, 1968. ((1972) Geological Survey Professional Paper. No.787. United States Geological Survey, Washington, DC, pp. 16-23. Available in http://resolver.caltech.edu/CaltechAUTHORS:20140917-144224324 Accessed 30th November 2017.

112 Joey Millar "Botswana rocked by strongest EVER earthquake - hours after South Africa struck" April 3rd 2017 Available in https://www.express.co.uk/news/world/787427/earthquake-botswana-largest-ever-south-africa Accessed 6th December 2017.

Examples of the earthquake causing disaster

Since 1138, earthquakes has resulted in disaster. The Aleppo Earthquake occurred on August 9th 1138.[113] The disaster was named after the city of Aleppo in Syria, now called Halab. Its death toll was approximated at 230,000 as the city was destroyed along with its surrounding areas.[114] The continent of Asia has recorded many cases of earthquake disasters. The Great Tokyo Earthquake which happened in Kanto, Japan, on September 1 1923 was equally devastating.[115] In China, the Haiyuan Earthquake also known as the 1920 Gansu Earthquake, occurred with a death toll of 240,000, on December 16th 1920.[116] In another recorded disaster in China, the Great Tangshan earthquake regarded as the deadliest earthquake of the 20th century, occurred on July 27th 1976 causing the death of 240,000 people and injuring 164,000 more.[117] In the Caribbean's, Haiti suffered from a disaster in 2010 where it was reported that according to official estimates, 300,000 were also injured, 1.3 million displaced, 97,294 houses destroyed and 188,383 damaged.[118] In Europe, earthquake struck, on May 11th 2011, and at least eight people were reported dead when the quake rocked southern Spain and 15,000 people were rendered homeless.[119] In Africa, a Nabro volcanic eruption happened in the southern red region of Eritrea on June 12th 2011 affecting about 5000 people.

113 NBC News.com "The top 10 deadliest earthquakes in historyAvailable in http://www.nbcnews.com/id/42029974/ns/world_news-asia_pacific/t/top-deadliest-earthquakes-history/#.Wh_Tpbc1jIU.

114 *Ibid*

115 *Ibid*

116 http://list25.com/25-worst-natural-disasters-recorded/

117 http://list25.com/25-worst-natural-disasters-recorded/

118 NBC News.com "The top 10 deadliest earthquakes in history Available inhttp://www.nbcnews.com/id/42029974/ns/world_news-asia_pacific/t/top-deadliest-earthquakes-history/#.Wh_Tpbc1jIU

119 "Spain earthquake: up to 15,000 left homeless The Telegraph Available in http://www.telegraph.co.uk/news/worldnews/europe/spain/8508784/Spain-earthquake-up-to-15000-left-homeless.htmlaccessed 30th November 2017

Mitigation/ Prevention of Earthquake

It may not be possible to prevent the occurrence of an earthquake. However, there can be ways to mitigate the effect. This may include early warning, putting in safety structures in buildings in earthquake-prone areas and proper planning.

XI. Famine

Meaning

A famine is a widespread scarcity of food. Nearly every continent in the world has experienced a period of famine throughout history.

Types/classification

There is no definite classification for famine. The classification available on famine is that leading to famine.[120] This is based on the level of food shortage. The classification is as follows – minimal stage is the phase 1, stressed stage is phase 2, crisis stage is phase 3, emergency stage is phase 4 and famine is the stage 5.[121]

Causes

Famine is caused by several factors including crop failure, population unbalance or government policies. Famine may be as a result of a drought. Long-term climate change is a contributing factor to the occurrence of drought. Famines result from a combination of "triple failure": production failure, access failure and response failure.[122]

120 "Integrated Phase Classification "FEWS NET Available in http://www.fews.net/IPC Accessed on 7th December 2017.

121 *Ibid*

122 "Famine in Somalia: causes and solutions" Oxfam International" Available in https://www.oxfam.org/es/node/3573 Accessed on 6th December 2017.

Effects of famine

This phenomenon is usually accompanied or followed by regional malnutrition, starvation, epidemic, and increased mortality.

Areas prone to famine

Some countries, particularly in sub-Sahara Africa and Ethiopia[123] continue to have extreme cases of famine.

Examples of the famine causing disaster

The famine that occurred in Ethiopia and Eritrea between 1983 and 1985 for a period of two years was reported to claim over 400, 000 lives.

Protective/Mitigating measures

These measures could be long term and short term. In a famine situation, humanitarian aid is very important as a short term measure. On the other hand, diversification of the economy and strengthening local food system resilience is a long term measure.[124]

XII. THUNDERSTORM

Meaning

Thunderstorms are simply electrical storms. They are characterized by the presence of lightning and its acoustic effect on the Earth's atmosphere.

Causes

Thunderstorms result from the rapid upward movement of warm moist air. They can occur inside warm, moist air masses and at fronts. As the warm, moist air moves upward, it cools, condenses,

123 "Africa's famine: Country by country" BBC News Available in http://news.bbc.co.uk/2/hi/africa/2027079.stm Accessed on 7th December 2017.

124 Suresh Babu and Paul Dorosh "Strategies for preventing recurring famines and building resilient food systems" April 6th, 2017International Food policy Research Institute Available inhttp://www.ifpri.org/blog/strategies-preventing-recurring-famines-and-building-resilient-food-systems. Accessed on 6th December 2017.

and forms cumulonimbus clouds (these are clouds that can reach heights of over 20km). Furthermore, as the rising air reaches its dew point, water droplets form and begin falling the long distance through the clouds towards the Earth's surface. As the droplets fall, they collide with other droplets and become larger. The falling droplets create a downdraft of air that spreads out at the Earth's surface and causes strong winds associated commonly with thunderstorms. Generally, thunderstorms require three conditions to form: moisture, an unstable air mass and a lifting force (heat).

The thunderstorm life cycle includes; The developing stage, maturing stage and the dissipating stage.

Type/Classification

There are 3 types of thunderstorms. They include the following:

i. Single-cell thunderstorms occur in temperate regions of the world.
ii. Multicell cluster thunderstorms: These types of storms are stronger than the single-cell storms.
iii. Multicell lines thunderstorms: This kind of storm is also known as "Wind of the Stony Lake." Supercell thunderstorms is the most powerful type of thunderstorm.

Effects

Thunderstorms result in heavy precipitation. Thunderstorms disrupt human life in more than one way. The felling of millions of trees, deaths due to lightning hazard and wind shear are just some of the manifestations. All thunderstorms produce lightning which often strikes away from the area where it is raining and is known to fall as far as 10 miles away from the rainfall area.

Areas Prone to Thunderstorm

Roughly, there are about 1800 thunderstorms occurring at any moment across the world and Kampala, Uganda, holds the record of about 240 thunderstorm per day, annually. Asia is also an area prone to thunderstorm.

Disasters caused as a result of Thunderstorm

Examples of disasters include a thunderstorm in 2007, in Ushari Dara, a remote mountain village in north-western Pakistan that killed 30 people.[125] The crash of LANSA Flight 508 which killed 91 people and a fuel explosion in Dronka, Egypt caused by lightning in 1994 which killed 469.[126]

Protective and Mitigating Measures

Thunderstorm cannot be prevented. However, the effects can be mitigated. Protective measures in a thunderstorm situation is best done and more effective before the occurrence. Avoiding the use of electronic gadgets during rainfall, use of proper roofing materials during construction, securing roof tops and doors, staying indoors and if necessary, evacuation and having an emergency survival kit ready are some measures to mitigate the effects of thunder.

B. Human/Man-made Disaster

They are as a result of human intent, error or as a result of failed systems. Often, these are as a result of neglect or ignorance and over the years, such disaster have claimed several lives. They are human-caused events that occur in or close to human settlements. Disasters having an element of human intent or negligence are called man-made disasters. They include, civil disorder, chemical/oil spill, toxic/hazardous waste, human stampede, war, mass murder, riots, terrorists accidents, fire: bush fire, wild fire, firestone, power outage, structural collapse, dam failure, telecommunication outage, plane, rail, and auto crashes. These man-made disasters can further be classified into technological disasters, sociological and transportation disasters. Some of these events can also occur as the result of a natural disaster. A few of these man-made disasters are discussed below.

125 "Lightning kills 30 people in Pakistan's north". Reuters. 2007-07-20. Accessed on 7th December 2017.

126 Evans, D. "An appraisal of underground gas storage technologies and incidents, for the development of risk assessment methodology" (PDF). British Geological Survey. Health and Safety Executive: 121. Retrieved 2008-08-14.

i. Arson/Fire

Meaning

A fire has been defined by the UCL Estate Department as "an undesirable event which emits heat, smoke and/or flame, which has the potential to cause damage, may require intervention either mechanical or human or has a cost implication."[127]

Setting fire with malice to property is known in legal parlance as Arson. It is when a person deliberately or intentionally set fire to a property in order to destroy it, with a criminal intent. A fire disaster has been defined as an event involving more than 25 deaths.[128]

The Uniform Crime Report (UCR) of the Federal Bureau of Investigation defines arson as: "any wilful or malicious burning or attempt to burn, with or without intent to defraud, a dwelling house, public building, motor vehicle or aircraft and personal property of another,"[129]

Types/Classification of Fire

Fire can be classified as natural, accidental, incendiary or origin not known.[130] Fires have different classification. They include forest fire, wild/bush fire and according to the intensity of the fire, fire is classified as Class A, B, C, D.

Causes of Fire

Lack of prevention and safety regulations and equipment like fire detecting system and fire safety equipment are known causes

127 UCL Estates Construction, Maintenance & Fire Safety Available at http://www.ucl.ac.uk/estates/maintenance/fire/incidents/ accessed on 8th December 2017.

128 M. Cavallini, M.F. Papagni, and F.W. Baruffaldi Preis "Fire Disasters in the Twentieth Century" 2007 Jun 30; 20(2): 101–103 Available in https://www.ncbi.nlm.nih.gov/pmc/articles/PMC3188063/ Accessed on 6th December 2017.

129 M. Cavallini, M.F. Papagni, and F.W. Baruffaldi Preis "Fire Disasters in the Twentieth Century" 2007 Jun 30; 20(2): 101–103 Available in https://www.ncbi.nlm.nih.gov/pmc/articles/PMC3188063/ Accessed on 6th December 2017.

130 Thomas J. Gardner, Terry M. Anderson, *Criminal Law Cengage Learning*, pg 418

of fire. Bad planning and defective installation in construction can cause fire. Explosion of flammable materials can cause a fire disaster. In recent times, terrorist criminal activities are causes of fire.

Effects of Fire

Apart from the tangible effect on man which include physical burns, there are mental and psychological effect, there is also the effect on property.[131] There is also the effect on the natural environment and ecosystem. Apart from the adverse effect of contamination of water, the effect may however be positive. The burning of forest habitat by fires can serve a number of important and beneficial ecological functions such as maintaining vegetation succession regimes and contributing to soil fertility [132]

Areas Prone to Fire

The areas include Trinidad and Tobago,[133] Australia, with an average of 2,000 wildfires a year,[134] and South of France.[135] In Nigeria, Kwara State in Northern Nigeria has been recognised as a fire-prone area.[136]

Disasters Caused as a Result of Fire

In Europe, the Barcelona fire in 1987, and in 1988, the fire disaster in London are examples of devastating fire disasters in Europe

131 Drew Martin, Mai Tomida, Brian Meacham "Environmental impact of Fire" Fire Science Reviews December 2016, 5:5 Available in https://www.ncbi.nlm.nih.gov/pmc/articles/PMC3188063/ Accessed on 6th December 2019.

132 De Bano, L.F., Neary, D.G. and Ffolhott, P.F. Fire Effects of Ecosystems. (1998) New York: John Wiley and Sons.

133 Fires Available in http://www.odpm.gov.tt/node/19accessed on 6th December 2017.

134 Natural Disaster Association http://www.n-d-a.org/fire.phpaccessed on 6th December 2017.

135 *Ibid*

136 "Kwara Leads in Fire Disasters | NEMA Nigeria" Available in nema.gov.ng/kwara-leads-in-fire-disastersaccessed on 6 December 2017

due to terrorist attacks.[137] In 1984, there was a LPG explosion in Mexico resulting in many deaths. In June 2017, the Grenfell Tower fire disaster occurred becoming one of the deadliest fire incidents in the 20th Century.[138] On December 7th 2017, the State of California was on fire resulting into loss of properties and life.[139]

Protection/Mitigating Measures

Regular inspection and servicing fire equipment, installation of fire alarms, quick evacuation are a few measures to mitigate or prevent fire or the effects. In the construction of buildings, the use of fire resistant material in construction, for example, concrete bricks as opposed to wood, is another preventive measure and the practice of storing flammable items such as fuel, wood and paint away from important structures such as dwelling spaces are some of the preventive measures identified.[140]

ii. Oil Spill

Meaning

An oil spill is a form of pollution and is a phenomenon common to oil producing areas around the globe. It is the release of a liquid petroleum hydrocarbon into the environment (especially the marine areas) mostly due to human activities.

Oil spills at sea are generally much more damaging than those on land, since they can spread for hundreds of nautical miles in a thin oil slick which can cover beaches with a thin coating of oil.

137 Available in https://www.ncbi.nlm.nih.gov/pmc/articles/PMC3188063/Accessed on 6th December 2017.

138 "The Latest: London Fire Deadliest in UK Since 20th Century" US News https://www.usnews.com/news/.../the-latest-moment-of-silence-for-london-fire-victimsLondon Fire Deadliest in UK Since 20th Century.

139 Richard Halsey "Why are California's homes burning? It isn't natural disaster it's bad planning" Los Angeles Times 7 December 2017 Available in http://www.latimes.com/opinion/op-ed/la-oe-halsey-socal-fires-why-20171207-story.html accessed on 8th December 2017.

140 http://www.odpm.gov.tt/node/19

This can kill sea bird, mammals, shellfish and other organisms it coats. Oil spills on ground are more readily containable if a makeshift earth dam can rapidly bulldoze around the spill site before most oil escapes.

Types/Classification

Oil spill may be classified according to geographical source. Oil spill may also be as a result of deliberate or routine or operational discharges of ships. Oil spill has also been classified as Class A, B, C, D and non petroleum/non fluid oil.[141]

Causes of Oil Spills

Oil spills can be caused by many factors. Carelessness or deliberate dumping by humans and natural causes such as floods, hurricanes, and storms are a few of them. The majority of the accidents occur due to human error or carelessness, which fall under controlled factors. High incidence of oil spills and marine accidents are caused by human error. Maritime transportation has grown steadily and the quantity of illegal oil discharges has grown with the volume of traffic.[142] The term 'oil spill' is usually applied to marine oil spills, where oil is released into the ocean or coastal waters; but spills may also occur on land. Oil spills may be due to the release/escape of crude oil from drilling rigs, oil wells, tankers and offshore platforms, as well as the spills or refined petroleum products (such as diesel gasoline etc.) or the spill of any refuse or waste oil.[143] Oil spills may result from intentional or unintentional releases into oceanic or coastal waters. According to Baker, other causes include blowout, pipeline failure, road and

141 Jennifer King "Types of Oil Spills" Sciencing April 25th 2017 Available in https://sciencing.com/types-oil-spills-6593214.html Accessed on 6th December 2017; See also"Types of Crude Oil" EPA Available in https://www.epa.gov/emergency-response/types-crude-oil Accessed on December 6th 2017.

142 http://ieeexplore.ieee.org/abstract/ document/6352042/?reload=true

143 Jennifer King "Types of Oil Spills" Sciencing April 25th 2017 Available in https://sciencing.com/types-oil-spills-6593214.html Accessed on 6 December 2017; See also "Types of Crude Oil" EPA Available in https://www.epa.gov/emergency-response/types-crude-oil Accessed on December 6th 2017.

rail accidents.[144] Lack of regular maintenance of the pipelines and storage tanks, oil production operations, engineering drill account for some cases of oil spill. Sabotage has also been identified as a cause of oil spill.[145] In Nigeria, oil bunkering, resulting in break of pipes to steal oil, is a major cause of this hazard.[146]

Human mistakes are not the only reason for such accidents. They may also occur due to break down of machineries or intentional spills due to vandalism, sabotage or wars. Illegal dumping of oil is also an important factor that needs to be mentioned.

Effect of Oil Spill

Oil spill is a major threat to ocean ecosystems and their health and has long-lasting effect on the natural environment. Pollution is another effect and may lead to death.

Areas Prone to Oil Spill

Areas along the coastal regions are prone to oil spill. The coastal zone of the Brazilian State of Maranhão is an area characterized by a large variety of human activities and services, in particular in the Itaqui-Bacanga port complex, and it is prone to oil spills resulting from the processes of transportation, storage, and tank cleaning.[147] Akwa Ibom State in the Delta Region of Nigeria is one such areas.[148]

---

144 Baker J.M. (1985) Causes and Classification of Oil Spills. In: Cairns J. (eds) Ecoaccidents. Springer, Boston, MA

145 Peter C. Nwilo and Olusegun T. Badejo "Impacts and Management of Oil Spill Pollution along the Nigerian Coastal Areas" Available in https://www.fig.net/resources/publications/figpub/pub36/chapters/chapter_8.pdf Accessed on December 5th 2017.

146 *Ibid*

147 "A socioeconomic and natural vulnerability index for oil spills in an Amazonian harbour: A case study using GIS and remote sensing" Science Direct Available in http://www.sciencedirect.com/science/article/pii/S0301479710001179

148 Available in http://www.stakeholderdemocracy.org/stockholm/wp-content/uploads/2015/04/Annex-8-OIL-SPILLS-IN-THE-NIGER-DELTA_-PROPOSALS-FOR-AN-EFFECTIVE-NONJUDICIAL-GRIEVANCE-MECHANISM_JEI-29-JULY-2014_FINAL-1.pdf Oil Spills In The Niger Delta: Proposals For An Effective Non-Judicial Grievance Mechanism

Disasters Caused by Oil Spill

GOCON's Escravos spill in 1978 of about 300,000 barrels, Idaho Oil Spill of January 1998, of about 40,000 barrels, Texaco Funiwa-5 blowout in 1980 of about 400,000 barrels[149] are examples of oil spill in Nigeria.[150]

Preventive and Mitigating Measures

Early detection of oil spill and oil spill monitoring by authorized agencies are effective ways of preventing oil spills and the attendant hazards. For example, since 2007, the European Maritime Safety Agency's oil spill monitoring and vessel detection service, 'CleanSeaNet' has been continuously monitoring the EU waters.[151]

iii. Toxic/Hazardous Waste

Meaning

Hazardous/toxic waste catastrophes are caused by people. Hazardous, a broader term, refers to all wastes that are dangerous for any reason, including those that are toxic. Toxic commonly refers to poisonous substances which cause death or serious injury to humans and animals by interfering with normal body physiology. The terms are often used interchangeably and the waste can pose a long-term risk to health or environment. There is a technical difference, and it is important, as well, to recognize that distinction. Waste includes all items that people no longer have any use for, which they either intend to get rid of or have already discarded. Additionally, wastes are such items which

149 Peter C. Nwilo and Olusegun T. Badejo "Impacts and Management of Oil Spill Pollution along the Nigerian Coastal Areas" Available in https://www.fig.net/resources/publications/figpub/pub36/chapters/chapter_8.pdf Accessed on December 5th 2017

150 J. C. Udoh and E. M. Ekanem

151 Suman Singha, Tim J. Bellerby, Olaf Trieschmann, "Satellite Oil Spill Detection Using Artificial Neural Networks", Selected Topics in Applied Earth Observations and Remote Sensing IEEE Journal of, vol. 6, pp. 2355-2363, 2013, ISSN 1939-1404 Available in http://ieeexplore.ieee.org/abstract/document/6352042/?reload=true Accessed 6th December 2017.

people are required to discard, for example by lay because of their hazardous properties. Many items can be considered as waste e.g., household rubbish, sewage sludge, wastes from manufacturing activities, packaging items, discarded cars, old electronic gadgets, garden waste, old paint containers etc. Thus, all our daily activities can give rise to a large variety of different wastes arising from different sources.

These are poisonous by-products of manufacturing, farming, city septic systems, construction, automotive garages, laboratories, hospitals, and other industries. The waste may be liquid, solid, or sludge and contain chemicals, heavy metals, radiation, dangerous pathogens, or other toxins. Even households generate hazardous waste from items such as batteries, used computer equipment, and leftover paints or pesticides.

The definition of hazardous waste varies from one country to another. One of the most widely used definitions, however, is contained in the U.S. Resource Conservation and Recovery Act of 1976 (RCRA).[152] RCRA considers wastes toxic and/or hazardous if they "cause or significantly contribute to an increase in mortality or an increase in serious irreversible, or incapacitating reversible illness; or pose a substantial present or potential hazard to human health or the environment when improperly treated, stored, transported, disposed of, or otherwise managed."[153]

Types/Classification

Hazardous waste can be classified into the following defined hazard classes: flammable liquid, combustible liquid, flammable solid, oxidizer, organic peroxide, corrosive, flammable gas, non-flammable gas, etiologic agents.

152 William C. Blackman, Jr. *Definition of Hazardous Waste Basic Hazardous Waste Management,* Third Edition CRC Press pp. 37.

153 *Ibid*

Causes of Hazardous Waste

Deposit of accumulation of toxic waste as a result of flooding can be a cause of a disaster of hazardous waste. Other causes of hazardous waste include nuclear waste, industrial waste, universal waste, medical waste, construction waste and electronic waste.

Effects of Hazardous Waste

When any of these classes of hazardous waste is disposed in our community or environment, it causes infection, injuries, loss of life of plants and animals and even cultural properties. Toxic waste is waste material that can cause death, injury or birth defects to living creatures. It spreads quite easily and can contaminate lakes, rivers, and the atmosphere. There are many types of hazardous waste that pose varying degrees of danger to public safety and environmental health and medical issues like cancer, physical anomalies and birth defects, respiratory illnesses and injuries. Disposal of these materials can be challenging and often needs to be handled by professionals. The effect of Hazardous waste can be classified as chemical risks, biological risks, physical risk and environmental risks.

Areas prone to Hazardous Waste

Due to the fact that developed countries tend to move their waste to less developed countries, and because these countries are not able to effectively handle the effects, developing countries are prone to hazardous waste.[154] The Koko Toxic Waste incidence is a good example. The Italian who was the key actor in the incidence, bought a piece of land in the small community with the intention of importing and disposing toxic waste from industrial nations who had strict laws on pollution.[155]

154 Grasso d, Kahn D, Kaseva, M. E and Mbuligwe S. E – "Hazardous Waste" HAZARDOUS WASTE MANAGEMENT Encyclopaedia of Life Support Systems (EOLSS) Available in http://www.eolss.net/sample-chapters/c09/e1-08-00-00.pdf

155 Segun Gbadegesin "Multinational Corporation Developed Nations and Environmental Racism: Toxic Waste, Exploration, and Eco-Catastrophe"in Faces of Environmental Racism: Confronting Issues of Global Justice. 2nd ed.' ed Laura Westra, Bill E. Lawson (Oxford: Rowman & Littlefield Publishers, Inc., 2001) 188.

Disasters as a result of Hazardous Waste

Love Canal hazardous waste disaster which occurred in Niagara Falls, New York in the late 20th century. The name "Love Canal" was derived from William T. Love, the industrialist who dug the canal that eventually led to the disaster. The cause of the disaster was the dumping of waste of all types ranging from chemical warfare materials, chemical waste and refuse by residents, companies and even the government.[156] In Nigeria, the Koko Toxic Waste incident in 1988 is a good example.[157] An Italian company registered in Nigeria imported toxic waste into a small town in Delta State of Nigeria under the disguise that it was industrial chemical.[158]

Preventive and Mitigating Measures

Hazardous waste disasters can be prevented by regulations and policies especially regulations restricting waste disposal in environmentally sensitive areas.[159] The active monitoring to be done by designated agencies in areas prone to risk. Inspection and maintenance of equipment and proper citing of industry with high risk of hazardous waste.[160]

156 John McNamara "The Love Canal Hazardous-Waste Disaster" The Gilder Lehrman Institute of American History Available in https://www.gilderlehrman.org/content/love-canal-hazardous-waste-disaster Accessed on 7th December 2017.

157 Toxic waste in Koko The Nation March 17, 2017 Available in http://thenationonlineng.net/toxic-waste-koko/ Accessed on 7th December 2017.

158 Segun Gbadegesin "Multinational Corporation Developed Nations and Environmental Racism: Toxic Waste, Exploration, and Eco-Catastrophe "in Faces of Environmental Racism: Confronting Issues of Global Justice. 2nd ed.' ed Laura Westra, Bill E. Lawson (Oxford: Rowman & Littlefield Publishers, Inc., 2001) 188.

159 Timothy J. Flynn, Stuart G. Walesh, James G. Titus, and Michael C. Barth "Implications of Sea Level Rise for Hazardous Waste Sites in Coastal Floodplains" Available in http://citeseerx.ist.psu.edu/viewdoc/download?doi=10.1.1.391.7715&rep=rep1&type=pdf Accessed 7th December 2017.

160 *Ibid*

iv. Human Stampede

Meaning

A stampede is an act of mass impulse amongst a group of people in which the crowd collectively begins running with no clear direction or purpose. Human stampede often occur during religious gatherings and sporting activities. There is a distinction between stampede and crushes. Crushes are very often and mistakenly referred to as stampedes but unlike true stampedes it causes a lot of death. It has been suggested that with crushes, crowd density rather than size is important, with a density of about four people per square meter beginning to be dangerous even if the crowd is not very large.

Types/classification of Stampede

Stampede can be classified according to the loss of lives in the incidence. The classification can then be as stated below:

Class I – mild stampedes with injuries and no death;
Class II – moderate stampedes with one to ten deaths;
Class III – severe stampedes with eleven to hundred deaths;
Class IV – devastating stampedes with 101 to 1000 deaths;
Class V – catastrophic stampedes with over 1000 deaths

In another classification, stampede can be according to who or what was involved in the stampede. There are therefore the following classification of human and non-human/cattle stampede. However, the discussion will be on human stampede.

Human Stampede

Human stampede is a man-made disaster which has not been completely understood, neither has it been fully documented, recorded or researched. The world population has continued to increase and be even more concentrated in urban settings. This has increased the gatherings of large crowds in various venues under different circumstances and with this comes the risk of various crowd disasters ranging from crowd bursting to human stampeding. A sudden rapid movement or reaction of a mass of people in response to a particular circumstance or stimulus is known as a human stampede. The Merriam-Webster dictionary defines it as a 'mass movement of people at a common impulse'.

Causes of Stampede

Acquisitive panic is said to be a cause of stampede. A stampede may occur when a group of people in a bid to lay hold or acquire something as on "black Friday" rush headlong in a panic. It may also occur where a large crowd as a result or panic occasioned by fright to get away from danger as when running away from an explosion or fire in a theatre move headlong. It usually occur during religious, social, musical, political, professional and sporting gatherings and other events that tend to involve a large number of people. They occur in times of panic as a result of fire or explosion as people try to get away urgently. It is also common when the crowd are trying to get towards something.

Effect of Stampede

Human stampede has been linked with dire consequences all over the world and has resulted in several degrees of human injuries and deaths. One key word in the definitions of stampede is the word 'impulse' and often it results in death and injuries. Other effects are structural damage of public facilities, loss of valuables etc. all of these have made it a veritable threat to public safety and a notable disaster among nations.

Areas Prone to the Disaster of Stampede

Any area where there is a sudden, rapid movement or reaction of a mass of people in response to a particular circumstance or stimulus is prone to stampede.

Examples of Disaster as a result of Stampede

On May 30th 1883, after the Brooklyn Bridge was declared open a few days before, there was a stampede on the bridge, after a false alarm that the bridge was collapsing. Hundreds of people ran for their lives and causing a stampede that led to many death.[161] An example of human stampede in Nigeria is the November 2013 stampede in a Church in Enugu town in Anambra State where more than 100,000, people gathered in a church built to contain 5,000 people and many people died as a result.[162] The March 2014 Nigerian Immigration Service interview stampede of more than 6.5 million applicants for 5000 jobs is another example. The Shanghai stampede occurred in December 2014 where 36 people were killed and 47 were injured during a New Year celebration.[163]

Preventive and Mitigating Measures

In a stampede involving livestock, cowboys attempt to turn the moving herd into itself so that it runs in circles rather than running off a cliff or into a river and also to avoid damaging human lives and property.[164] Tactics used to do this include firing a pistol which creates noise causing the leaders of the stampede to turn.[165]

Animals that stampede like cattle are less likely to do so after having eaten and spread out in smaller groups to digest.[166] To

161 Jaya Saxena"130 Years Ago, Elephants Solved Panic On the Brooklyn Bridge" New York Historical society May 29, 2014 Available in http://behindthescenes.nyhistory.org/elephants-panic-brooklyn-bridge-1883/ Accessed on 8th December 2017.

162 Christopher Isiguzo and Emanuel Ugwu "Nigeria: Anambra Stampede - CAN Bans Political Campaigns in Churches" This DayNov 5, 2013 Available in allafrica.com/stories/201311050824.htmlAccessed on 8th December 2017.

163 Jonathan Kaiman"Shanghai: dozens killed and injured in stampede at New Year celebrations" The Guardian I January 2015 Available in https://www.theguardian.com/world/2014/dec/31/shanghai-35-people-killed-42-injured-new-year-crush. Accessed on 7th December 2017.

164 https://ipfs.io/ipfs/QmXoypizjW3WknFiJnKLwHCnL72vedxjQkDDP1mXWo6uco/wiki/ Stampede.html.

165 *Ibid*

166 *Ibid*

further reduce the risk of stampedes, in animals, it is said that cowboys sometimes sing or whistle to calm the herds which are disquieted by nightfall.[167] Those on watch at night avoid doing things which could startle the herd and even distance themselves before dismounting a horse or lighting a match.

Improved understanding of crowd behaviour and more systematic research will help design mitigation strategies that can be implemented across the world to prevent human stampedes

v. War

Meaning

A war can be defined as a state of open, armed, often prolonged conflict carried on between nations, states, or parties or the period of such conflict or the techniques and procedures of war; military science. War is the waging of armed conflict against the enemy or a legal state created by a declaration of war and ended by official declaration during which the international rules of war apply. In 2003, Richard Smalley identified war as the sixth (of ten) biggest problems facing humanity in the next fifty years.

Types/Classification of War

The first classification of the types of war is discussed below:[168] asymmetric war, biological war, chemical war, civil war, conventional war, cyber war, nuclear war, total war, unconventional war, war of aggression.

Wars have also been classified as border wars, cold war, frontline war, proxy war, range war, religious war, total war and nuclear war. [169]

167 *Ibid*

168 "What are the types of wars that exist?" Quora Available in https://www.quora.com/What-are-the-types-of-wars-that-exist Accessed on 7th December 2017.

169 Types of war https://en.wikipedia.org/wiki/Outline_of_war

Causes of War

There is no single cause of war. There are many factors that cause war.[170] They include power tussle, ethnicity, sociocultural differences, elimination of resurgent rivals, economic factors, marginalization, hatred. Generally, poverty and political, social, and economic inequalities between groups may cause war.[171]

Effect of War

Civilian casualties, military personnel suffer mental and physical injuries, and depression, negative socio-economic effects, forced migration and refugee development. Health challenge, brain drains, malnutrition and hunger.

Areas Prone to War

The Middle East has the reputation of being one of the most conflict-prone areas in the world.[172] Other areas include North Africa, Ethiopia, Angola, Israel, and Vietnam.[173]

Example of Disaster as a Result of War

The deadliest war in history in terms of cumulative number of deaths since its start is the second world war from 1939 – 1945 with 60 – 80 million deaths. Proportionally speaking, the most destructive war in modern history is the War of the Triple Alliance which took the lives of over 60% of Paraguay's population. According to Steve Pinker, examples of wars in the world are the Nigerian civil war 1966-1970 also known as the Biafra war, the Iraq Afghanistan war, the Gulf war, the US-Iraq war, Korean war, Vietnam war, and so on. The US-Iraq war of 2003 is another

170 R.J. Rummel "Causes And Conditions Of International Conflict And War" Understanding Conflict And War: Vol. 4:War, Power, Peace Available in https://www.hawaii.edu/powerkills/WPP.CHAP16.HTM.

171 Root causes of violent conflict in developing countries Available in https://www.ncbi.nlm.nih.gov/pmc/articles/PMC1122271/ Accessed on 7th December 2017.

172 Dec 13, 2015 –Available in https://history105.libraries.wsu.edu/.../12/.../the-start-of-the-conflict-prone-middle-eas....

173 Conflict-prone countries The Economist Available in http://www.economist.com/node/5084685.

example of war causing disaster. Officially, the war on Iraq ended in 2011 with the withdrawal of virtually all of America's combat troops. But the havoc wreaked by the US invasion and regime change goes on. Over a hundred thousand Iraqis were killed in the war itself, but many more died in the aftermath from sectarian violence and the obliterated infrastructure. On the American side, the deaths approached 4,500, with tens of thousands shattered in body and spirit. For the U.S. taxpayer, the price is over a trillion dollars, with billions lost to sheer corruption in the so-called rebuilding.

Preventive and Mitigating Measures

These include dialogue, tolerance, planning ahead, establishing peace institutions, involvement of religious groups e.g. the Second Vatican Council, the involvement of anti-war movements, awareness, lobbying and peaceful demonstrations.

Similarities/Differences between Man-Made and Natural Disasters

Natural disasters are brought about by change in natural phenomenon or what is known as acts of God. The extent of loss experienced is dependent on the vulnerability of the population. As such, this means that this can only occur in areas that are susceptible to vulnerability. On the other hand, man-made disasters are influenced by humans and they are often as a result of negligence and human error among other factors.

Despite the difference between these two, it is ideal to note that they can cause irrevocable damage if the right measures are not put in place to avoid the same. This is where the need for disaster preparedness comes in. It goes a long way to cushion people from the after effects of such happenings. There are several sources that provide useful resources that make it possible to meet this end.

Whether the disaster is natural or man-made, the manner in which action is taken goes a long way to determine how people fare from the experience. In both instances, casualties should be treated immediately and the best way to meet this end is placing the necessary measures in place that counteract this. Note that the costs associated with handling of the man-made and natural disasters are enormous every year and this negatively affects the economy.

vi. Terrorism

The act of terrorism is a man-made disaster and it is a key security threat all over the world.

Meaning and Definition

The definition of terrorism has proved controversial and even the United Nations attempt failed.

Terrorism comes from French word *'terrorisme'* in turn derived from the Latin word *'terrere'* meaning 'I frighten', use of violence and treats to intimidate or coerce, especially for political purposes.[174] Common definitions of terrorism refer only to those violent acts which are intended to create fear (terror) are perpetuated for a religious, political or ideological goals and deliberately target or disregard the safety of civilians. All terrorists' acts involve violence or the threat of violence. Terrorists attempt not only to cause panic but also to undermine confidence in the government and political leadership of their target country. Terrorism, is designed to have psychological effects that reach far beyond its impact on the immediate victims or object of an attack.

Types of Terrorism

There are different forms of terrorism, with their own characteristics. Below are some of the common types of terrorism.

174 Wilkinson P. (1974) Concepts of Terror and Terrorism. In: *Political Terrorism. Studies in Comparative Politics.* Palgrave, London Available in https://link.springer.com/chapter/10.1007/978-1-349-15550-7_1#citeas. Accessed on 14th December 2017.

i) *State terrorism* which is original form of terrorism is the systematic use of terror by a government in order to control its population. It is carried out entirely by the group holding power in a country and not a non-governmental organisation.

ii) *Religious terrorism* motivated by religious ideologies and grievances which may be dangerous due to the fanaticism of those who practice it and their willingness to sacrifice themselves for the cause.

iii) *Right wing terrorism* aims to combat liberal governments and preserve traditional social orders and is commonly characterised by militants and gangs.

iv) *Left wing terrorism* seek to overthrow capitalist democracies and establish socialist or communist governments in their place.

v) *Pathological terrorism* is the use of terrorism by individuals who utilize such strategies for the sheer joy of terrorizing others as seen in school shootings and serial killing scenarios.

vi) *Issue-oriented terrorism* is carried out for the purpose of advancing a specific issue. Commonly, these issues are social in nature or deal with the environment.

vii) *Separatist terrorism,* typical of minorities within a nation state that desire their own state, seek to cause fragmentation within a country and establish a new state.

viii) *Narco-terrorism* referred to organisations that gain funds through the sale of drugs and or using violence to make the sale of drugs easier.

Causes of Terrorism

The causes of terrorism can be historical, cultural, political, social, psychological, economic or religious — or any combination of these. However, these causes can be traced down to poverty and marginalization.

Characteristics of Terrorism

Terrorism is characterised by the different stages including planning and organisation stages. All terrorists share one characteristic: they never commit actions randomly and senselessly. It is often targeted at innocent civilians in order to create an atmosphere of fear, intimidation and insecurity and there are weapons of terrorism of which the use of bomb is the most common while other weapons include firearms, chemical and biological weapons.

Areas prone to Terrorism

It is reported that more than 72% of terrorist deaths last year occurred in just five countries mainly in Middle East, Asia and Africa.[175] The countries are Libya, Egypt, Somalia, India, Yemen, Syria, Pakistan, Nigeria, Afghanistan and Iraq.

Examples of Terrorism Causing Disaster

In Africa, recent acts of terrorism caused by groups like Boko Haram in Nigeria, Al–shabaab in Somalia, and Al Qaeda in Algeria are stated to be recent. However, act of terrorism have been recorded before this era. Examples of terrorism include the Sousse beach resort killing of 38 people in Tunisia on June 26th 2015,[176] and the killing of about 150 people in a University campus in Kenya on April 2nd 2015. The 2011 bombing of the United Nations Building in Abuja resulted in the death of 23 people and many injured on August 26.[177]

175 Dominic Dudley "The Ten Countries Most Affected By Terrorism" Forbes Nov 18, 2016 Available in https://www.forbes.com/sites/dominicdudley/2016/11/18/countries-most-affected-by-terrorism/#78b6de5430d9 Accessed on 13th December 2107.

176 Wycliffe Muga"Recent terrorists attacks in Africa" Star Mar. 26th, 2016, Available in https://www.the-star.co.ke/news/2016/03/26/recent-terrorists-attacks-in-africa_c1320034 accessed 14th December 2017.

177 Christopher Bartolotta"Terrorism in Nigeria: the Rise of Boko Haram" World Policy September 19, 2011 Available in http://www.worldpolicy.org/blog/2011/09/19/terrorism-nigeria-rise-boko-haram Accessed on 14th December 2017.

In the United States of America, one of the significant and dramatic terrorist attack took place on September 11th 2001 when the Islamic extremist group al-Qaeda hijacked four airplanes and carried out an attack killing almost 3,000 people.[178]

In the Middle East, one of the deadliest terrorist attack took place in Egypt, in November, when an Islamic militant bombed and sprayed gun fire in a mosque leaving 350 people dead.[179]

The Effect of Terrorism

The effect of terrorism covers a wide area and cannot be over-emphasised. The act of terrorism paralyzes almost all sections, sectors and activities of any community. Terrorism impedes peace and progressive development and hinders political development. It affects rapid economic growth and it distorts socio-cultural equilibrium which may lead to environmental deterioration.

The effect of terrorism results in conflict and leads to dislocation of people from their base resulting in disruption of economic activities. On the social sphere, family welfare services, health, housing, community development are affected.

Other areas affected by terrorism include education, especially in the Boko Haram terrorism in Nigeria where the major target of the boko haram is to frustrate the western education which they regarded as sinful. Agriculture and food supply, security, social amenities and the environment are also affected.

The environment has always formed a central part of the development focus. Conflict lead to environmental degradation, poverty and hindered rapid economic growth and development. Other effects are pollution by war equipment on the environment by gases, chemical and toxics, radio-actives, which may scare away foreign investors and the refugee problems to neighbouring countries and cities.

---

178 Peter L Bergen "September 11 Attacks" Encyclopaedia Britannica available in https://www.britannica.com/event/September-11-attacks Accessed on 14th December 2017

179 Declan Walsh and NourYoussef "Militants Kill 305 at Sufi Mosque in Egypt's Deadliest Terrorist Attack." The New York Times November. 24, 2017, Available in https://www.nytimes.com/2017/11/24/world/middleeast/mosque-attack-egypt.html Accessed on 14th December 2017.

Protective/Mitigating Measures

There are different protective measures. Some are physical or structural in the sense that structures such as protective design of houses, vehicle barriers, perimeter walls, strengthened perimeter columns, blast resistant glazing, are in place. On the other hand, awareness and vigilance in the environment and reporting suspicious activities to the relevant agencies are other measures. There should be adequate response and recovery measures in place

C. Complex Disaster

This is a disaster occurring where there is no single root cause. This is more common in developing countries. A specific disaster may spawn a secondary disaster that increases the impact.[180] A classic example is an earthquake that causes a tsunami, resulting in coastal flooding. One specific event may spawn another type of disaster. For example, a hurricane may cause flooding by excessive rain or by pulling seawater onshore, also called a storm surge. A volcanic eruption is particularly prone to spawning additional events like avalanche, earthquakes, structural fires and wildfires, flooding from melting of mountain snow and ice, landslides, mudslides, thunderstorms and, if the volcano is located in or near the ocean, tsunamis.

Examples of Disasters resulting in Complex Disaster

Indian Ocean Earthquake: The Sumatra-Andaman earthquake, it was estimated to have a magnitude of 9.15, and occurred on December 26th 2004. While the earthquake itself lasted for only 10 seconds, it caused a tsunami that killed an estimated 200,000 to 310,000 people along the shores of Indonesia, Sri Lanka, South India, and Thailand with one death even occurring at Port Elizabeth in South Africa, 8,000 miles away from the epicenter.[181]

180 Ian Davis "Disaster Risk Management in Asia and the Pacific" Available in https://books.google.com.ng/books?isbn=1317644875 Accessed 10th December 2017.

181 http://list25.com/25-worst-natural-disasters-recorded/

Another example of a complex disaster is hurricane Andrew. This disaster was formed by a tropical wave that moved to the West Coast of Africa on August 14th 1992. During the next few days, it developed into a destructive tropical cyclone. The storm was responsible for 65 fatalities and US $26 billion in property damages becoming the 5th costliest hurricane in US history after hurricanes Katrina, Wilma, and Ike.[182] Another Example was Tōhoku Earthquake and Tsunami in 2011. A quake followed by a tsunami was reported to have hit the East Coast of Japan on March 11th 2011 killing 15,894 people with 6,152 injured and 2,562 people missing. It was also said to have caused a near nuclear disaster when there was a partial meltdown in 3 reactors of the Fukushima Daiichi nuclear power plant, which is the 2nd largest nuclear disaster after Chernob.[183]

Food-security emergencies may also result in complex disasters with multiple root causes.[184] Severe drought and/or conflict can produce an acute food emergency, whereas chronic food insecurity is often a reflection of poverty, a worsening debt crisis, and the economic effects at household level of the HIV/AIDS pandemic or mismanagement or abuse of water resources.[185]

Recommendations and Conclusion

The havoc caused by natural and man-made hazards that has been recorded in history has been enormous. On daily basis, a part of the world, experiences a disaster. This points to the fact that disasters are inevitable. However, the magnitude of the effect of disaster can be greatly reduced. There is therefore the need for proper awareness of causes of disasters and what to do in the

182 *Ibid*

183 *Ibid*

184 International Federation of Red Cross and Red Crescent Societies "Complex/ manmade hazards: famine, food insecurity" http://www.ifrc.org/en/what-we-do/disaster-management/about-disasters/definition-of-hazard/famine-food-insecurity/

185 International Federation of Red Cross and Red Crescent Societies (IFRC) "Complex/manmade hazards: famine, food insecurity" http://www.ifrc.org/en/what-we-do/disaster-management/about-disasters/definition-of-hazard/famine-food-insecurity/ Accessed 30th November 2017.

event of a particular hazard as each has its peculiarities. Adequate planning to mitigate against occurrence and effects are germane. Very sound preparation is vital to forestall or reduce the impact of hazards

To aid the recovery process in an event of a disaster, humanitarian aid from not only the big agencies are important. Community response and engagement go a long way especially with respect to essential clothing, food and temporary accommodation to lessen the effect of the disaster. Governmental effort at the planning/preparedness stages before the occurrence of a disaster and the response/ recovery stages at the occurrence and after the disaster is very important in determining the impact of such disaster in a given community.

In any given situation, whether before, during or after a disaster, all hands must be on deck to find a solution either on a short term or long term basis to handle the present and future problems.

2

HUMAN RIGHT-BASED APPROACH TO DISASTER RISK MANAGEMENT

Deborah D. Adeyemo

Abstract

Since the beginning of the 21st century, there has been an increasing trend of disasters caused by natural hazards such as earthquake, tsunami, hurricane, flood etc., and trends in climate change show increasing vulnerability and exposure of people to such disasters. Disasters have huge economic, environmental and social impacts ranging from loss of life, injury, disease displacement and other negative effects on human, physical, mental and social wellbeing, to damage to property, destruction of assets, loss of services, social and economic disruption, and environmental degradation.

While disasters predispose victims (especially vulnerable groups such as children, women and persons with disabilities) to human rights violations emanating from disaster impacts, disaster risk management requires a holistic approach which integrates the rights of victims. Disaster risk management encompasses a systematic process of implementing strategies for preparing, responding and recovery from disaster in order to lessen the adverse impacts of hazard and reduce the possibility of disaster. Thus, human rights protection forms an important part of disaster risk management. The United Nations presupposes that human rights principles should guide disaster risk management, ranging from pre-disaster mitigation and preparedness measures to emergency relief and rehabilitation efforts. Even though there are a number of international soft law instruments concerning the rights of victims in disaster and State obligations in disaster prevention and management activities, the operational context of these rules should be explored in order to ensure an effective disaster risk management regime. Human rights may also provide appropriate mechanisms for compensation for preventable disaster losses. This article considers the subject of disaster risk management from a human rights perspective.

Keywords: Disaster, Disaster Risk Management, Human Rights, Victims

Introduction

Disasters are unpleasant and hazardous events which may take place in moments but have impacts that could last for months, years or even decades. Disasters may be natural-occurring or man-made, associated with factors such as climate change, environmental degradation etc. Winser *et al* stated that the risk involved in disasters are created by vulnerabilities caused by everyday living conditions.[1] United Nations (UN) reports show increasing trends in the frequency and severity of disasters

1 B. Winser, P. Blaikie, T. Cannon, & I. Davis, *At Risk: Natural Hazards, People's Vulnerability and Disasters* (London: Routledge 2004), 18.

globally over the last decade.[2] Recognising that no state is immune from disaster risk, the international community has consistently adopted frameworks on action plans towards disaster risk reduction and strengthening states resilience to reduce disaster risk since 2005. Regional strategies and platforms are also created in realising the goals of these frameworks. These frameworks are to serve as guidelines to state parties in formulating their disaster risk management system. National action plans, laws and policies on Disaster risk management however are viewed from a development perspective with focus on the environment, climate change and less focus on human rights issues involved in disaster management and disaster risk reduction. This chapter examines the concept of disaster risk management from a human rights perspective. It interrogates incorporating human rights as an operational guideline for disaster risk management and to what extent human rights considerations can be implemented in the different phascs of disaster risk management.

This chapter is divided into six parts. The first part is the introductory section. Part two addresses the concept of disaster, disaster impacts and disaster risk management. Part three examines legal instruments concerning disaster risk management. Part four examines human rights issues in disaster. Part six analyses a human approach to disaster risk management while part six presents the conclusion.

Disaster and its Impacts

Disaster

Disaster generally refers to a sudden or calamitous event which has unfortunate consequences. It has been defined as 'a serious disruption of the functioning of a community or a society at any scale due to hazardous events interacting with conditions of

2 United Nations Office for Disaster Risk Reduction (UNISDR) 'The Economic and Human Impact of Disasters in the Last Ten years' Available at <http://www.unisdr.org/files/42862_economichumanimpact20052014unisdr.pdf> last accessed 21st October 2017.

exposure, vulnerability and capacity leading to human, material, economic and environmental losses and impacts'[3] Disasters may happen with a sudden onset such as in the case of natural disasters like earthquakes, cyclones, volcanic eruption, landslides and Tsunamis; it could also have a slow onset as in the case of droughts and environmental degradation.[4] Disaster could also occur as a result of technological or man-made hazards such as industrial accidents and transport accidents. Disasters could be small scale or large scale.[5] In the context of this chapter, disasters are limited to naturally occurring hazards and exclude such man-made disasters occasioned by armed conflicts.

Disaster Impacts

Disaster impacts is the total effect of a hazardous event or disaster which could be positive or negative.[6] The effect of disaster could be immediate and localised[7], but is often widespread and could last for a long period of time. Since disasters are influenced by both natural hazards and human actions/inactions, the impacts of disaster depend largely on vulnerabilities of the victims and its interplay with other social, economic and political factors. Disaster impacts is often measured in terms of the negative effects on the environment, human beings and the society.[8] The level of preparedness and resilience of the state determines how well it is able to absorb the intensity of the impacts of disaster. According to the United Nations, disasters are as a result of the combination

3 United Nations Office for Disaster Risk Reduction, (UNISDR) 'Terminology' (2009) Available at <https://www.unisdr.org/we/inform/terminology> Last Accessed 20th October 2017. UNISDR opine that disaster impact could be positive in terms of economic gains.

4 Elretha Louw & Simon Van Wyk, 'Disaster Risk Management: Planning for Resilient and Sustainable Societies' [2011] *South Africa Civil Engineering* 16.

5 UNISDR Terminology (2009). Available at <https://www.unisdr.org/we/inform/terminology 20th October 2017>. Last Accessed 20th October, 2017.

6 UNISDR Terminology (2009) *Ibid* (n. 3).

7 This is measured in terms of disaster damage.

8 Evaluation of disaster impacts in terms of the negative outcomes/results provides the contextual perspective of the concept of disaster risk.

of conditions of vulnerability; insufficient capacity or measures to reduce or cope with the potential negative consequences; and exposure to a natural hazard.[9] Thus, 'Disaster impacts may include loss of life, injury, disease and other negative effects on human physical, mental and social wellbeing, together with damage to property, destruction of assets, loss of services, social and economic disruption, and environmental degradation.'[10] Broadly, the impacts of disaster could be direct or indirect; or at individual or community levels; economic, physical or social etc. While economic impacts of disasters are usually measured in terms of monetary cost, social expenditure and capital investment or capital stock lost, other factors such as damage to homes, services, public infrastructure, jobs loss, decreased productivity, land degradation etc. also constitute important parameters for measuring economic impacts of disasters.[11]

Physical impacts are the most obvious and easily measured. Physical impacts of disaster include death and injuries, property damage and displacement of victims. One of the most immediate effects of disasters is the displacement of victims. Disasters such as floods, earthquakes, tsunamis etc. may force people out of their residences and homes. Displacement refers to forced movement of persons from their locality or environment and occupational activities.[12] Victims of disasters may be forced to move from areas that have been hit by disaster to areas of relative safety which

9 UNISDR WMO UN System Task team on the Post-2015 UN Development Agenda. 'Disaster Risk and Resilience' (2012) p. Available at <http://www.un.org/en/development/desa/policy/untaskteam_undf/thinkpieces/3_disaster_risk_resilience.pdf> Last accessed 17th October 2017.

10 Dewald Van Niekerk 'Introduction to Disaster Risk Reduction' (United States Agency for International Development (USAID), August 2011) 8. Available at <http://www.preventionweb.net/files/26081_kp1concepdisasterrisk1.pdf > Last accessed 21st November 2017.

11 Internal Displacement Monitoring Centre, 'Global Disaster Displacement Risk: A Baseline for Future Work' (October 2017) (Report) 13

12 United Nations Educational Scientific and Cultural Organisation (UNESCO) 'Glossary of Migration Related Terms': Displacement' Available at <http://www.unesco.org/new/en/social-and-human-sciences/themes/international-migration/glossary/displaced-person-displacement/> last accessed 20th November 2017.

could be within the boundaries of a state (internal displacement)[13] or across international borders of a state. However, most of the disaster displaced victims remain within the borders of their country.[14] Displacement in this context may not necessarily include pre-emptive and voluntary movement or evacuation carried out based on early warnings. Apart from disrupting their normal course of life, disasters also affect victims' access to basic necessities of life such as food, clothing and shelter.[15] Victims stand health risk in the outbreak of disasters especially when their access to basic supplies and infrastructure which are essential to health is seriously affected. The health impacts may be direct and immediate or may be a consequence of the disruption of health system and services. Physical impacts of disaster such as injuries, illness have psychological effect on victims especially women and children.[16]

Disaster-induced displacement may also increase the economic impacts of disasters. Displacement increases the need for emergency measures, costly evacuation and resettlement plans which have huge monetary cost and economic implications.[17] Internal Displacement Monitoring Centre (IDMC) reports that disaster-induced displacement is the least reported impacts of

13 Victims of displacement within a State are referred to as internally displaced persons (IDPs) while victims who have moved out of their States are referred to as refugees.

14 Displacement in the context of this chapter is restricted to internal displacements as opposed to displacement across borders.

15 IDMC suggests that while a high number of displacement may be an issue of concern, displacements may not necessarily constitute negative disasters impact. Pre-emptive movement of people based on timely early warnings may have saved many lives. Internal Displacement Monitoring Centre, 'Global Disaster Displacement Risk: A Baseline for Future Work' (October 2017) (Report) 13.

16 'Emergency Risk Management Health: Overview' (Emergency Risk Management for Health Fact Sheets) (Global Platform May 2013). Available at <http://www.who.int/hac/techguidance/preparedness/risk_management_overview_17may2013.pdf> Last Accessed 20 November 2017; 'Protecting People's health from the Risk of Disasters' (March 2015) Available at <http://www.who.int/hac/techguidance/preparedness/protecting_peoples_health_march2015.pdf?ua=1> Last Accessed 23rd November 2017.

17 IDMC (2017) *op. cit.* (n.11).

sudden-onset disasters.[18] The focus on displacement whether internal or across borders is usually on violence-induced displacement occasioned by armed conflicts. While economic disaster impacts are included in the United Nations Global Assessment Report (GAR), there is little or no report on the human impacts of disaster in terms of disaster-induced displacement.[19] Overall, disaster impacts whether physical, economic or social are ultimately surmised in its impacts on human lives.

DISASTER RISK MANAGEMENT

Risk is an inevitable part of life which applies to many aspects of daily lives at both individual and national levels.[20] The concept of risk broadly covers an assessment of the likelihood of negative outcomes and the efforts to mitigate them.[21] Mitigating efforts could involve individuals, communities, states or regions and international community at large.[22] Disaster risk is the potential or probability of harmful consequences resulting from disaster. Disaster risks are determined by the probability of the combination of hazards, exposure, vulnerability and coping capacity.[23] Disaster risk management describes policies, strategies and plans aimed at reducing existing or prospective disaster risk in order to strengthen state's resilience and reduce disaster impacts.

Disaster risk management (DRM) refers to the 'systematic process of using administrative directives, organisations and operational skills and capacities to implement strategies,

18 Internal Displacement Monitoring Centre, 'Global Disaster Displacement Risk: A Baseline for Future Work' (October 2017) (Report) 5. Sharona Hoffman, Preparing for Disaster: Protecting the Most Vulnerable in Emergencies (2009) 42, University of California Davis 1491, 1494. Hoffman reports that disaster preparedness plans of various States across six different countries reveal that there usually little or no provision for the needs of vulnerable groups in disasters.

19 IDMC (2017) *op. cit.* (n. 11).

20 IDMC *op. cit.* (n. 11) 8.

21 UNISDR (2009) defines risk as the combination of the probability of an event and its negative consequences. USAID (2011) *op. cit.* posits that risk is associated with the coping capacity of humans with a particular situation e.g. natural hazards.

22 IDMC. *op. cit.* (n. 11) 8.

23 UNISDR 'Terminology' (2009) *op. cit.* (n. 3)

policies and improved coping capacities in order to lessen the adverse impacts of hazards and the possibility of disaster.'[24] It encompasses both preventive and responsive approach to disaster management. It is preventive by aiming at lessening the possibility of disaster and responsive in reducing the adverse impact of disasters. While it may be impossible to entirely prevent the occurrence of disasters especially naturally occurring disasters, DRM recognises measures which aim at disaster risk reduction and play an important role in effectively managing the frequency and severity of such disasters.[25]

Thus, disaster risk management incorporates both 'disaster management (DM) which includes preparedness, response and post-disaster recovery processes, and disaster risk reduction (DRR) which principally involves processes that attempt to reduce or control the levels of existing disaster risk and which can also be incorporated into post-disaster recovery.'[26] Consequently, 'Disaster risk management systems refer to the legal, policy, administrative and institutional frameworks established within a country for coordinated and systematic DRM.'[27]

DRM involves activities and measures against disaster hazards aimed at prevention, mitigation, preparedness, response, recovery and rehabilitation.[28] DRM essentially involves these six components which reflect three key phases of disaster risk management. The pre-disaster phase incorporates activities relating to prevention and mitigation while disaster phase includes measures relating to response and the post-disaster phase addresses recovery, rehabilitation and reconstruction measures. The United Nations Office for Disaster Risk Reduction

24 International Federation of Red Cross and Red Crescent Societies (IFRC) and United Nations Development Programme (UNDP) 'The Handbook on Law and Disaster Risk Reduction' (2015) 7.

25 IDMC posits that natural hazard disasters are not entirely attributable to an 'act of God' occurring in vacuum and that mitigation effort can reduce disaster risk and loss. Internal Displacement Monitoring Centre, 'Global Disaster Displacement Risk: A Baseline for Future Work' (October 2017) (Report) 8.

26 IFRC 'The Handbook on Law and Disaster Risk Reduction' *op. cit.* (n. 24) 7.

27 UNISDR 'Terminology' (2009) *op. cit.* (n 3).

28 IFRC 'The Handbook on Law and Disaster Risk Reduction' (2015) *op. cit.* (n. 24), 7.

(UNISDR) broadly categorises disaster risk management actions into prospective, corrective or compensatory.[29] Prospective DRM activities are preventive measures which address disaster risks reduction policies to avoid prospective or increased disaster risk. Corrective activities address prevailing disaster risks which are required to be managed and reduced. Compensatory activities focus on strengthening the social and economic resilience of individuals and societies in the face of residual risk that cannot be effectively reduced.

Disaster risk management is therefore a multifaceted approach which cuts across various fields and sectors.[30] DRM infuses issues on development, climate change, environment and health etc. In line with the multi-faceted approach to DRM, human rights considerations must maintain high priority in disaster risk reduction as much as in disaster management. Displacement risk is an inevitable element of disaster risks which must be adequately incorporated in designing effective disaster risk management regime in any state. Climate change in combination with other factors, is projected as an imperative risk driver of displacement.[31]

The United Nations reports that globally, 14 million people are likely to be displaced by weather related and metrological disasters yearly.[32] Therefore disaster risk management measures must envisage and adequately provide for displacement risk and other forms of

29 UNISDR 'Terminology' (2009) *op.cit.* (n. 3).

30 'Disaster Risk Reduction: A Preparedness Approach' in Pardeep Sahni, Madhavi Ariyabandu (Eds.) *Disaster Risk Reduction in South Asia* (PHI Learning Pvt Ltd. 2003) 132- 144; The South African Disaster Risk Management Act 57 of 2002 recognises disaster management as a continuous and integrated multi-sectoral and multi-disciplinary process.

31 The Assessment Report of the IPCC Working Group II reports that climate change over the 21st century is projected to increase the displacement of people. Intergovernmental Panel on Climate Change (IPCC) Working Group II AR 5 Climate Change 2014: Impacts, Adaptation and Vulnerability Summary for Policy Makers (IPCC 2014) Available at <http://www.ipcc.ch/report/ar5/wg2/docs/WGIIAR5_SPM_Top_Level_Findings.pdf> Last accessed 23 November 2017.

32 UN News Centre 'On International Day: UN forecasts 14 million people made homeless each year by disasters' (13 October 2017) Available at <http://www.un.org/apps/news/story.asp?NewsID=57881#.Wey6sDtrzIU> Last accessed 21st October 2017.

physical disaster impacts such as death, injuries in the three main phases of disaster risk management. The risk of disaster-induced displacement is largely determined by the exposure of people in terms of 'the situation of people, infrastructure, housing, production capacities and other tangible human assets located in hazard-prone areas'[33] to vulnerabilities.[34] Vulnerabilities are measured in terms of 'physical, social, economic and environmental factors or processes which increases the susceptibility of an individual or a community, asset or systems to impacts of hazards.'[35] A multi-faceted approach to disaster-induced displacement risk, must incorporate a human rights approach beyond the conventional humanitarian response to disaster-induced displacement as a post-disaster recovery measure.

Disaster Risk Management Laws

International Legal Instruments on Disaster Risk Management

In 2005, states globally under the auspices of the United Nations adopted Hyogo Framework for Action (2005-2015): Building the Resilience of Nations and Communities to disasters (HFA) in Hyogo Japan.[36] The Hyogo Framework was designed as a ten-year plan aimed at reducing vulnerabilities to hazards which cause disasters. The HFA outlined five priorities for action under which States committed themselves towards disaster risk reduction by developing a legal framework (policy, legislative and institutional) for their states and implementing such framework.

33 United Nations General Assembly, Report of the Open-ended Intergovernmental Expert Working Group on Indicators and Terminology relating to Disaster risk Reduction A/71/644 I December 2016 Available at <https://reliefweb.int/sites/reliefweb.int/files/resources/50683_oiewgreportenglish.pdf> Last Accessed 20 November 2017.

34 Intergovernmental Panel on Climate Change (IPCC) Summary for Policymakers in Managing the Risk of Extreme Events and Disasters to Advance Climate Change Adaptation, (Cambridge University Press 2012) Available at <https://www.ipcc.ch/pdf/special-reports/srex/SREX_Full_Report.pdf> Last accessed 20 November 2017.

35 UNISDR 'Terminology' (2009) *op. cit.* (n 3).

36 Adopted by the UN General Assembly in Resolution A/RES/60/195.

At the expiration of the HFA in 2015, the Sendai Framework for Disaster Risk Reduction (2015-2030) was adopted as a fifteen-year plan with four action plans and seven global targets. The framework aims primarily at substantial reduction of disaster risk and disaster impacts. It recognises state parties as the primary duty bearers in disaster risk reduction. State parties agreed to implement the Sendai Framework with a clear focus of shifting from managing disaster to managing disaster risk.[37] The Sendai framework global targets are reflective of the importance of recognising human rights in disaster risk management. While global targets 1 and 2 focus on reducing global disaster mortality rate and the number of persons affected by disasters generally, global targets 4 and 7 also aimed at reducing disaster damage to basic health and educational services and providing access to risk information.[38] Regional strategies and platforms are created based on the framework to strengthen commitment and implementation. These frameworks are however soft laws which are non-binding and create only voluntary obligations without specific obligation in recognising human rights.

National Disaster Risk Management Laws

National Disaster Risk Management laws capture the whole body of institutional and policy framework of the disaster management system. These laws could be in form of a single dedicated law or a collection of interacting laws which set out different aspects of disaster risk management system. In Nigeria, the National Emergency Management (Establishment) Act 1999 is the main disaster management law. It established the National Emergency Management Agency (NEMA) at the federal level and makes

37 UN News Centre 'With focus on natural disaster UN Risk Reduction forum open in Mexico' (25th May 2017) Available at <http://www.un.org/apps/news/story.asp?NewsID=56843#.Wey6CTtrzIU> Last Accessed 21st October 2017.

38 Available at <http://www.unisdr.org/we/coordinate/sendai-framework>Last accessed 20th October 2017.

provision for equivalent agencies and laws at the state[39] and local[40] government levels. Apart from this federal law, there is no single comprehensive law which addresses disaster risk management in Nigeria, especially as it concerns disaster risk reduction. There are however other specific laws which are designed to address specific areas relating to disaster risk reduction. These laws focus on specific aspect such as the environment, climate and land use etc.

There are also national policies on disaster management such as the National Disaster Response Plan (NDRP) 2010 and the National Disaster Management Framework (NDMF) 2010 which were set out to incorporate policies on disaster risk reduction as set out in the Hyogo Framework priorities.[41] The lingering questions however are: do national laws and policies integrate a human right approach to disaster risk management? How well do they incorporate the main concerns of individual disaster victims in all the phases of disaster risk management? While it can be argued that national laws significantly address disaster risk, vulnerabilities to such risk, coping capacities of the state and disaster impacts on the community, there is a need to look beyond community response and focus on individuals who are affected by disaster impacts. Even though the international framework create non-obligatory duties, states may be obliged to implement the guidelines of the framework where the priorities are viewed from a human rights standpoint.

39 State Emergency Management Agency (SEMA). Some states of the federation have equivalent provisions adapted from the federal Act.

40 Local Government Emergency Agency (LEMA).

41 International Federation of Red Cross and Red Crescent Societies, 'Background Report Law and Regulation for the Reduction of Risk from Natural Disasters in Nigeria: A National Desk Survey' (International Federation of Red Cross and Red Crescent Societies October 2012). Available at <http://drr-law.org/resources/Nigeria-Desk-Survey.pdf> Last accessed 21st October 2017.

Disaster and Human Rights

Human Rights in Disasters

The frequency and magnitude of disasters have increased over the years and climate change has had an increasing influence on disaster trends in the recent decades.[42] In 2015, the Centre for Research on Epidemiology of Disasters (CERD) Annual Disaster Statistical Review shows an increasing trend of natural disasters globally with 376 natural disasters with over 110 million victims globally.[43] IDMC reports that since 2008, it had recorded more than 3800 disaster events in over 170 countries and territories.[44]

Disaster impacts, whether economic, social or physical, are borne by people either as direct victims or indirect victims. While disaster may be a development issue, it is as much a human rights issue because human rights are at risk generally at disaster times. Beyond the impacts already identified above, disasters pose great humanitarian challenge and problem and this has increased the need to ensure human rights protection in times of disasters. Disasters impacts impinge on the human rights of victims. Direct victims suffer violation of their basic human rights through death,

42 Cyclone in Myanmar (2008), earthquakes in Japan (2010), Haiti 2010, and Nepal (2015) are only a few of the high impact disasters which have attracted global attention; Worldwatch Institute quoting Munich Re data reports that the frequency of water-related disasters such as flood and wind storms attributed to climate change have increased six-folds since the 1950s' while the frequency of other types of disasters have risen only slightly. Available at <http://www.worldwatch.org/node/4127> Last accessed 23rd November 2017; U.S. Climate Resilience Toolkit reports that increased extreme weather and climate conditions have increased the risk and vulnerability of certain categories of human population disaster risk and eventually disaster impacts. 'Disaster Risk Reduction' (Excerpts from the Report on Climate Change Impacts in the United States: The Third National Climate Assessment Available at <https://toolkit.climate.gov/topics/tribal-nations/disaster-risk-reduction> Last accessed 23rd November 2017.

43 Debarati Guha-Sapir Philippe Hoyois and Regina Below 'Annual Disaster Statistical Review 2015: The Number and Trends' (ZCERD October 2016) Available at <https://reliefweb.int/sites/reliefweb.int/files/resources/ADSR_2015.pdf> Last Accessed 23rd November 2017. The report shows that in China, there were 36 incidences of natural disaster consisting of 17 storms, 13 floods and landslides, 5 earthquakes and an incident of drought. In the United States, 28 disaster incidents, 21 in India 15 in Philippines and 10 in Indonesia.

44 IDMC (2017) op. cit. (n. 11) 6.

injuries, displacement, loss of body parts or means of livelihood etc. Victims of disasters, especially vulnerable groups face huge human rights challenges and their rights are often overlooked during disasters.

Vulnerable Groups in Disasters

Certain groups are regarded as vulnerable in times of disaster. These group of persons may have specific needs which render them more vulnerable than others. Women, children, older people, internally displaced persons (IDPs), people living with disabilities or HIV/AIDS, ethnic minorities etc. have been recognised as vulnerable. This is because they have less physical strength and weaker immune system. The elderly are also regarded as vulnerable because they are more dependent on others for survival. These group suffer the impacts of disaster more than any other and are more likely to suffer discrimination in the post recovery measures of DRM. In post-disaster periods, IDPs especially women and children are highly susceptible to gross violations of their human rights. IDPs move from places of disaster to places of relative safety while some are lodged in camps established in areas where they have restricted access to food, shelter and health services. The Guiding Principles on Internal Displacement[45] recognise that these set of people have a right to protection and assistance required by their condition and treatment which takes into account their special needs.[46] States have the obligation to provide such protection and assistance.

Internal displacement of disaster victims, especially the vulnerable groups poses a great risk of violations of human rights. Victims often face issues such as enforced relocation, discrimination in aid provision, sexual and gender based violence, unsafe and involuntary return to settlement etc. Since 2008, disasters have displaced an average of 26.4 million people annually and in 2014, more than 19.3 million people in 100

45 Adopted by UN General Assembly in 2005 by resolution A/60/L.1.
46 *Ibid.*

countries were displaced from their homes mostly within their own countries.[47] The United Nations reports that in 2016, 445 million people were affected by disasters globally out of which 8000 people lost their lives.[48] IDMC reports that 24.2 million new disaster-induced displacement were recorded in 2016 and as at September 2017, more than 8 million people were displaced by disaster. Internally displaced persons (IDPs) as a result of disasters usually do not have access to basic economic, social and cultural rights. Disasters may deprive the rights of IDPs to access health care, education, employment and social amenities. While such human rights violations are often unintended and attributed to lack of sufficient resources and capacity to prepare and respond to disasters and their aftermath, such violations could be a result of failure to recognise the role of human rights in national policies and laws on DRM.

International Legal Rules on Human Rights and Disasters

Human rights principles of international human rights law and international humanitarian laws are applicable to persons generally, irrespective of whether they are affected by disasters or not. During disasters, human rights principles should be made applicable to disaster victims especially in spite of their plight. The fact that they are victims of disaster does not detract from their human rights guaranteed under the various legal instruments. These essentially include their basic civil, political, social, economic and cultural rights guaranteed by the principal international human rights documents such as the Universal Declaration of Human Rights,[49] International Covenant on Civil and Political Rights,[50]

47 World Disaster Report: 'Resilience: Saving Lives today and Investing Tomorrow' (IFRC 2016) 21. Available at <http://www.ifrc.org/Global/Documents/Secretariat/201610/WDR%202016-FINAL_web.pdf> Last accessed 23rd November 2017.

48 UN News Centre 'With focus on natural disasters, UN risk reduction forum opens in Mexico' (15th May 2017) Available at <http://www.un.org/apps/news/story.asp?NewsID=56843#.We0IbDtrzIV> Last accessed 21st October 2017.

49 Adopted by the UN General Assembly on 10th Dec. 1948

50 Adopted by the Un General Assembly on 16th Dec. 1966

International Covenant on Economic Social and Cultural Rights[51] and African Charter on Human and Peoples' Rights.[52] Victims in fact, may have specific protection and assistance needs, distinct from persons not directly affected by the disaster. International humanitarian law and human rights law have limited provisions in the context of disaster displacement and IDPs do not enjoy the same human rights provisions as refugees.

In 2007, the United Nations International Law Commission (ILC) began working on draft article on the protection of persons in the event of disasters.[53] The draft articles aim among other things 'to facilitate adequate and effective response to disaster that meets the essential need of persons concerned with respect to their rights.'[54] While this draft articles are still a work in progress, draft articles 5 specifically provides for the respect and protection of the rights of victims of disasters.

The operational guidelines on human rights and natural disasters adopted by the United Nations Inter-Agency Standing Committee (IASC) in 2006 and subsequently revised in 2011 are meant to complement existing humanitarian standards on situations of natural disaster to ensure protection of victims, principles of non-discrimination and accountability in disaster response, recovery and reconstruction efforts.

International human rights should therefore form part of the guidelines for disaster risk management. The Guiding Principles on Internal Displacement[55] while recognising the role of non-state actors, imposes legal obligation on states in respect of persons displaced by natural disaster. The guiding principles, drawing on basic principles of international human rights and humanitarian law protects the rights of victims to compensation, participation and return.[56]

51 *Ibid.*

52 Adopted 1st June 1981.

53 UN General Assembly Resolution 62/66 of 6th December 2007.

54 Draft Article 2.

55 Adopted by UN General Assembly in 2005 by resolution A/60/L.1.

56 Principle 4 of 'the Guiding Principles.'

Disaster Risk Management and Human Rights

Whether as development, health or environmental issue, DRM can be viewed from the rights perspective. Individuals have basic civil rights such as; the right to life, right to human dignity, or economic, cultural and social rights such as; the right to housing, the right to a safe and healthy environment, the right to health or more controversial rights such as the right to development. Generally, States have the sacrosanct three fold duty to respect, protect and fulfil these rights. Specifically, states where such disasters occur are the primary duty bearers and they have obligations under international human rights principles to discharge these obligations. In respect of disaster risk management and all its components, human rights are essential in the three phases of disaster risk management. It is important to incorporate a right based approach to disaster risk management so as to focus on the needs of the individuals before, during and after disasters. Disaster impacts have more severe and intense human impact than any other element that make up the society. While disaster impacts may affect humans, the environment, economic situation etc. Overall, these impacts eventually have greater human impacts over all other forms of impacts. Thus, human rights of individual should form the underlying principles of disaster risk management whether in disaster risk reduction or disaster management. The focus should not be restricted to disaster management and the humanitarian approach to the plight of disaster victims in terms of reliefs offered as a component of response and post-disaster recovery process of DRM.

Human rights principles could form the underlying basis for pre-disaster DRM activities in respect of prevention, preparedness and mitigation. In this respect, the human rights may be broadly categorised into two groups; the right to protection and the right against discrimination while other rights are subsumed in the two broad categories.[57] Recognition of certain basic and fundamental

57 This is in line with the provision of the United Nations Inter-Agency Standing Committee's Operational Guidelines on Human Rights and Natural Disaster

individual rights in DRM activities is an important consideration in fulfilling states' obligation towards disaster risk reduction. In states where certain categories of human rights especially economic, social and cultural rights as provided by international and regional human rights documents are either not recognised or non-justiciable, human rights principles guaranteeing the right to food, shelter, health and work; the right to access information relevant to disaster risk and the right to safe environment could be implied from broader and basic human rights such as; the right to life and human dignity. Human rights considerations create pre-disaster responsibility on the part of states to ensure that they regard their obligations towards risk reduction in terms of prevention, mitigation and preparedness. State will not only ensure that they put measures in place towards building risk resilience but also ensure that such measures are implemented to protect human rights of persons within their territory.

In pre-disaster situations, the right to life may serve as a legal basis for a pre-disaster risk DRM measure toward reducing the risk of exposure and vulnerabilities of individuals. The right to life places a corresponding obligation on States to ensure that persons within their territory are protected from disaster risks and hazards and their right to access information on early warning signs regarding metrological and hydrological hazards. While it may be impossible to absolutely safeguard against disaster-induced displacement, prospective disaster risk management must make adequate provisions for the protection of displaced persons and permanent relief for resettlement. Temporary, planned evacuation may be a pre-disaster management measure which may ensure that the human rights are not grossly violated in disaster period, whereas sudden displacement is chaotic, disorganised and uncertain.[58]

During disasters, mitigation, preparedness and rehabilitation measures should also incorporate human rights principles aimed

58 World Disaster Report: 'Resilience: Saving Lives today and Investing Tomorrow' (IFRC 2016) 22. Available at <http:// www. ifrc. org /Global / Documents/Secretariat /201610/WDR%202016-FINAL_web.pdf> Last accessed 23rd November 2017.

at protecting disaster victims and re-integrating them into the society.

First, disaster victims have the right to protection which includes their right to survival and physical security.[59] This covers victims' right to life and human dignity. Their right to survival incorporates the right to basic necessities of life such as food, water, shelter, adequate clothing and health services etc. [60] Victims' right to survival and physical security will only be fully realised where other basic rights such as the right to education, the right to work, right to receive restitution or compensation for lost property and such other related rights are guaranteed. Thus, relief and assistance extended to victims displaced by disasters are not just humanitarian reliefs but they are entitlements of the victims.

Post-disaster human rights concerns have been linked to inadequate disaster risk prevention and mitigation in terms of planning and preparedness which overlooks human rights guarantees.[61] In post-disaster situations, states also have an obligation to prevent re-occurrence of the violations of these rights and to ensure reparations and full rehabilitation.[62] Victims should be protected from secondary hazards which may predispose them to further disaster risk.

Second, victims have a right to non-discrimination. Victims must be protected against any kind of discrimination on the basis of their sex, colour, religion, language, age, disability etc. Disaster victims should enjoy the same rights as those who are not affected. The right to non-discrimination protects vulnerable groups as much as other persons.

59 Draft Article 4 of the Protection of Persons in the Event of Disasters provides for the protection of human dignity in the event of disaster; Operational Guidelines on Human Rights Manual (2011) Ibid., 8.

60 *Ibid.*

61 'Human Rights and Natural Disaster Operational Guidelines and Field Manual on Human Rights Protection in Situations of Natural Disaster' (Brookings-Bern Project on Internal Displacement March 2011), 2. Available at

62 *UN Guiding Principles on Internal Displacement 2005.*

National Human Rights Laws and Disaster Risk Management

While DRM laws do not necessarily enumerate or provide for human rights, national constitutions and specific human rights laws generally mirror international human rights obligations of states. Constitutional provisions on human rights may serve as the basis for legal responsibility and liability of national government in respect of failure to fulfil their obligations on disaster risk reduction. As regards recovery and rehabilitation, constitutional provisions regarding economic and social rights may provide the essential legal framework for DRM activities such as compensation and re-integration of disaster victims. Sendai Framework priority 4 recognises recovery, rehabilitation and reconstruction as important phase in DRM. Specifically, such rights as the right to good food and healthy environment, right to access risk information are pivotal to DRM in realising priority 4. National DRM focus on regulating such issues as land use and urban planning, construction, building, environmental and resource management, climate change, health and safety standards etc. The laws regulating these sectors are aimed at reducing disaster risk and ensuring that people are safer.

The principle of *Ubi jus ibi remedium (where there is a right, there is a remedy)* ensures that where there is a breach of these right, there should be a remedy for such breach. This would provide a mechanism for compensation for preventable disaster losses and also ensure accountability and legal responsibility on the part of states.

Conclusion

Disasters and disaster risks are increasing phenomena globally. The focus of international and national actions and policies have been disaster risk reduction by addressing the various factors that predispose societies to disasters with little or no consideration for human rights especially of those directly affected by disasters.

Victims' protection are considered as entirely humanitarian in terms of assistance and relief aids to victims, however the protection of victims is not just humanitarian but right based. Thus, the right to protection should be reflected in pre-disaster and much as in post disaster phase. While no human rights principle is absolute and derogation of these rights is permitted in exceptional and unforeseen circumstances, with the frequency and severity of the trend of disasters over the years, disasters are no longer considered exceptional. As such, adequate measures towards disaster risk reduction and prospective disaster risk management is an imperative in any society. DRM laws and policies should adopt an integrated approach which recognises human rights principles before, during and after disasters as guiding principles for disaster risk reduction and disaster management.

3

INTERNATIONAL LAW AND DISASTER MANAGEMENT

Idiat F. Akande

Abstract

This chapter, entitled International Law and Disaster Management is aimed at studying the relevant institutions responsible for the management of natural disasters in international law with a view to understanding the adequacy or otherwise of such institutions which include among others, the African Union Convention for the Protection and Assistance of Victims of Disaster; International and Regional Instruments on Human Rights; United Nations High Commission for Refugees (UNHR); International Committee of Red Cross (ICRC), and the current Sendai Framework for Disaster Risk Reduction 2015-2030. The management of natural disaster is of great significance in international law of recent due to the overwhelming increase of situation of disaster constituting serious challenge of national and global implications. Such implications manifest themselves in the violations of human rights of persons that are particularly affected by the disaster. Persons particularly affected by disasters are vulnerable and often victims of violations of human rights because they suffer from many things such as loss of homes, employment, threat to life and liberty, deprivation of food, adequate health care and education. All these implications call for the need for disaster risk management not only at the local level but also at the international and regional levels which is the concern of this chapter. Therefore, the objective of this chapter is to define institutions that are measurable, flexible and adaptable to coordinating structures which deals with key roles so that in the final analysis, the efficacy or otherwise of the studied institutions will serve as a lesson for domestic implementation in Nigeria. Cumulatively, this chapter justifies itself by providing a basis as a way forward to managing and addressing the challenges of victims of disaster in all its ramifications at all levels of governance, that is international, regional and local levels.

Keywords: International Law, Human Rights, Disaster, Victims and Management

Introduction

This chapter provides an international perspective that focuses on disaster risk management and its evolution. It highlights some of the main aspects of humanitarian disaster response training in relation to health specialists' approach which is intended to reduce the health consequences linked to disasters globally. The first section explores how disaster management has evolved to its present status, while the latter section explores avenues for reducing disaster negative impact. Examples of developments in emergency medicine education and research are also discussed.

Literarily, disaster is perceived as hazard. Hazard is defined in the Hyogo Framework for Action as:

> A potentially damaging physical event, phenomenon or human activity that may cause the loss of life or injury, property damage, social and economic disruption or environmental degradation. Hazards can include latent conditions that may represent future threats and can have different origins: natural (geological, hydrometeorological and biological) or induced by

human processes (environmental degradation and technological hazards).[1]

In specific terms, the Internal Displacement Monitoring Centre (IDMC), Norwegian Refugee Council defines disaster as a serious disruption of the functioning of a community or a society causing widespread human, material, economic or environmental losses which exceed the ability of the affected community or society to cope using its own resources. It results from the combination of hazards, conditions of vulnerability and insufficient capacity or measures to reduce the potential negative consequences of risk.[2] Section 2 (2) of the National Emergency Management Act[3] defines disasters to include any disaster arising from any crisis, epidemic, drought, floods, tornado, earthquake, train, aircraft or other accidents and mass deportation or repatriation of Nigerians from abroad. Thus, in broad categories, some of these disasters may further be categorised as natural or man-made in the global contexts.

The Development of Disaster Risk Management in International Law

This section shows the evolution of disaster risk management over the past years. Thus, the following stages of developments in international practice show commitment in disaster management and the research gaps that occur at intervals of development process:

a. *The Late 20th Century Disaster and 1976 Management Measure*

One of the most distressing phenomena in the post-cold war (of the 20th century) is the substantial occurrence of disaster

1 World Conference on Disaster Reduction 2005-2015, UNGA Resolution 46/182 (accessed 4th October, 2017).

2 Internal Displacement Monitoring Centre (IDMC),www.internal-displacement.org, (accessed 4th October, 2017)

3 National Emergency Management Act, 1999

resulting in the displacement of persons within their homelands.[4] Today, there have been monumental increase in the report of disaster events. While some of these disasters are rapid, others are slow-onset resulting in catastrophic situations leading to loss of lives, property and degradation of environment. This disaster occur in form of drought, desertification, flooding, epidemics, coastal erosion, building collapse, oil spillage, volcanic erosions, tornados, hurricanes, wildfire, maritime collision and earthquakes. In 1976, a major shift took place in the field of humanitarian disaster response. Several disasters occurred in a relatively short period of time in the same geographical region. The most significant of these were earthquakes impacting Peru in 1970, Nicaragua in 1972, and Guatemala in 1976. These caused significant devastation and loss of life, and the ministers of health in Latin America and the Caribbean subsequently called for changes in the international humanitarian response mechanism. Until that point, disaster response was mostly reactive, both at the national and international levels. In the 2010 Haitian earthquake, more than 400 health and medical groups provided services. While several were excellent, many of them were of questionable skills and efficiency, and some may have actually inflicted harm.[5]

More recently, the approval of the International Health Regulations in 2005 by the World Health Assembly empowered public health officers, infectious disease specialists, and epidemiologists to manage evolving epidemics with the potential to reach catastrophic levels such as seen in the 1918 "Spanish flu." The disciplines of public health and emergency medicine have both made substantial contributions to the field and are now intimately linked. Subsequently, a growing number of professionals have systematically investigated disasters from a multidisciplinary and multihazard perspective.[6]

---

4 Protecting Internally Displaced Persons: A Manual for Law and Policymakers Brookings Institution University of Bern, Project of Internal Displacement.

5 *Ibid*

6 *Ibid*

b. The Period between 1987-1989

In 1987, the UN Assembly adopted a resolution launching the International Decade for Natural Disaster Reduction.[7] Its goal was to reduce loss of life, property damage, and social and economic disruption caused by disasters, especially in developing countries. The resolution establishing the International Decade for Natural Disaster Reduction was implemented in 1990.[8] The concept of mitigation was born.

Later, the mitigation approach helped produce the concept of risk reduction, which recognises the importance of moving beyond preparedness. Risk can be defined as a function of the hazard and vulnerability in which the hazard is an environmental (e.g., earthquake or hurricane), technological (e.g., chemical or radiological accident), or political (e.g., war or civil strife) event. The essential idea of mitigation focuses on separating the hazard (an earthquake or biological agent) from the vulnerability of the institution or the system. If a building collapses, it is not attributed to the earthquake; rather, it is a consequence of poor building design or failure to use appropriate shake-resistant construction techniques.

Since the late 1980s, an ongoing effort has existed in the health sector, especially in Latin America and the Caribbean, to protect health facilities so that life-saving functions can continue after a disaster. In the beginning, efforts centred on mitigating and refurbishing health facilities. Currently, the approach includes a more comprehensive vision, not just focusing on the construction aspects (structural and nonstructural dimensions), but also considering the functional aspects of a hospital.[9] In this context, "functional" refers to all organisational components needed to provide service.

7 *Ibid*
8 *Ibid*
9 *Ibid*

c. *The Period between 1990 to date*

In the late 1990s and early 2000s, the increasing engagement of financial institutions in risk reduction opened the door to considering new incentives and justifications for such activities in addition to typical health-centred metrics such as lives saved. Participating institutions included the World Bank through its Global Facility for Disaster Reduction and Recovery and regional banks in Asia and Latin America, so as to assist in financial considerations.[10]

Additional medical tools have also been developed to improve the quality of patient care. These instruments provide guidance on how to ensure quality care in disaster situations. For example, the U.S. Institute of Medicine within the National Academies has proposed a series of protocols referred to as "Crisis Standards of Care: A System Framework for Catastrophic Disaster Response."[11] Many organisations have succeeded at improving the quality of care under catastrophic conditions. In 2012 and 2013, groups such as the ICRC (the International Committee of the Red Cross), MSF (Medecins Sans Frontieres), and AusAID (Australian Agency for International Development) have developed their own guidelines, especially for the delivery of care during armed conflicts.

The UN launched a 2-year campaign in 2008 called "Hospitals Safe from Disasters" to ensure that these institutions are prioritised in reducing their vulnerability to hazards. Spearheaded by WHO and the UN International Strategy for Disaster Reduction (UNISDR), "the campaign focused on structural safety of hospitals and health facilities, on keeping health facilities functioning during and after disasters, and on making sure health workers are prepared when natural hazards strike."[12]

10 *Ibid*

11 *Ibid*

12 *Ibid*

An Overview of the Rights of Victims of Disaster in International Law

Victims of disaster are governed by the same human rights instrument applicable to every individual regardless of the position he/she finds himself/herself. Human rights instrument guaranteed the enjoyment of the following rights namely, right to life, right to human dignity, right to personal liberty, right to fair hearing, right to private and family life, right to freedom of thought, conscience and religion, right to freedom of expression and the press, right to peaceful assembly and association, right to freedom of movement, right to freedom from discrimination, right to acquire and own immovable property, and the right to access to court.

The above rights are promoted and protected by the following international instrument at various levels namely, Universal Declaration of Human Rights (UDHR),[13] International Covenant on Civil and Political Rights (ICCPR),[14] the International Convention on Economic, Social and Cultural Rights (ICESCR),[15] the First Optional Protocol to the ICCPR,[16] the Second Optional Protocol to the ICCPR,[17] International Convention on the Elimination of all Forms of Violence Against Women (CEDAW),[18] the International Convention on the Rights of the Child (CRC),[19] International Humanitarian Law,[20] International Refugee Law, the Guiding Principles on Internal Displacement, AU Convention for the Protection and Assistance of Internally Displaced Persons

13 Adopted by the United Nations General Assembly in 1948.

14 Adopted by the United Nations General Assembly in 1966 and enter into force in 1976.

15 *Ibid*

16 *Ibid*

17 Adopted by the United Nations General Assembly in 1989 and entered into force in 1991.

18 Adopted by the United Nations General Assembly in 1979.

19 Adopted by the United Nations General Assembly in 1989 and enter into force 1990.

20 Comprising of the four Geneva Conventions of 1949 and the two additional protocol to the Geneva Conventions. This aspect of the law is principally classify into two viz, the Hague and the Geneva Law.

in Africa (Kampala Convention),[21] African Charter on Human and Peoples Rights,[22] the United Nations High Commission for Refugees (UNHCR), United Nations Convention Relating to the Status of Refugees[23] and the International Committee of the Red Cross (ICRC).[24]

At the domestic level, human rights of person are promoted and protected by the following instruments namely, the Constitution Federal Republic of Nigeria, the National Human Rights Commission,[25] National Disaster Framework,[26] National Commission for Refugees (NCR)[27], National Emergency Management Agency (NEMA),[28] National Policy on Internally Displaced Persons (IDPs)[29] and the Nigerian Red Cross Society.

By virtue of the above instruments, victims of disaster are entitled to the same right like any other individual in the society. However, the conditions they find themselves due to the prevailing disaster on them make them to be vulnerable to enforce their rights. For this reason, there is a gap in the law for the promotion and protection of their rights. Therefore, the attempt to address such gaps in the law consequently brings about another area of study which is the study of disaster management aiming at enforcing the protection and the promotion of the rights of victims of disaster, examined below.

---

21 Adopted by Member States of Africa Union (AU) in Oct. 2009.

22 Adopted by Heads and Government of Africa State in Kenya, Nairobi in 1981 and entered into force in 1986.

23 Adopted by the United Nations General Assembly in 1951.

24 Established in 1864 by Henry Dunant.

25 Established in 1995.

26 Established in 2011 to provide mechanism that serves as a regulatory guideline for effective and efficient disaster management in Nigeria.

27 Established in 1989 by virtue of Decree 52 of 1989. Currently cited as National Commission for Refugees Act Cap 21 LFN 2004.

28 Cap N34 LFN 2004.

29 Established July 2012.

International Legal Framework for Disaster Risk Management

International Legal Framework for Disaster Management includes conventions, charters, guidelines, treaties/agreements, code of conducts, action plan, resolutions and declarations all of which aimed at promoting welfare and development, saving human lives and alleviating suffering, and safeguarding the dignity and rights of the vulnerable. Persons particularly affected by disasters are vulnerable and often victims of violations of human rights because they suffer from many things such as loss of homes, employment, threat to life and liberty, deprivation of food, adequate health care and education, all of which the aforesaid legal framework sets out to address not only at the local level but also at the international level.

i. International Decade for Natural Disaster Reduction 1990 - 1999[30]

The Natural Disaster Reduction calls on states to institute disaster reduction policies and strategies.This further calls for an effective coordinating policies and practices for disaster risk management by states. By so doing, an effective orientation based on an understanding of disaster risk in all its dimensions of vulnerability, capacity, exposure of persons and assets, hazard characteristics and the environment are enhanced. Such knowledge can be leveraged for the purpose of pre-disaster risk assessment, for prevention and mitigation and for the development and implementation of appropriate preparedness and effective response to disasters. This framework calls on the United Nations Office for Disaster Risk Reduction, in particular, to support the implementation, follow-up and review of the present framework by: preparing periodic reviews on progress, in particular for the Global Platform for Disaster Risk Reduction, and, as appropriate, in a timely manner, along with the follow-up process at the United Nations, supporting

30 GA Res 45/185 of 1990.

the development of coherent global and regional follow-up and indicators.

ii. *Yokohama Strategy for a Safer World: World Conference on Natural Disaster Reduction 1994*

World Conference on Natural Disaster Reduction 1994: Yokohama Strategy for a Safer World resolved as follows:

> The international community accepts the need to share the necessary technology to prevent, reduce and mitigate disaster; this should be made freely available and in a timely manner as an integral part of technical cooperation.

Drawing from the principles contained in the Yokohama Strategy for a Safer World: Guidelines for Natural Disaster Prevention, Preparedness and Mitigation and its Plan of Action[31] and the Hyogo Framework for Action, the implementation of the present framework will be guided by the following principles, while taking into account national circumstances, and consistent with domestic laws as well as international obligations and commitments:

a. Each state has the primary responsibility to prevent and reduce disaster risk, including through international, regional, sub-regional, trans-boundary and bilateral cooperation. The reduction of disaster risk is a common concern for all states and the extent to which developing countries are able to effectively enhance and implement national disaster risk reduction policies and measures in the context of their respective circumstance and capabilities can be further enhanced through the provision of sustainable international cooperation;

b. Disaster risk reduction requires that responsibilities be shared by central governments and relevant national authorities, sectors and stakeholders, as appropriate to their national circumstances and systems of governance;

31 A/CONF. 172/9, Chap. I, Resolution 1, Annex I.

c. Managing the risk of disasters is aimed at protecting persons and their property, health, livelihoods and productive assets, as well as cultural and environmental assets, while promoting and protecting all human rights, including the right to development;

d. Disaster risk reduction requires an all-of-society engagement and partnership. It also requires empowerment and inclusive, accessible and non-discriminatory participation, paying special attention to people disproportionately affected by disasters, especially the poorest. Gender, age, disability and cultural perspectives should be integrated in all policies and practices, and women and youth leadership should be promoted. In this context, special attention should be paid to the improvement of organised voluntary work of citizens;

e. Disaster risk reduction and management depends on coordination mechanisms within and across sectors and with relevant stakeholders at all levels, and it requires the full engagement of all state institutions of an executive and legislative nature at national and local levels and a clear articulation of responsibilities across public and private stakeholders, including business and academia, to ensure mutual outreach, partnership, complementarity in roles and accountability and follow-up;

f. While the enabling, guiding and coordinating role of national and federal state governments remain essential, it is necessary to empower local authorities and local communities to reduce disaster risk, through resources, incentives and decision-making responsibilities, as appropriate;

g. Disaster risk reduction requires a multi-hazard approach and inclusive risk-informed decision making based on the open exchange and dissemination of disaggregated data, including by sex, age and disability, as well as on easily accessible, up-to-date, comprehensible, science-based, non-sensitive risk information, complemented by traditional knowledge;

h. The development, strengthening and implementation of relevant policies, plans, practices and mechanisms need to aim at coherence, as appropriate, across sustainable development and growth, food security, health and safety, climate change and variability environmental management and disaster risk reduction agendas. Disaster risk reduction is essential to achieve sustainable development;

i. While the drivers of disaster risk may be local, national, regional or global in scope, disaster risks have local and specific characteristics that must be understood for the determination of measures to reduce disaster risk;

j. Addressing underlying disaster risk factors through disaster risk-informed public and private investments is more cost-effective than primary reliance on post-disaster response and recovery, and contributes to sustainable development;

k. In the post-disaster recovery, rehabilitation and reconstruction phase, it is critical to prevent the creation of and to reduce disaster risk by "Building Back Better" and increasing public education and awareness of disaster risk;

l. An effective and meaningful global partnership and the further strengthening of international cooperation, including the fulfilment of respective commitments of official development assistance by developed countries, are essential for effective disaster risk management.

Developing countries, in particular the least developed countries, small island developing states, landlocked developing countries and African countries, as well as middle-income and other countries facing specific disaster risk challenges, need adequate, sustainable and timely provision of support, including through finance, technology transfer and capacity-building from developed countries and partners tailored to their needs and priorities, as identified by them.

iii. United Nations Office for Disaster Risk Reduction (UNISDR)[32]

The United Nations Office for Disaster Risk Reduction (UNISDR) was created in December 1999 to facilitate the implementation of the International Strategy for Disaster Reduction (ISDR). It was created to be the focal point in United Nation system for the coordination of disaster risk reduction and to ensure synergies in international practice. It built on the work conducted for the Integrated Research on Disaster Risk (IRDR) assessment.

iv. The United Nation Commission on Sustainable Development[33]

The 11th session of the Commission of Sustainable Development 2003 adopted the themes of disaster management for the protection of the need of both the present and future generation without compromising the interest of the former against the latter. This further creates an obligation on the United Nations Office for Disaster Risk Reduction, in particular, to support the implementation, follow-up and review of the present framework by: preparing periodic reviews on progress, in particular for the Global Platform for Disaster Risk Reduction, and, as appropriate, in a timely manner, along with the follow-up process at the United Nations, supporting the development of coherent global and regional follow-up and indicators and in coordination, as appropriate, with other relevant mechanisms for sustainable development and climate change. Sustainable indicators require collaboration with other relevant institutions as follows:

i. updating the existing web-based Hyogo Framework for Action Monitor accordingly;

32 Established 1999/ https://sustainabledevelopment.UN.org,(accessed 10th Oct., 2017).

33 Adopted by the UNGA on the 28th April – 9th May 2003, New York. https://sustainabledevelopment.UN.org,(accessed 10th Oct., 2017).

ii. participating actively in the work of the Inter-Agency and Expert Group on Sustainable Development Goal Indicators;

iii. generating evidence-based and practical guidance for implementation in close collaboration with states and through the mobilisation of experts;

iv. reinforcing a culture of prevention among relevant stakeholders through supporting development of standards by experts and technical organisations, advocacy initiatives and dissemination of disaster risk information, policies and practices, as well as by providing education and training on disaster risk reduction through affiliated organisations;

v. supporting countries, through national platforms or their equivalents, in their development of national plans and monitoring trends and patterns in disaster risk, loss and impacts; convening the Global Platform for Disaster Risk Reduction and supporting the organisation of regional platforms for disaster risk reduction in cooperation with regional organisations;

vi. leading the revision of the United Nations Plan of Action on Disaster Risk Reduction for Resilience;

vii. facilitating the enhancement of, and continuing to service, the United Nations Office for Disaster Risk Reduction Scientific and Advisory Group in mobilising science and technical work on disaster risk reduction;

viii. leading, in close coordination with states, the update of the publication entitled "2009 UNISDR Terminology on Disaster Risk Reduction," in line with the terminology agreed upon by states; and maintaining the stakeholders' commitment registry.

It is recalled that the outcome document of the United Nations Conference on Sustainable Development, held in 2012, entitled "The future we want,"[34] called for disaster risk reduction and the

34 Resolution 66/288, annex.

building of resilience to disasters to be addressed with a renewed sense of urgency in the context of sustainable development and poverty eradication and, as appropriate, to be integrated at all levels. The conference also reaffirmed all the principles of the Rio Declaration on Environment and Development.[35]

v. Hyogo Framework for Action[36]

Hyogo Framework for Action is focused on building the resilience of nations and communities to disasters. The international community recognised that "a culture of disaster prevention and resilience, and associated pre-disaster strategies, which are sound investments, must be fostered at all levels, ranging from the individual to international levels."[37] Based on international perspective, Hyogo Framework for Action focuses on risk prevention as follows:

i. in order to reduce the impact of disasters, there should be increased awareness of the need for establishing disaster mitigation strategies, particularly in disaster-prone countries.

ii. any action undertaken by the Civil Defence Service of a state for the benefit of another state, with the objective of preventing, or mitigating the consequences of disasters.[38]

iii. Cooperation among states, as well as with non-state entities and intergovernmental organisations, to facilitate the use of telecommunications for, inter alia, measures designed to prevent, predict, prepare for, respond to, monitor and/or mitigate the impact of, disasters.[39]

---

35 Report of the United Nations Conference on Environment and Development, Rio de Janeiro, 3-14 June 1992, vol. I. Resolutions Adopted by the ConferenceUnited Nations publication, Sales No. E.93,1.8 and corrigendum, resolution 1. annex I.

36 World Conference on Disaster Reduction 2005-2015, UNGA Resolution 46/182,(accessed 4th October, 2017).

37 *Ibid.* Preamble

38 FCCDA 2000

39 Tampere Convention on the provisions of Telecommunication Resources for Disaster Mitigation and Relief Operations adopted on June 1998 and enter into force on 8 January, 2005.

iv. States Parties are required to "give priority to prevention and mitigation, and thus taking precautionary measures to prevent, monitor and mitigate disasters."[40]

Cumulatively, the above provisions create emphasis on response and relief, recognition of prevention and pre–disaster preparedness. In the long run, this has the tendency of transforming into resilience development of prevention and management based on specific targets, including legislative and judicial system targets.

Since the adoption of the Hyogo Framework for Action in 2005, as documented in national and regional progress reports on its implementation as well as in other global reports, progress has been achieved in reducing disaster risk at local, national, regional and global levels by countries and other relevant stakeholders, leading to a decrease in mortality in the case of some hazards.[41] Reducing disaster risk is a cost-effective investment in preventing future losses. Effective disaster risk management contributes to sustainable development. Countries have enhanced their capacities in disaster risk management. International mechanisms for strategic advice, coordination and partnership development for disaster risk reduction, such as the Global Platform for Disaster Risk Reduction and the regional platforms for disaster risk reduction, as well as other relevant international and regional forums for cooperation, have been instrumental in the development of policies and strategies and the advancement of knowledge and mutual learning. Overall, the Hyogo Framework for Action has been an important instrument for raising public and institutional awareness, generating political commitment and focusing and catalysing actions by a wide range of stakeholders at all levels.

40 *Ibid*

41 Hazard is defined in the Hyogo Framework for Action as: "A potentially damaging physical event, phenomenon or human activity that may cause the loss of life or injury, property damage, social and economic disruption or environmental degradation. Hazards can include latent conditions that may represent future threats and can have different origins: natural geological, hydrometeorological and biological or induced by human processes environmental degradation and technological hazards."

It is urgent and critical to anticipate, plan for and reduce disaster risk in order to more effectively protect persons, communities and countries, their livelihoods, health, cultural heritage, socioeconomic assets and ecosystems, and thus strengthen their resilience. Enhanced work to reduce exposure and vulnerability, thus preventing the creation of new disaster risks, and accountability for disaster risk creation are needed at all levels.

On the whole, the Hyogo Framework for Action has provided critical guidance in efforts to reduce disaster risk and has contributed to the progress towards the achievement of the Millennium Development Goals. Its implementation has, however, highlighted a number of gaps in addressing the underlying disaster risk factors, in the formulation of goals and priorities for action,[42] in the need to foster disaster resilience at all levels and in ensuring adequate means of implementation. The gaps indicate a need to develop an action-oriented framework that governments and relevant stakeholders can implement in a supportive and complementary manner, and which helps to identify disaster risks to be managed and guides investment to improve resilience. Ten years after the adoption of the Hyogo Framework for Action, disasters continue to undermine efforts to achieve sustainable development.The present framework applies to the risk of small-scale and large-scale, frequent and infrequent, sudden and slow-onset disasters caused by natural or man-made hazards, as well as related environmental, technological and biological hazards and risks. It aims to guide the multi-hazard management of disaster risk in development at all levels as well as within and across all sectors.

42 The Hyogo Framework priorities for action 2005-2015 are: (1) to ensure that disaster risk reduction is a national and a local priority with a strong institutional basis for implementation; (2) to identify, assess and monitor disaster risks and enhance early warning; (3) to use knowledge, innovation and education to build a culture of safety and resilience at all levels; (4) to reduce the underlying risk factors; and (5) to strengthen disaster preparedness for effective response at all levels.

vi. *Sendai Framework for Disaster Risk Reduction 2015-2030*

This resolution was adopted by the General Assembly on June 3rd 2015.[43] The Sendai Framework for Disaster Risk Reduction 2015-2030 was adopted at the Third United Nations World Conference on Disaster Risk Reduction, held from March 14th to 18th 2015 in Sendai, Miyagi, Japan, which represented a unique opportunity for countries:[44]

a. To adopt a concise, focused, forward-looking and action-oriented post-2015 framework for disaster risk reduction;
b. To complete the assessment and review of the implementation of the Hyogo Framework for Action 2005-2015: Building the Resilience of Nations and Communities to Disasters;[45]
c. To consider the experience gained through the regional and national strategies/institutions and plans for disaster risk reduction and their recommendations, as well as relevant regional agreements for the implementation of the Hyogo Framework for Action;
d. To identify modalities of cooperation based on commitments to implement a post-2015 framework for disaster risk reduction;
e. To determine modalities for the periodic review of the implementation of a post-2015 framework for disaster risk reduction.

During the World Conference, states also reiterated their commitment to address disaster risk reduction and the building of resilience[46] to disasters with a renewed sense of urgency within

43 United Nations 69/283. Sendai Framework for Disaster Risk Reduction 2015-2030.

44 *Ibid.* Preamble to Sendai Declaration

45 *Ibid*

46 Resilience is defined as: "The ability of a system, community or society exposed to hazards to resist,absorb, accommodate to and recover from the effects of a hazard in a timely and efficient manner,including through the preservation and restoration of its essential basic structures and functions" see www.unisdr.org/we/inform/terminology.

the context of sustainable development and poverty eradication, and to integrate, as appropriate, both disaster risk reduction and the building of resilience into policies, plans, programmes and budgets at all levels and to consider both within relevant frameworks.

Obligations of State on Disaster Risk Management in International Law

In the management of disaster risk reduction, states, regional and international organisations and other relevant stakeholders should take into consideration the key activities listed under each of these four priorities and should implement them as appropriate, taking into consideration respective capacities and capabilities, in line with national laws and regulations. Cumulatively, existing legal framework and policy guidelines create the following obligations (among others) on state in relation to disaster risk management in international law:[47]

i. To enhance access of states, in particular developing countries, to finance, environmentally sound technology, science and inclusive innovation, as well as knowledge and information-sharing through existing mechanisms, namely bilateral, regional and multilateral collaborative arrangements, including the United Nations and other relevant bodies.

ii. To promote the use and expansion of thematic platforms of cooperation, such as global technology pools and global systems to share know-how, innovation and research and ensure access to technology and information on disaster risk reduction.

47 See generally guiding policies creating similar obligations with on states; A/CONF. 172/9, Chap. I, Resolution 1, Annex I; World Conference on Disaster Reduction 2005-2015, UNGA Resolution 46/182,(accessed 4th October, 2017) and United Nations. Treaty Series, vol. 1771, No. 30822. The climate change issues mentioned in the present Framework remain within the mandate of the United Nations Framework Convention on Climate Change under the competences of the States Parties to the Convention, (accessed 17th September, 2017).

iii. To incorporate disaster risk reduction measures into multilateral and bilateral development assistance programmes within and across all sectors, as appropriate, related to poverty reduction, sustainable development, natural resource management, the environment, urban development and adaptation to climate change.

iv. To mainstream and integrate disaster risk reduction within and across all sectors and review and promote the coherence and further development, as appropriate, of national and local frameworks of laws, regulations and public policies, which, by defining roles and responsibilities, guide the public and private sectors in: (i) addressing disaster risk in publically owned, managed or regulated services and infrastructures; (ii) promoting and providing incentives, as relevant, for actions by persons, households, communities and businesses; (iii) enhancing relevant mechanisms and initiatives for disaster risk transparency, which may include financial incentives, public awareness-raising and training initiatives, reporting requirements and legal and administrative measures; and (iv) putting in place coordination and organisational structures.

v. To adopt and implement national and local disaster risk reduction strategies and plans, across different timescales, with targets, indicators and time frames, aimed at preventing the creation of risk, the reduction of existing risk and the strengthening of economic, social, health and environmental resilience.

vi. To encourage the establishment of necessary mechanisms and incentives to ensure high levels of compliance with the existing safety-enhancing provisions of sectoral laws and regulations, including those addressing land use and urban planning, building codes, environmental and resource management and health and safety standards, and update them, where needed, to ensure an adequate focus on disaster risk management.

vii. To develop and strengthen, as appropriate, mechanisms to follow-up, periodically assess and publicly report on progress on national and local plans; promote public scrutiny and encourage institutional debates, by parliamentarians and other relevant officials, on the progress reports of local and national plans for disaster risk reduction.

viii. To assign, as appropriate, clear roles and tasks to community representatives within disaster risk management institutions processes and decision-making through relevant legal frameworks, and undertake comprehensive public and community consultations during the development of such laws and regulations to support their implementation.

ix. To establish and strengthen government coordination forums composed of relevant stakeholders at the national and local levels, such as national and local platforms for disaster risk reduction, and a designated national focal point for implementing the Sendai Framework for Disaster Risk Reduction 2015-2030. It is necessary for such mechanisms to have a strong foundation in national institutional frameworks with clearly assigned responsibilities and authority to, inter alia, identify sectoral and multisectoral disaster risk, build awareness and knowledge of disaster risk through sharing and dissemination of non-sensitive disaster risk, coordinate public awareness campaigns on disaster risk, facilitate and support local multisectoral cooperation (e.g. among local governments) and contribute to the determination of and reporting on national and local disaster risk management plans and all policies relevant for disaster risk management. These responsibilities should be established through laws, regulations, standards and procedures.

x. To empower local authorities, as appropriate, through regulatory and financial means to work and coordinate with civil society, communities and indigenous peoples and migrants in disaster risk management at the local level.

xi. To encourage parliamentarians to support the implementation of disaster risk reduction by developing new or amending relevant legislation and setting budget allocations.

xii. To formulate public policies, where applicable, aimed at addressing the issues of prevention or relocation, where possible, of human settlements in disaster risk-prone zones, subject to national law and legal systems.

xiii. To guide action at the regional level through agreed regional and sub-regional strategies and mechanisms for cooperation for disaster risk reduction, as appropriate, in the light of the present framework, in order to foster more efficient planning, create common information systems and exchange good practices and programmes for cooperation and capacity development, in particular to address common and trans-boundary disaster risks.

xiv. To actively engage in the Global Platform for Disaster Risk Reduction, the regional and subregional platforms for disaster risk reduction and the thematic platforms in order to forge partnerships, periodically assess progress on implementation and share practice and knowledge on disaster risk-informed policies, programmes and investments, including development and climate issues, as appropriate, as well as to promote the integration of disaster risk management in other relevant sectors. Regional intergovernmental organisations should play an important role in the regional platforms for disaster risk reduction, for example, provision of free food and medical care.

xv. To promote trans-boundary cooperation to enable policy and planning for the implementation of ecosystem-based approaches with regard to shared resources, such as within river basins and along coastlines, to build resilience and reduce disaster risk, including epidemic and displacement risk.

xvi. To allocate the necessary resources, including finance and logistics, as appropriate, at all levels of administration for the development and the implementation of disaster risk reduction strategies, policies, plans, laws and regulations in all relevant sectors.

xvii. To promote mechanisms for disaster risk transfer and insurance, risk-sharing and retention and financial protection, as appropriate, for both public and private investment in order to reduce the financial impact of disasters on governments and societies, in urban and rural areas.

xviii. To strengthen, as appropriate, disaster-resilient public and private investments, particularly through structural, non-structural and functional disaster risk prevention and reduction measures in critical facilities, in particular schools and hospitals and physical infrastructures: building better from the start to withstand hazards through proper design and construction, including the use of the principles of universal design and the standardisation of building materials; retrofitting and rebuilding; nurturing a culture of maintenance; and taking into account economic, social, structural, technological and environmental impact assessments.

xix. To protect or support the protection of cultural and collecting institutions and other sites of historical, cultural heritage and religious interest.

xx. To promote the disaster risk resilience of workplaces through structural and non-structural measures such as evacuating persons in disaster–prone areas and providing psychological support and mental health services for all persons in need.

xxi. To promote the mainstreaming of disaster risk assessments into land-use policy development and implementation, including urban planning, land degradation assessments and informal and non-permanent housing, and the use of guidelines and follow-up tools informed by anticipated demographic and environmental changes.

xxii. To promote the mainstreaming of disaster risk assessment, mapping and management into rural development planning and management of, inter alia, mountains, rivers, coastal flood plain areas, dry lands, wetlands and all other areas prone to droughts and flooding, including the identification of areas that are safe for human settlement, and at the same time preserving ecosystem functions that help to reduce risks.

xxiii. To encourage the revision of existing or the development of new building codes and standards and rehabilitation and reconstruction practices at the national or local levels, as appropriate, with the aim of making them more applicable within the local context, particularly in informal and marginal human settlements, and reinforce the capacity to implement, survey and enforce such codes through an appropriate approach, with a view to fostering disaster-resistant structures.

xxiv. To enhance the resilience of national health systems, including by integrating disaster risk management into primary, secondary and tertiary health care, especially at the local level; developing the capacity of health workers in understanding disaster risk and applying and implementing disaster risk reduction approaches in health work; promoting and enhancing the training capacities in the field of disaster medicine; and supporting and training community health groups in disaster risk reduction approaches in health programmes, in collaboration with other sectors, as well as in the implementation of the International Health Regulations (2005) of the World Health Organisation.

xxv. To strengthen the design and implementation of inclusive policies and social safety-net mechanisms, including through community involvement, integrated with livelihood enhancement programmes, and access to basic health-care services, including maternal, newborn and child health, sexual and reproductive health, food security and nutrition, housing and education, towards the eradication of poverty, to find durable solutions in the post-disaster phase and to empower and assist people disproportionately affected by disasters; people with life-threatening and chronic disease(s), due to their particular needs, should be included in the design of policies and plans to manage their risks before, during and after disasters, including having access to life-saving services.

xxvi. To promote, as appropriate, the integration of disaster risk reduction considerations and measures in financial and fiscal instruments.

xxvii. To strengthen the sustainable use and management of ecosystems and implement integrated environmental and natural resource management approaches that incorporate disaster risk reduction.

xxviii. To increase business resilience and protection of livelihoods and productive assets throughout the supply chains, ensure continuity of services and integrate disaster risk management into business models and practices.

xxix. To strengthen the protection of livelihoods and productive assets, including livestock, working animals, tools and seeds.

xxx. To promote and integrate disaster risk management approaches throughout the tourism industry, given the often heavy reliance on tourism as a key economic driver.

xxxi. To promote the development and strengthening of disaster risk transfer and sharing mechanisms and instruments in close cooperation with partners in the international community, business, international financial institutions and other relevant stakeholders.

xxxii. To promote cooperation between academic, scientific and research entities and networks and the private sector to develop new products and services to help to reduce disaster risk, in particular those that would assist developing countries and their specific challenges.

xxxiii. To encourage the coordination between global and regional financial institutions with a view to assessing and anticipating the potential economic and social impacts of disasters.

xxxiv. To strengthen and promote collaboration and capacity-building for the protection of productive assets, including livestock, working animals, tools and seeds.

xxxv. To strengthen and broaden international efforts aimed at eradicating hunger and poverty through disaster risk reduction.

xxxvi. To promote and support collaboration among relevant public and private stakeholders to enhance the resilience of business to disasters. The steady growth of disaster risk, including the increase of people and assets exposure, combined with the lessons learned from past disasters, indicates the need to further strengthen disaster preparedness for response, take action in anticipation of events, integrate disaster risk reduction in response preparedness and ensure that capacities are in place for effective response and recovery at all levels. Empowering women and persons with disabilities to publicly lead and promote gender equitable and universally accessible response, recovery, rehabilitation and reconstruction approaches is key. Disasters have demonstrated that the recovery, rehabilitation and reconstruction phase, which needs to be prepared ahead of a disaster, is a critical opportunity to "Build Back Better", including through integrating disaster risk reduction into development measures, making nations and communities resilient to disasters.

xxxvii. To prepare or review and periodically update disaster preparedness and contingency policies, plans and programmes with the involvement of the relevant institutions, considering climate change scenarios and their impact on disaster risk, and facilitating as appropriate, the participation of all sectors and relevant stakeholders.

xxxviii. To invest in, develop, maintain and strengthen people-centred multi-hazard, multisectoral forecasting and early warning systems, disaster risk and emergency communications mechanisms, social technologies and hazard-monitoring telecommunications systems.

xxxix. To promote the resilience of new and existing critical infrastructure, including water, transportation and telecommunications infrastructure, educational facilities, hospitals and other health facilities, to ensure that they remain safe, effective and operational during and after disasters in order to provide live-saving and essential services.

xl. To establish community centres for the promotion of public awareness and the stockpiling of necessary materials to implement rescue and relief activities.

xli. To ensure the use of traditional, indigenous and local knowledge and practices, as appropriate, to complement scientific knowledge in disaster risk assessment and the development and implementation of policies, strategies, plans and programmes of specific sectors, with a cross-sectoral approach, which should be tailored to localities and to the context.

Role of Stakeholders in Disaster Risk Management in International Law

States have the overall responsibility of reducing disaster risk, yet, it is a shared responsibility between governments and relevant stakeholders. In particular, non-state stakeholders play an important role as enablers in providing support to states, in accordance with national policies, laws and regulations, in the implementation of the present framework at local, national, regional and global levels. Their commitment, goodwill, knowledge, experience and resources will be required. When determining specific roles and responsibilities for stakeholders, and at the same time building on existing relevant international instruments, states should encourage the following actions on the part of all public and private stakeholders:

a. Civil societies, volunteers, organised voluntary work organisations and community-based organisations to participate, in collaboration with public institutions, to, inter alia, provide specific knowledge and pragmatic guidance in the context of the development and implementation of normative frameworks, standards and plans for disaster risk reduction; engage in the implementation of local, national, regional and global plans and strategies; contribute to and support public awareness, a culture of prevention and education on disaster risk; and advocate for resilient communities and an inclusive and all-of-society disaster risk management that strengthen synergies across groups, as appropriate. On this point, it should be noted that:

i. Women and their participation are critical to effectively managing disaster risk and designing, resourcing and implementing gender-sensitive disaster risk reduction policies, plans and programmes; and adequate capacity building measures need to be taken to empower women for preparedness as well as to build their capacity to secure alternate means of livelihood in post-disaster situations;

ii. Children and youth are agents of change and should be given the space and modalities to contribute to disaster risk reduction, in accordancc with legislation, national practice and educational curricula;

iii. Persons with disabilities and their organisations are critical in the assessment of disaster risk and in designing and implementing plans tailored to specific requirements, taking into consideration, inter alia, the principles of universal design;

iv. Older persons have years of knowledge, skills and wisdom, which are invaluable assets to reduce disaster risk, and they should be included in the design of policies, plans and mechanisms, including for early warning;

v. Indigenous peoples, through their experience and traditional knowledge, provide an important contribution to the development and implementation of plans and mechanisms, including for early warning;

vi. Migrants contribute to the resilience of communities and societies, and their knowledge, skills and capacities can be useful in the design and implementation of disaster risk reduction.

b. Academia, scientific and research entities and networks to focus on the disaster risk factors and scenarios, including emerging disaster risks, in the medium and long term; increase research for regional, national and local application; support action by local communities and authorities: and support the interface between policy and science for decision making.

c. Business, professional associations and private sector financial institutions, including financial regulators and accounting bodies, as well as philanthropic foundations, to integrate disaster risk management, including business continuity, into business models and practices through disaster-risk-informed investments, especially in micro, small and medium-sized enterprises; engage in awareness-raising and training for their employees and customers; engage in and support research and innovation, as well as technological development for disaster risk management; share and disseminate knowledge, practices and non-sensitive data; and actively participate, as appropriate and under the guidance of the public sector, in the development of normative frameworks and technical standards that incorporate disaster risk management;

d. Media to take an active and inclusive role at the local, national, regional and global levels in contributing to the raising of public awareness and understanding and disseminate accurate and non-sensitive disaster risk, hazard and disaster information, including on small-scale disasters, in a simple, transparent, easy-to-understand and accessible manner, in close cooperation with national authorities; adopt specific disaster risk reduction communications policies; support, as appropriate, early warning systems and life-saving protective measures; and stimulate a culture of prevention and strong community involvement in sustained public education campaigns and public consultations at all levels of society, in accordance with national practices.

With reference to General Assembly resolution,[48] commitments by relevant stakeholders are important in order to identify modalities of cooperation and to implement the present framework. Those commitments should be specific and time-bound in order to support the development of partnerships at local, national, regional and global levels and the implementation of local and national disaster risk reduction strategies and plans. All stakeholders are encouraged to publicise their commitments and their fulfilment in support of the implementation of the existing framework, or of the national and local disaster risk management plans, through the website of the United Nations Office for Disaster Risk Reduction.

Domestic Implementation

For the purposes of this domestic implementation, the following factors should be put into consideration as lessons from other jurisdictions:

i. **Disaster Preparedness:** The first pillars of health preparedness began in the late 1970s with simulations or tabletop exercises, and drills or live exercises that included the participation of several institutions through a multidisciplinary approach. Following suit, a number of countries started preparedness planning in their hospitals and later expanded activities to other institutions such as water systems.[49] Presently, a wealth of guidelines exist on the web covering a variety of topics, from establishing an Emergency Operations Centre to describing the amount of water in liters that should be distributed to displaced populations or those in shelters.[50] This

48 68/211 of 20 December 2013
49 *Ibid*
50 *Ibid*

and other information can now be found at virtual knowledge centres such as the Knowledge Centre on Public Health and Disasters,[51] and may be found in various jurisdictions globally but subject to local circumstances.

ii. **Health and Disaster Legislation:** Health and disaster legislation are greatly needed as a prevention measure in order to reduce disaster risk. In many countries, the progressive expansion of health and disaster-related standards and legislation resulted largely in response to the occurrence of disasters. These events helped governments identify problems and propose solutions. Based on these experiences, research activities and field work, new standards and legislation were eventually created.[52] Some examples of these are found in countries where they form the basis of establishing hospital emergency committees and defining hospital construction standards. On a broader scale, sub-regional institutions — such as the Ministries of Health for Central America, South American Andean Countries, and South East Asia — also pass resolutions providing standards and defining the scope of regulations that form and implement a national disaster relief and prevention system (which is a combination of disaster preparedness and health legislation).

Although great progress has been made in preparedness, issues first identified in the 1980s remain as significant challenges. For example, nongovernmental organisations (NGOs), representatives from governments, and United Nations (UN) agencies met in Costa Rica in 1986 and established a series of specific recommendations to guide direct international donations. The following areas of concern (in relation to the myth and reality of local circumstance) must be given priority in the event of international donations, the summary of which is shown in the table below.[53]

---

51 *Ibid*
52 *Ibid*
53 *Ibid*

Table 3.1: Disaster Myths and Realities[54]

Myth	Reality
Foreign medical volunteers with any kind of medical background are needed. Donations of cash or credit provided to health authorities or international agencies should be used whenever possible.	The local population almost always covers immediate life-saving needs. Only medical personnel with skills that are not available in the affected country may be needed.
Any kind of international assistance is needed, and it is needed now! Donations should be aimed at restoring the quality of health care to pre-disaster levels.	A hasty response that is not based on an impartial evaluation only contributes to the chaos. It is better to wait until genuine needs have been assessed.
Epidemics and plagues are inevitable after every disaster. Perishables or short-life donations should only be made on request from, and with prior approval by, the National Health Disaster Coordinator or other Ministry of Health authority.	Epidemics do not spontaneously occur after a disaster and dead bodies will not lead to catastrophic outbreaks of exotic diseases. The key to preventing disease is to improve sanitary conditions and educate the public.
The affected population is too shocked and helpless to take responsibility for their own survival.	On the contrary, many find new strength during an emergency, as evidenced by the thousands of volunteers spontaneously united to sift through the rubble in search of victims after the 1985 Mexico City earthquake.

54 *Ibid*

Disasters are random killers. The World Health Organisation's list of essential drugs and supplies should be used as a guideline by those wishing to donate.	Disasters strike hardest at the most vulnerable group, the poor, and especially women, children and the elderly.
Locating disaster victims in temporary settlements is the best alternate. Recipient countries should improve their distribution systems to ensure the best utilisation of donated resources.	Temporary settlements should be the last alternate. Many agencies used funds normally meant for tent to purchase building materials, tools and other construction-related support in the affected country.
Conditions are back to baseline within a few weeks	The effects of a disaster last a long time. Disaster-affected countries deplete much of their financial and material resources in the immediate post-impact phase. Successful relief programmes gear their operations to the fact that international interest wanes as needs and shortages become more pressing.

iii. **Disaster resilience:** Resilience is defined as "the ability of a system, community or society exposed to hazards to resist, absorb, accommodate to and recover from the effects of a hazard in a timely and efficient manner, including through the preservation and restoration of its essential basic structures and functions." [55]

While some progress in building resilience and reducing losses and damages has been achieved, a substantial reduction of disaster risk requires perseverance and persistence, with a

55 Policy guideline updated over time, PAHO/WHO.17 https://aneskey.com/international-perspectives-on-disaster-management

more explicit focus on people and their health and livelihoods, and regular follow-up.

Challenges

In view of what has been discussed in this chapter, the following challenges are identified:

1. Lack of concerted effort at the international level through the establishment of a viable machinery to coordinate and monitor the usages of the resource designated for disaster management at the local levels. For example, during the 2017 Muslim Ramadan fasting, the Kingdom of Saudi Arabia donated 200 tons of dates to the federal government of Nigeria for distribution to the victims of disaster (that is, the internally displaced persons as the sole beneficiaries of the gifts). Unfortunately, the dates were reported to be mismanaged and found in the open market for sale to the general public contrary to the intent of the donor (the Kingdom of Saudi Arabia). This event generated a lot of regrets by the donor and even the federal government of Nigeria due to lack of proper monitoring and coordination.[56]
2. Financial constraint militate against effective implementations of policies and guidelines for disaster risk management at the local levels.
3. Lack of uniform coordination strategy within the institutions responsible for implementing policies and guidelines for disaster risk management at all levels (that is at the international, regional and local levels). As a result, there is inconsistences in implementation strategy. For example, international instrument and policy provide high standard for logistic implementation of disaster risk management which is difficult to attain at the regional levels and particularly at the local levels. However, simply having

56 Kingdom of Saudi Arabia Premium Times Reports; See also Naij.com https://www.naija.com. (Accessed 17th September, 2017).

safe structure and processes may not be comprehensive enough or may not apply to a given situation. For example, countries with very weak economies could not expect to have a state of the art safe society in a decade.

4. Lack of adequate cooperation within the local mechanism for implementation. This is because their roles seems to be interwoven and there is no specification for limitation of duties on disaster risk management.

5. There is inadequate practical initiative by the United Nations for engagement with private sectors and business on disaster risk management and reduction. Currently, at best the policy guideline on the role of organisations is not educative enough because their role is greatly undermine by government agencies.

Recommendations

In view of the challenges identified above, the following recommendations are made:

1. Policies and practices for disaster risk management should be based on an understanding of disaster risk in all its dimensions of vulnerability, capacity, exposure of persons and assets, hazard characteristics and the environment. Such knowledge can be leveraged for the purpose of pre-disaster risk assessment, for prevention and mitigation and for the development and implementation of appropriate preparedness and effective response to disasters. The United Nations Office for Disaster Risk Reduction, in particular, to support the implementation, follow-up and review of the present framework by: preparing periodic reviews on progress, in particular for the Global Platform for Disaster Risk Reduction, and, as appropriate, in a timely manner, along with the follow-up process at the United Nations, supporting the development of coherent global, and regional follow-up and indicators to be in coordination as

appropriate, with other relevant mechanisms for sustainable development and climate change, and updating the existing web-based Hyogo Framework for Action Monitor accordingly.

2. Developing countries need enhanced provision of coordinated, sustained and adequate international support for disaster risk reduction. This is more needed by the least developed countries, small island developing states, landlocked developing countries and African countries, as well as middle-income countries facing specific challenges, through bilateral and multilateral channels, including enhanced technical and financial support and technology transfer on concessional and preferential terms, as mutually agreed, for the development and strengthening of their capacities. International financial institutions, such as the World Bank and regional development banks, to consider the priorities of the present framework for providing financial support and loans for integrated disaster risk reduction to developing countries.

   Other international organisations and treaty bodies, including the Conference of the Parties to the United Nations Framework Convention on Climate Change, international financial institutions at the global and regional levels and the International Red Cross and Red Crescent Movement to support developing countries, at their request, in the implementation of the present framework, in coordination with other relevant frameworks. The overall capacity of the United Nations system to assist developing countries in disaster risk reduction should be strengthened by providing adequate resources through various funding mechanisms, including increased, timely, stable and predictable contributions to the United Nations Trust Fund for Disaster Reduction and by enhancing the role of the Trust Fund in relation to the implementation of the present framework.

3. The United Nations and other international and regional organisations, international and regional financial institutions

and donor agencies engaged in disaster risk reduction are requested, as appropriate, to enhance the coordination of their strategies in this regard; the entities of the United Nations system, including the funds and programmes and the specialised agencies, through the United Nations Plan of Action on Disaster Risk Reduction for Resilience, United Nations Development Assistance Frameworks and country programmes, to promote the optimum use of resources and to support developing countries, at their request, in the implementation of the present framework, in coordination with other relevant frameworks, such as the International Health Regulations (2005), including through the development and the strengthening of capacities and clear and focused programmes that support the priorities of states in a balanced, well-coordinated and sustainable manner, within their respective mandates.

4. State, local government, organisation and other relevant bodies in the society must give supporting cooperation and mutual learning among themselves for disaster risk reduction and management in order to achieve an efficient implementation of the present framework and guiding policies.

5. The United Nations Global Compact, as the main United Nations initiative for engagement with the private sector and business, to further engage with and promote the critical importance of disaster risk reduction and management for an effective sustainable development and resilience. The Inter-Parliamentary Union and other relevant regional bodies and mechanisms for parliamentarians, as appropriate, to continue supporting and advocating disaster risk reduction and management in order to strengthen national legal frameworks.

CONCLUSION

Disaster risk governance at the national, regional and global levels is of great importance for the effective management of disaster risk. Clear vision, plans, competence, guidance and coordination within and across sectors, as well as participation of relevant stakeholders, are needed. Strengthening disaster risk governance for prevention, mitigation, preparedness, response, recovery and rehabilitation is therefore necessary and fosters collaboration and partnership across mechanisms and institutions for the implementation of instruments relevant to disaster risk reduction and management so as to achieve an effective, sustainable development. Thus, international, regional, sub-regional and trans-boundary cooperation remains pivotal in supporting the efforts of states, their national and local authorities, as well as communities and business must also be safety streamlined in order to reduce disaster risk. Existing mechanisms may require strengthening in order to provide effective support and achieve better implementation. Developing countries, in particular the least developed countries, small island developing states, landlocked developing countries and African countries, as well as middle-income countries facing specific challenges, need special attention and support to augment domestic resources and capabilities through bilateral and multilateral channels in order to ensure adequate, sustainable, and timely means of implementation in capacity-building, financial and technical assistance, and technology transfer, in accordance with international commitments.

4

INDIVIDUAL AND CORPORATE LIABILITY FOR DISASTERS CAUSED BY CRIMINAL NEGLIGENCE

Akeem O. Bello

Abstract

Disasters with devastating consequences resulting in loss of human lives, destruction of properties and environmental degradation are common phenomenon in all human societies. Where disasters are caused wholly by natural factors, the victims usually pick their losses unless the losses are covered by insurance. This chapter covers liabilities for disasters caused wholly or partly by criminal negligence attributable to an individual or a corporate entity or where an on-going disaster is aggravated by criminal negligence or failing to take steps to mitigate it. The chapter examines the basis of individual and corporate criminal liability for negligence and the emergent concept of corporate manslaughter. Statutory provisions on criminal negligence in Nigeria, England and Australia are examined. The chapter also interrogates whether the criminal law has any preventive role to play in disaster prevention and reduction. The author finds that while there are existing provisions which provide a platform to hold individuals and corporate entities liable for disasters caused or aggravated by criminal negligence, there is the need to fill some identifiable gaps in existing statutory provisions in Nigeria. The chapter concludes by offering some suggestions to strengthen existing statutory provisions on criminal negligence in Nigeria.

Keywords: Disaster, Criminal Negligence, Corporate Criminal Liability

INTRODUCTION

Disasters are a common occurrence in human societies. Disasters are caused by a variety of factors including natural factors, industrial accidents, and wholly or partly by criminal negligence. Disasters often result in loss of human lives, destruction of properties and environmental degradation. Disasters caused wholly or partly by criminal negligence may result in criminal prosecution of persons responsible for the disaster. This chapter examines the case law and statutory provisions in Nigeria, England and Australia on liability in criminal negligence for disasters attributable to the fault of individual or corporate entities. The chapter also interrogates whether the criminal law has any preventive role to play in disaster prevention and reduction.

The chapter is divided into four parts. Part one is the introduction. In part two the author clarifies the meaning of concepts such as disaster, criminal liability, negligence, individual criminal liability, and corporate criminal liability. Part three covers issue of individual and corporate criminal liability at common law, the theory of negligence, distinction between intentional and negligent conduct, basis of liability for negligence, and concept of corporate manslaughter in England and Australia, and whether there is any lesson for Nigeria from the concept. Part three also examines liability for causing disaster by criminal negligence in Nigeria against the background of recent disasters in Nigeria and whether there is any preventive role for criminal law in disaster management. Part four concludes the chapter.

Clarification of Concepts

Generally, a disaster can be defined as a calamitous event, especially one occuring suddenly and causing great loss of life, damage, or hardship, as a flood, airplane crash, or business failure.[1] In this chapter, a disaster is defined as a calamitous event, especially one occurring suddenly and causing great loss of life, damage to property, or hardship wholly or partly caused by the fault of an individual or a corporate entity or where an on-going disaster is aggravated by the fault of an individual or corporate entity or failure to take steps to mitigate it. Criminal liability is the burden or punishment imposed by the criminal law on a person after the person has been found liable for contravening the provisions of the law. Such a person is said to be criminally responsible because he is liable to be punished for an offence.[2] In the context of this chapter, a person may be criminally liable when he is found to be responsible for causing, failing to prevent or mitigate a disaster.

Negligence in the law of torts governing civil liability simply means breach of duty of care which has resulted in damage to another, attributable to the fault of a person responsible for the breach.[3] This type of negligence results in the award of monetary damages against the person at fault. Conversely, criminal negligence involves a breach of duty of care which in the words of Lord Hewart must show a "disregard for the life and safety of others as to amount to a crime against the state and conduct deserving punishment."[4] In this chapter, criminal negligence involves not just a breach of duty of care, but such breach of duty that shows a disregard for human life that justifies the intervention of the criminal law and the imposition of punishment. The Criminal Law imposes liability on "any person." A person in law

1 http://www.dictionary.com/browse/disaster (last accessed on 15th of September 2017).

2 See section 1 of the Criminal Code, Schedule to the Criminal Code Act, Cap. C38 Laws of the Federation of Nigeria 2010.

3 Winfield & Jolowicz, Tort, W.V.H. Rogers, ed. Sweet & Maxwell, 18th ed. 2010, at p. 150.

4 R v. Bateman (1925) 133 LT. 730 at 732.

usually consists of a natural human being and corporate entities in the form of private limited liability companies, public limited liability companies or statutory corporations. In this chapter, whenever the law imposes criminal liability on a natural human being and corporate entities for causing or failing to prevent disaster, the law is said to impose individual criminal liability and corporate criminal liability respectively.

Individual and Corporate Criminal Liability at Common Law

The liability for infractions of the criminal law has always been imposed on an individual usually a natural human being. The imposition of criminal liability for the acts or omissions of corporate entities at common law was however a gradual process. The initial position in England was that a corporation could not be convicted of an offence at all.[5] Despite the judgment in 1842 in the case of *R v. Birmingham and Gloucester Rly Co.*[6] convicting a corporation for failing to fulfil a statutory duty, "true" corporate criminal liability was not firmly established in the UK until the 20th century.[7]

The common law courts have been able to surmount the initial obstacle preventing imposition of criminal liability on corporations for offences involving *mens rea*. A corporation is a separate legal entity from its shareholder.[8] A company, being a legal fictional entity, was initially perceived to be incapable of

5 Anon (1701) 12 Mod Rep. 560, per Holt CJ. For some of the reasons advanced for the position that a corporation could not be convicted

6 (1842) 3 QB 233.

7 See generally M. Wenger, "Corporate Criminal Liability National and International Responses" (Background Paper for the International Society for the Reform of Criminal Law,13th International Conference on Commercial and Financial Fraud: A comparative Perspective held at Malta, 8-12th July 1999, available at http://www.icclr.law.ubc.ca/Publications/Reports/CorporateCriminal.pdf (last accessed on 15th of September 2017) It is therefore not surprising that the Criminal Code that was introduced into Nigeria at the turn of the 20th precisely in 1904 did not make any deliberate effort to provide for the criminal liability of corporations

8 See Salomon v Salomon Ltd (1897) A.C. 22.

possessing the mental element of an offence.[9] The courts later developed a doctrine through which, they view the acts and thoughts of certain officers of a company as those of the company itself. The minds and bodies of officers of the corporation are then taken to supply its lack of mental and physical faculties. The doctrine is known as the "alter ego doctrine or identification theory"[10] (hereinafter "identification doctrine"). It enables the court to hold a corporation liable for the criminal conduct of its servants. The corporation is deemed liable not for the acts of its servants, but its own act.[11]

The application of the identification doctrine involves the problem of the identification of officers who represent the directing mind and the will of the company for whose acts the company would be liable as distinct from those officers who merely represents its hands.[12] An attempt was made in the leading case of *Tesco Supermarkets Ltd v. Natrass* [13] to identify the category of such officers whose acts would be taken to be the acts of the company. According to Lord Reid, the company would only be liable for the acts of "the board of directors, the managing directors and perhaps other superior officers of a company [who] carry out the functions of management and speak and acts as the company."[14] In the words of Lord Dilhorne, a person

> ...who is in actual control of the operations of a company or of part of them and who is not responsible to another person in the company for the manner in which he discharges his duties in the sense of being under orders...[15]

---

9 See G Williams, *Textbook of Criminal Law The General Part* (London, Stevens & Sons Limited, Second Edition, 1961) 856.

10 G Williams, *Textbook of Criminal Law,* (London: Stevens & Sons, 1978) 946.

11 See P. K. Fogam, "Company Law and the Corporate Polluter; Who to Blame" in Environmental Laws Including Compensation in J A Omotola, (ed), (Lagos, Faculty of Law, University of Lagos, 1999) 87, 90, M J Allen, Text Book on Criminal Law, (Blackstone Press Limited, London, 1999) pp. 222-223.

12 See the comments of Lord Denning in Bolton (Engineering) Co Ltd v. T.J. Graham & Sons Ltd[1957] 1 QB 159, 172.

13 [1972] AC 153.

14 *Ibid*, 171.

15 *Ibid,* 187.

It is thus clear that persons representing the directing mind and will of the company cover directors, managing director, the company secretary and other superior officers responsible for managing the affairs of the corporation.[16]

The application of the identification doctrine has however proved problematic and has consequently attracted criticisms. A person may not qualify as "directing mind and will of a company" merely because he exercises some managerial discretion.[17] The followings have been held not to qualify as "directing mind and will" of a corporation: (i) the manager of a supermarket owned by a company with hundreds of chain of supermarkets;[18] (ii) a depot engineer;[19] and (iii) a weighbridge operator.[20] One of the major shortcomings of the identification doctrine is that it fails to reflect the reality of the modern-day large multinational corporations.[21] How do we identify the directing and mind will of a large multinational corporations with different levels of management and control? The identification doctrine renders it disproportionately easy to prosecute smaller companies in which the controllers are more readily identifiable and allows larger corporations to avoid liability by disguising their true organisational structure.[22] Another objection to the test is that it is unclear which individuals will qualify, and whether the same approach is applicable in all contexts.[23]

It is important to note that a statute may impose corporate liability in respect of an employee who could not be said to be the "directing mind and will" of the corporation. In M*eridian Global Funds Management Asia Ltd v. Securities Commission* [24]

16 M J Allen, supra note 11, at p. 223.

17 D. Ormerod, Smith and Hogan's Criminal Law, 13th Edition, (Oxford University Press, London, 12th Edition, 2008)p. 260.

18 Tesco Supermarkets Ltd. v. Nattrass [1972] AC 153.

19 Magna Plant Ltd v Mitchel [1966] Crim L R 394.

20 John Henshall (Quarries) Ltd v Harvey [1965] 2 QB 233.

21 Ormerod, supra note 17, at p. 260.

22 *Ibid*, at p. 261.

23 *Ibid*.

24 [1995] 2 AC 500.

the Privy Council noted that whether an act is to be attributed to a corporation is a question of the construction of the particular statute under which proceedings are brought, so that the statute may impose corporate liability in respect of an employee who could not be said to be the "directing mind and will" of the corporation.

Liability for Criminal Negligence

The first section of this part examines theoretical issues relating to criminal negligence which encompasses the distinction between negligent and intentional conduct and the basis for criminalising negligence. The second section relates to evaluation of statutory provisions imposing individual and corporate liability for criminal negligence and the emergence of the offence of corporate manslaughter in England and Australia.

Theory of Criminal Negligence

Negligence has been described as an inferior and an aberrant basis for imposing criminal liability. This is because every interest protected by criminal law is protected against intentional violations with only a few such as bodily integrity and sometimes property protected against negligent risks.[25] The reluctance to punish negligence generally is not unconnected with the distinction often drawn between negligence and intentional conduct and the questioning of the justness of punishing for unintentional conduct.

The most obvious difference between intentional and negligent conduct is that in the former case, the actor chooses to do the harm, while in the latter, he is unaware that he is causing harm.[26] Intentional conduct involves the offender foreseeing the consequences of his act and desiring it.[27] Conversely, to establish

25 G. P. Fletcher, "The Theory of Criminal Negligence: A Comparative Perspective" (1971) 119 (No. 3) University of Pennsylvania Law Review, 401, at p. 402.

26 *Ibid*, p. 408.

27 C.O. Okonkwo ed., Okonkwo and Naish on Criminal Law in Nigeria, 2nd Edition, (Spectrum Books Limited, Abuja, 1980) at p. 51.

liability for negligence, the prosecution must prove that the defendant's conduct failed to conform to what would be expected of the reasonable person in those circumstances. The defendant may not have foreseen what was going to happen, yet the law sometimes holds that he ought to have foreseen it and was guilty of negligence.[28] The relevant question is whether the defendant's conduct falls below the reasonable standard. This might be because he has seen the risk of the proscribed consequences or circumstance but reacted unreasonably to it, or because he has not seen the risk when he ought to. If the reasonable person ought to have seen the risk of the proscribed harm or consequence, the defendant can be held liable for criminal negligence even if he personally failed to see the risk.[29] Negligence therefore involves a failure by the defendant to comply with an objective standard of behaviour set by the law.[30]

The imposition of criminal liability for negligence has not always been accepted without questioning the basis for its imposition. Questions have been raised about the basis of culpability for negligent conduct. If it is accepted that negligence involves liability for causing harm even when the defendant is unaware of the consequences of his act, how can it be fair to subject him to criminal sanctions?[31] This takes us to examine the question of the basis for criminalising negligence.

Before examining the basis for criminalising negligence, a subsidiary and associated issue is the questions of whether negligence is a form of *mens rea*. The question of the relevance of *mens rea* to criminal liability is often expressed in the latin maxim *actus non facit reum nisi mens sit rea*. This is interpreted generally to mean that an act does not make a man guilty unless his mind is guilty. The problem with this maxim is that its real import in criminal law is beclouded because of the different senses in which the terms

28 *Ibid.*

29 Ormerod, supra note 17, p. 146.

30 *Ibid.*

31 *Ibid,* at p. 408.

is used. Three meanings have been ascribed to *mens rea*.[32] The first is the conception of *mens rea* as the subjective mental element of an offence in relation to the definition of an offence. The second meaning is the emphasis on historical requirement of intentional conduct: the choosing to do harm coupled with the foresight of harm as basis of common law of crimes. The third meaning is a normative approach that merges the analysis of *mens rea* with the question of whether negligent conduct ought to be punished as crimes.[33] It is when *mens rea* is used in this third sense that we can say that negligence involves a form of *mens rea*. Ormerod agrees with the view thus:

> If *mens rea* is used simply as a compendious expression for the varieties of fault that may give rise to criminal liability, then it does, of course, include negligence.[34]

This point was also stressed by Fletcher thus:

> We require *mens rea* as an essential condition for criminal liability, not because we suppose that mental states are essential for criminality, but because we realise intuitively that the condemnatory sanctions should apply only to those who are justly condemned for their conduct. And men are not justly condemned and deprived of their liberty unless they are personally culpable in violating the law.[35]

The punishment of negligence as a crime is not an arbitrary exercise of imposition of criminal liability. The basis of punishing negligence is not the culpability of choice, but rather of failing to bring to bear one's faculties to perceive the risks that one is taking.[36] A negligent actor, has the capacity of doing otherwise; he could have brought to bear his faculties to perceive and to avoid the risk he created.[37]

32 This can be contrasted with two different meaning ascribed to the term by Okonkwo and Naish, see supra note 27, at p.49. The three different meanings encompass the said two meanings.

33 Fletcher, supra note 25, at p. 410.

34 Ormerod, supra note 17, at p. 147.

35 Fletcher, supra note 25, at p. 414.

36 *Ibid*, at p. 413.

37 *Ibid*, at p. 416.

MANSLAUGHTER AT COMMON LAW AND CORPORATE MANSLAUGHTER IN ENGLAND AND AUSTRALIA

Liability for criminal negligence at common law is founded in the offence of involuntary manslaughter. At common law, there are three types of involuntary manslaughter offences. The first category is unlawful act manslaughter which involves the defendant committing a criminal act in dangerous circumstances and this causes the death of the victim. For unlawful act manslaughter, the test of liability is objective — a reasonable person in the defendant's position would have recognised a risk of some harm to the victim.[38] The second category is gross negligence manslaughter where the defendant caused the death of the victim through criminal negligence.[39] The gross negligence manslaughter is designed to catch cases where, despite the lack of conscious choice to risk harms, the defendant's conduct is seriously negligent (unreasonably careless or inattentive) and where that negligence causes death: the most serious harm.[40] In *R v Adomako*[41] Lord Mackay set out the test of gross negligence in cases of manslaughter by criminal negligence involving a breach of duty thus:

> ...the ordinary principles of the law of negligence apply to ascertain whether or not the defendant has been in breach of a duty of care towards the victim who has died. If such breach of duty is established, the next question is whether that breach of duty caused the death of the victim. If so, the jury must go on to consider whether that breach of duty should be characterised as gross negligence and therefore as a crime. This will depend on the seriousness of the breach of duty committed by the defendant in all the circumstances in which the defendant was

38 J. Child & D. Ormerod, Smith and Hogan's Essentials of Criminal Law, Oxford University Press, United Kingdom, 2015, at p. 182.

39 The third category is reckless manslaughter where the defendant caused the victims death, being reckless as to causing death or grievous bodily harm. The first two categories are the commonest cases of manslaughter because cases coming within the third category will almost always satisfy the elements of the first and second categories.

40 *Ibid*, at p. 191.

41 [1994] 3 All ER 79.

> placed when it occurred. The jury will have to consider whether the extent to which the defendant's conduct departed from the proper standard of care incumbent upon him, involving as it must have done a risk of death to the patient, was such that it should be judged criminal.The essence of the matter which is supremely a jury question is whether having regard to the risk of death involved, the conduct of the defendant was so bad in all the circumstances as to amount in their judgment to a criminal act or omission...[42]

Where death results from a disaster caused by a breach of duty of care, the person responsible for the breach will be liable for manslaughter by criminal negligence if the conduct of the defendant was bad in all the circumstances as to amount to a criminal act. As noted earlier it was possible to convict a corporate entity for an offence involving a fault element under the identification doctrine. The problem of identifying officers of a corporate entity who qualifies as the controlling mind and will, whose conduct would be ascribed to the entity however proved problematic with large corporations. Consequently, it was difficult for the law to hold big corporations liable for manslaughter by gross negligence. The unsatisfactory state of the law led to the enactment of the Corporate Manslaughter and Corporate Homicide Act 2007 (CMA 2007).

The CMA 2007 shifts the basis of holding a company liable for corporate manslaughter from the liability of individual officers to the existence of a culture of negligence within an organisation that creates the environment for the breaches resulting in death. An organisation is guilty of an offence if the way in which its activities are managed or organised-

(a) causes a person's death, and

(b) amounts to a gross breach of a relevant duty of care owed by the organisation to the deceased.[43]

42 *Ibid*, at pp 86-87.

43 Section 1(1) (a) 7 (b) of CMA 2007.

The organisation is liable only if the way in which its activities are managed or organised by its senior management is a substantial element in the breach of duty.[44] A "senior management" in relation to an organisation, means the persons who play significant roles in- (i) the making of decisions about how the whole or a substantial part of its activities are to be managed or organised, or (ii) the actual managing or organising of the whole or a substantial part of those activities.[45] Commenting on this requirement, it has been noted in contrast to the old identification doctrine, that the breach does not have to be identified in the conduct of a single representative of the organisation, but can be found through an aggregation of the organisation's activities.[46] In dispensing with the need to identify a specific individual within the organisation who has committed manslaughter, the CMA 2007 represents a major improvement over the identification doctrine.[47] The definition of senior management under the Act is more encompassing and realistic than the identification doctrine's restriction to persons who comprise the "directing mind and will" of a company although it has also been argued that linkage of senior management to persons who play significant role in the formulation and or implementation of organisation policy may indirectly continue the identification doctrine's preoccupation with individual rather than systemic fault.[48]

Another problem with the CMA 2007 is that its scope is limited only to cases of homicide. It does not address corporate liability for causing grievous bodily harm, a statistical far more prevalent problem than corporate manslaughter.[49] The Australian Criminal Code (Code) is better than the CMA 2007 in this respect.[50] Section 12.1(2) provides that a body corporate may be found guilty of any offence,

44 Section 1(3) of CMA 2007.

45 Section 1(4(c) of CMA 2007.

46 J. Child & D. Ormerod, supra note at p. 200.

47 J. Gobert, "The Corporate Manslaughter and Corporate Homicide Act 2007: Thirteen Years in the Making But Was it Worth the Wait? (2008) 71(3) Modern Law Review, 413 at p.p. 427-428.

48 *Ibid.*

49 *Ibid*, at p. 420.

50 Act. No. 12 of 1995 as amended.

including one punishable by imprisonment. This potentially extends the provision of the code on corporate criminal liability to liability for causing grievous bodily harm. Another attractive aspect of the Australian approach is that one of the grounds for corporate criminal liability is holding a company liable based on corporate culture. A company can be liable for any offence if it is proved that a corporate culture existed within the body corporate that directed, encouraged, tolerated or led to non-compliance with the relevant provision;[51] or that the body corporate failed to create and maintain a corporate culture that required compliance with the relevant provision.[52]

The overall purpose of CMA 2007 is further strengthened by sections 9 and 10 empowering the court to make appropriate remedial order and adverse publicity order respectively. After conviction for corporate manslaughter, the court may make an order requiring the organisation to take specified steps to remedy-

(a) the breach that caused the death

(b) any matter that appears to the court to have resulted from the relevant breach and to have been a cause of the death;

(c) any deficiency, as regards health and safety matters, in the organisation's policies, systems or practices of which the relevant breach appears to the court to be an indication.

The making of the order can only be on the application of the prosecution specifying the terms of the proposed order. A remedial order- (a) must specify a period within which the steps ordered by the court are to be taken; (b) may require the organisation to supply to an enforcement authority within a specified period, evidence that those steps have been taken. These orders ensure that the court is empowered to take steps that would prevent the same or similar negligence resulting in deaths in the future. These provisions are potent weapons in the hands of the court to prevent future disasters.

51 Section 12.3(2)(c) of the Code.

52 Section 12.3(2)(d) of the Code

Another interesting feature of the CMA 2007 is section 10 which empowers the court convicting an organisation for corporate manslaughter to make an order requiring it to publicise in a specified manner –

(a) the fact that it has been convicted of the offence;
(b) specified particulars of the offence;
(c) the amount of any fine imposed;
(d) the terms of any remedial order made.

This provision enables the court to make what has been described as adverse publicity order. The assumption underlying this kind of order is that corporate entities are sensitive to the negative effect that such orders may have on their bottom line-profits. The combination of remedial orders and adverse publicity orders would stimulate corporate compliance and the taking of measure(s) that would prevent future disasters.

Criminal Liability for Disasters in Nigeria- Any Preventive Role for Criminal Law?

Statutory provisions in Nigeria impose criminal liability on "any person"[53] or "whoever"[54] commits an infraction of the law. "Any person" or "whoever" includes a natural human being and by the provisions of the Interpretation Act "person" includes a body of persons corporate or unincorporate."[55] These provisions would arguably impose criminal liability on corporations. The question of how to hold a company liable for offences involving *mens rea* would however remain. While the common law courts evolved

53 The Criminal Code Schedule to the Criminal Code Act now applicable in States in Southern Nigeria as Criminal Code Law except Lagos which enacted a Criminal Law of Lagos State in 2011, section 7 of the Harmful Waste (Special Criminal Provisions, etc.) Act Cap. H1 Laws of the Federation of Nigeria 2010.

54 The Penal Code Cap. 89 Laws of Northern Nigeria, 1963 applicable in the States in Northern Nigeria as Penal Code Law.

55 Section 18 of the Interpretation Act Cap. I23 Laws of the Federation of Nigeria. See also section 5 of the Penal Code Law defining a "person" to include any company or association or body of person whether incorporated or not.

the identification doctrine to address the problem, the applicability of the doctrine in Nigeria is not a straightforward matter. This is because with the introduction of the criminal code and the penal code the doctrines of common law no longer automatically apply in the application of the law of crimes in Nigeria.[56] The criminal and the penal codes did not make any special provisions concerning the criminal liability of corporations.[57] Consequently, the extent of corporate criminal liability has been said to be of some doubt.[58]

Statutory interventions have however imposed criminal liability on corporations in a manner like the identification doctrine. The statutes usually creates offences and impose criminal liability on corporations where the contravention of the statutes is caused by "a director, manager, secretary or other similar officer of the body corporate." These statutes in addition impose personal liability on the designated officers unless the officer can prove that the offence was committed without his consent or connivance and that he exercised all such diligence to prevent the commission of the offence.[59] Corporate entities have been convicted of criminal offences in Nigerian law.[60]

Section 65(1) of the Companies and Allied Matters Act[61] (hereafter "CAMA") imposes criminal liability on corporations for any act of its members in a general meeting, the board of directors or the managing director while carrying on in the usual way the business of the company. Section 66(1)(a) of CAMA deems a company liable only for acts of any of its officer or agent provided

56 On the relationship between the Nigerian Criminal Law and English Law see Okonkwo, ed. supra note 27, at pp 12-17.

57 The Criminal Code is a Schedule to the Criminal Code Act, Cap. C38 Laws of the Federation of Nigeria 2010.

58 Okonkwo ed., supra note 27, at p. 124.

59 See section 49 of the Banks and Other Financial Institutions Act, Cap. B3 Laws of the Federation of Nigeria 2010, see also section 19 of the Money Laundering (Prohibition) Act No.11 of 2011 as amended by the Money Laundering (Prohibition) (Amendment) Act 2012, 10 of the Advance Fee Fraud and Other Fraud Related Offences Act Cap. A6 Laws of the Federation of Nigeria, 2010.

60 In FRN v Olowe {2006] 7 Q.C.C.R. 185 aa company was convicted of conspiracy to defraud with the managing director who acted for the company. In FRN v Sir Kingsley Ikpe & Anor {2005) 2 Q.C.C.R. 162 a company was convicted of offences including stealing, uttering and forgery along with its managing director.

61 Cap. C20 *Laws of the Federation of Nigeria* 2010.

the company acting through its members in a general meeting, the board of directors, or managing director expressly or impliedly authorised such officer or agent to act in the matter. Consequently, it would therefore appear to be irrelevant to a company's liability that an employee is not the directing mind under the identification doctrine if there is supporting evidence that the company's principal organs expressly or impliedly authorised such officer or agent. Whether an employee is authorised impliedly or expressly is usually a question of fact and depends on the circumstances of the case.[62] Where an officer is impliedly or expressly authorised to do an act which amounts to a crime, the company can be held liable under the provision of section 7 of the Criminal Code[63] as a party for aiding, counselling or procuring the commission of the crime.[64]

Under CAMA, the members in a general meeting, the board of directors, or managing director constitute the "brain and nerve" or "directing mind and will" of the company. Consequently, the acts of any of these organs of the company shall for all practical purposes be deemed as the act of the company and their *mens rea* "personifies the culpability of the corporation for the purpose of criminal liability."[65] If those persons who are responsible for the general management of the company delegate their duties to another, then the acts of that other person will be the acts of the company.[66]

The recently enacted Criminal Law of Lagos State[67] also contains provisions on the criminal liabilities of corporations. Section 20(1) puts the question of criminal liability of corporations

62 C. Okoli, "Criminal Liability of Corporations in Nigeria: A Current Perspective" (1994) 38(1) *Journal of African Law,* 35, 38.

63 Schedule to the Criminal Code Act 1965, Cap C38 *Laws of the Federation of Nigeria* 2010.

64 J. Orojo, *Company Law and Practice in Nigeria,* (South Africa, Lexis Nexis Butterworths, 2006) 121.

65 L. Ali, *Corporate Criminal Liability in Nigeria,* (Lagos, Malthouse Press Limited, 2008) at p. 115

66 (1972) AC 153, 170, 173, per Per Lord Reid.

67 Cap C17 Laws of Lagos State of Nigeria, 2015.

beyond dispute by providing that where the context admits, a company shall be criminally responsible for an offence under the law. A company is held liable for the acts or omission of its officer.[68] In determining whether a company should be held liable for the acts or omissions of its officers, the court is to consider all the circumstances including the fact that the person has apparent or real authority to bind the company.[69] The provisions ostensibly attempted to address some of the problem associated with the identification theory. A combination of subsection (2) and (3) would seems to suggest that the legislator has moved away from the identification doctrine as a basis of ascribing criminal liability to corporations. This legislator has not adopted the limited confines of the identification doctrine for ascribing criminal liability to corporations. The ascription of liability to a corporation is not limited to the officers qualifying as "directing mind and will." A company is simply held liable for any crime committed by its officers. It is arguable that the legislators cognisant of the identification doctrine directly allows the courts to consider all the circumstances including the fact that the officer has apparent or real authority to bind the company in deciding whether the corporation should be held liable for the acts or omission of the officer. This requirement seems to emphasise that where the officer does not qualify as "directing mind and will" it is sufficient that the officer has apparent or real authority to bind the company. Where the offence is one involving *mens rea* the officers would thus supply the *mens rea* required to hold the company liable. The approach adopted by the legislator in simply holding a company liable for the acts or omissions of its officer is however problematic. Without defining who is an "officer" the provision leaves open holding a company liable for the acts or omissions of any category of officers including not only the "directing mind and will" but also officers who are the "hands and legs" of the company. What is the basis for simply holding a company liable for the acts or omissions of its officers? The

68 Section 20(2) of the Criminal Law of Lagos State.

69 Section 20(3) of the Criminal Law of Lagos State.

identification doctrine justifiably sought to restrict the category of officers whose conduct can be attributed to the company. Without such a restriction or other criteria for ascribing fault to companies, the law would invariably impose liability on a company for all acts or omissions of its employees. This certainly would result in a general imposition of vicarious criminal liability on companies for the acts or omissions of their officers. It is still generally accepted today that vicarious criminal liability is the exception rather than the rule.[70] It would be interesting to see how the courts in Nigeria would construe the provisions of the statute.

A company is also liable for acts or omission of an employee (who is not an officer), if it fails to take steps to prevent the act or omission.The provisions implicitly draw a distinction between an "officer" of a company and an "employee" without defining who qualifies as an "officer." The distinction between an "officer" and an employee of a company would have been meaningful in the context of an officer who qualifies as "directing mind and will" within the identification theory and an employee. But without defining who an officer is and with the appearance of the statute jettisoning the identification doctrine as basis for imposing liability on companies, the distinction between an "officer" and an "employee" is not helpful.

Where a disaster is caused or aggravated in Nigeria by the acts or omissions of an individual or corporate entity, liability for criminal negligence may arise under the following statutory provisions for breach of statutory duties imposed by the criminal law. Section 300 of the Criminal Code imposes a duty on every person having charge of another who is unable because of age, sickness, unsoundness of mind, detention or any other cause to withdraw himself from such charge and who is unable to provide himself necessaries of life, whether the charge is undertaken under a contract, or is imposed by law, or arises because of any act, whether lawful or unlawful, to provide the person with necessaries of life. Where the duty bearer fails to discharge that

70 Ormerod, supra note 17, at pp. 274-275.

duty, he is liable for any consequences which result to the life or health of the other person because of failure to perform the duty. A similar duty is imposed on the head of a family to provide necessaries of life for a child under the age of fourteen years who is a member of his household and he is responsible for any consequences to the life or health of the child for omission to perform that duty.[71] A duty is imposed on a master or mistress under contract to provide necessary food, clothing, or lodging for any servant or apprentice under the age of sixteen and he or she is liable for any consequences which result to the life or health of the servant or apprentice for omission to perform the duty.[72] A duty to have reasonable skill and to use reasonable care is imposed on any person administering surgical or medical treatment and is responsible for any consequences which result to the life or health of any person for omission to observe or perform that duty.[73] The law imposes a duty on every person in charge of dangerous things to use reasonable care and take reasonable precautions to avoid such danger and he is liable for any consequences which may result to the life or health of any person by reason of any omission to perform that duty.[74] When a person undertakes to do any act, the omission to do which is or may be dangerous to human life or health, the law imposes a duty on him to do that act; and he is held liable to have caused any consequence which result to the life or health of any person for omission to perform the duty.[75]

The provisions of the criminal code discussed above does not create any specific offence, it merely provides liability where the duty holder omits to perform the duty. Where the breach of duty or omission to observe the required standard set by the law occurs, the duty holder may be liable in criminal negligence. Where death is caused by omission to perform any of duties outlined above, the duty holder may be liable for involuntary manslaughter by criminal negligence. Involuntary manslaughter occurs where a

71 Section 301.
72 Section 302.
73 Section 303.
74 Section 304.
75 Section 305.

person causes death under circumstances that he did not intend to kill and did not foresee death as a probable consequence of his conduct but there is some blameworthiness, such as gross negligence in his conduct.[76]

Before the enactment of the Criminal Law of Lagos State in 2011, the criminal code did not clearly set out what constitutes manslaughter and did not use the term voluntary or involuntary manslaughter. Manslaughter was merely defined as any unlawful killing "in circumstances as not to constitute murder."[77] The criminal code provided for killings arising from provocation covered by section 318. Involuntary manslaughter refers to killing caused by an unlawful and dangerous act. In *Amayo v. The State*[78] the Supreme Court quoted with approval the statement of Humphrey J in R. v. Larkin thus:

> Where the act which a person is engaged in performing is unlawful, if at the same time it is a dangerous act, that is, an act which is likely to injure another person and quite incidentally the doer of the act causes the death of that other person by that act then he is guilty of manslaughter.

Section 224 of the Criminal Law of Lagos State has now clearly sets out the circumstances that would constitute involuntary manslaughter. It provides that a person who unlawfully kills another in circumstances which does not constitute murder commits manslaughter if he causes death- (a) by an unlawful and dangerous act; or (b) with gross negligence or reckless disregard for human life. The first leg of the provisions deals with an unlawful and dangerous act. The breach of duties outlined in sections 300 to 305 where it is caused by gross negligence may result in conviction for causing death by gross negligence.

The penal code applicable in states in northern Nigeria does not contain a detailed list of duties the breach of which may result in loss of human life. It imposes in section 196 criminal liability on whoever does any rash or negligent act that endangers human

76 Okonkwo, supra note 27, at p. 253.

77 Section 317.

78 (2002) 5 WRN 1.

life or is likely to cause hurt or injury to property or person. It is the offence of culpable homicide under section 220(c) of the Penal Code to cause death by a rash or negligent act that equates the common law offence of manslaughter by gross negligence[79] earlier discussed.

The breach of some of the provisions of the Federal Factories Act[80] (FA) may give rise to liability for manslaughter by an unlawful and dangerous act or manslaughter by gross negligence. The FA is designed to ensure the health and safety of workers in the work place. It contains several provisions regulating cleanliness,[81] overcrowding,[82] ventilation,[83] lighting,[84] fencing of dangerous parts of machinery[85] amongst others.

Where a breach of any of the provisions results in a disaster causing death of workers, such a breach may amount to an unlawful and dangerous act giving rise to liability for involuntary manslaughter under the Criminal Code. The FA provides for liability of the occupier or owner of the factory where a person dies or suffers any bodily injury arising from the breach of any provisions of the FA. The penalty for the resulting death or injury is however a fine not exceeding N5,000 or to imprisonment for a term not exceeding two years or both fine and imprisonment.[86] The occupiers or owners of the factory may be prosecuted under the criminal code.

There have been several reported disasters in Nigeria over the years implicating liability for manslaughter by gross negligence.

---

79 S.S. Richardson, Notes on the Penal Code Law (Ahmadu Bello University, 1993 4th Edition) p.171. See also State v Gari (1977) N.N.L.R. 235 where Wheeler C.J. held that the degree of negligence under section 220 to establish the offence of culpable homicide by causing death by a negligent act is the same as that required to establish a charge of manslaughter at common law, that is to say a high degree of negligence must be proved.

80 Cap F1 *Laws of the Federation of Nigeria* 2010.

81 Section 7.

82 Section 8

83 Section 9.

84 Section 10.

85 Section

86 Section 71 of the Factories Act.

One of such disasters which killed several hundreds was the Ikeja military cantonment bomb blast of January 27th, 2002. Following various allegations of negligence by military authorities for failing to take proper care of the armoury, the government promised an enquiry to unearth the cause of the incident. It was reported in 27th of January, 2015 (thirteen years after), that a "Proper forensic report of the event is yet to be seen and the outcome of the panel of inquiry set up after the incident has not been made public."[87] A proper inquiry into the incident would probably have revealed whether any individual or the Nigerian Army as a corporate entity is liable for criminal negligence in causing the explosion. Without holding persons who cause disasters resulting in loss of human lives responsible, there would be no possibility of personal accountability to stimulate greater vigilance and responsibility thereby promoting disaster prevention.

Another significant disaster involved 45 workers who lost their lives on the night of September 15th, 2002 when a fire swept through a Nigerian plastics factory, West Africa Rubber Products Limited Ikorodu, Lagos. The fire gutted the factory and the adjacent Super Engineering Limited, both of which are owned by a conglomerate based in Shanghai and Hong Kong. There were speculations then that the staff were locked up in the factory and thus causing their deaths when they were unable to escape. If it was true that staff were locked up in the factory that could have given rise to possible liability in criminal negligence. There were no prosecutions that followed the incident which might suggest that the Police could not substantiate the allegations made against the factory owners.[88]

There have been recent developments which appeared to signal that the days of impunity for causing disaster by criminal negligence are over. On 12th of September 2014, a six-storey building owned by the Synagogue Church of All Nations collapsed in Ikotun area of Lagos resulting in the death of 116 persons. The Lagos State

87 "Reminiscence: 13 Years After Ikeja Cantonment Bomb Blasts, The Tears Have Not Dried up"https:// www.the cable.ng/reminiscence-13-years-ikeja-cantonment-bomb-blast-tears-not-dried

88 "Nigeria Factory Fire Kills 45" www.wsws.org/en/articles/2002/09/fire-s25.html (last accessed on 15th September 2017).

Government subsequently charged the trustees of the church, and two engineers who constructed the structure to court on 111 counts of involuntary manslaughter and failure to obtain building approval before constructing the building.[89] The construction companies owned by the two engineers were also charged along with their owners. A similar building collapse occurred in Lekki on 10th of March 2016 killing thirty-four persons. The owner of the building, engineers and companies involved in the construction of the building were charged to court for involuntary manslaughter and failure to obtain required building approvals.[90] A successful prosecution of these cases and any other similar case that may emerge will be a wakeup call that the government will no longer treat cases of disasters caused by criminal negligence with levity.

The forgoing discussion has revealed that the criminal law has a limited role to play in disaster prevention and management. The criminal law is limited because it only applies after a disaster had occurred to bring persons at fault to justice. The existence of statutory provisions criminalising negligence cannot without more prevent the occurrence of disasters. Where a disaster occurs resulting in death and the disaster is attributable to the fault of individuals or corporate entities, the due enforcement of the criminal law and imposition of sanction may achieve some positive outcomes. The due enforcement of criminal laws will signal that the government is willing and able to prosecute persons responsible for causing disasters by negligence. The publicity that would accompany finding a big corporation liable for such disasters may create negative public image that is generally not good for business and profit of corporations.

89 "Collapsed Building, Synagogue's Trustees, Engineers arraigned for Manslaughter… As Court Sends Accused Person to Prison" http://globalpatriotnews.com/collapsed-building-synagogues-trustees-engineers-arraigned-for-manslaughter-as-court-sends-accused-persons-to-prison/(last accessed on 15th September 2017).

90 "Building Collapse: Court Dismisses Lekki Gardens Director's Motion Seeking to Quash Charge" https://www.vanguardngr.com/2017/06/building-collapse-court-dismisses-lekki-gardens-directors-motion-seeking-quash-charge/ (last accessed on 15th September 2017).

RECOMMENDATIONS AND CONCLUSION

Some of the statutory provisions imposing corporate criminal liability in Nigeria still adopts language suggesting the importation of the identification doctrine and the problems already identified in relation to the application of the doctrine. The approach adopted in England and Australia is recommended for adoption in Nigeria. That approach as earlier noted shifts the basis of holding a corporation liable away from the fault of specified individuals to the way the activities of corporation is conducted (CMA 2007) or the existence of a corporate culture that created the environment for an offence to be committed (Australia). One advantage the Australian approach has over that of England is that it is not limited to gross negligence resulting only in homicide. The Australian Criminal Code extends corporate liability to causing grievous bodily harm. This would ensure that corporations can also be held liable for grievous bodily harm especially those occurring in the work place. It is recommended that relevant statutes in Nigeria be amended to reflect the Australian position. It is also recommended that in reforming our laws relating to gross negligence involving corporations, we should also incorporate provisions empowering the courts to make remedial consequential orders and adverse publicity orders as provided under CMA as obtains under the CMA 2007.

5

DISABILITY, DISASTER MANAGEMENT AND THE LAW IN NIGERIA

Bukola R. Akinbola

Abstract

A strong link exists between disaster and disability. Disasters frequently cause disabilities for persons who had none and worsen or cause additional disabilities for persons with disabilities (PWDs). PWDs are usually more at risk in the event of a life-threatening human induced disaster such as bomb explosion or natural disasters like earthquake, tsunami, landslide. Disasters can lead to harm that can be prevented or reduced by including disability in disaster management. PWDs are disproportionately represented among casualties of disaster due to failure to mainstream disability in disaster risk reduction programmes. Incidental or charitable intervention for PWDs in the event of disaster needs to be replaced by a right-based approach to disaster risk management, which mainstreams disability and the peculiar needs of PWDs. While disaster management should cover all segments of society, PWDs are viewed as recipients of charity rather than subjects of rights. Article 11 of the CRPD provides for the right of PWDs to all available measures which the state should provide, to ensure protection and safety of PWDs in situations of risk, including armed conflict, humanitarian emergencies and the occurrence of natural disasters. This chapter discusses the intersections between disability and disaster management in Nigeria. It focuses on PWDs and disaster, due to the high number of persons affected by disability either as PWDs or as caregivers, the existing level of discrimination against PWDs in access to social benefits, as well as the potential of disasters to either aggravate disability or increase the number of PWDs by causing injuries, damages, displacements, health hazards due to poor hygiene, food and water shortages. The paper concludes with recommendations including the need to mainstream disability in disaster management plans, implementation and post disaster recovery activities, reflecting the principles of equality and non-discrimination in all disaster management programmes, enacting disaster management laws at all levels, creating awareness and preparedness drills for everyone, and for PWDs in the language that they understand, like Braille and sign language.

INTRODUCTION

The United Nations Office for Disaster Risk Reduction has stated that vulnerability and exposure to disasters is increasing as more people and assets locate in areas of high risk. Since 1970, the world's population has grown by 87 per cent. During the same time, the proportion of people living in flood-prone river basins increased by 114 per cent and on cyclone-exposed coastlines by 192%.[1] More than half of the world's large cities, with populations ranging from 2 to 15 million, are located in areas of high earthquake risk. Rapid urbanisation will increase exposure to natural hazards, especially in coastal zones. In terms of its impact, it has had severe consequences.[2]

1 United Nations Office for Disaster Risk Reduction, Global Assessment Report 2011: Revealing Risk, Redefining Development (Geneva, 2011)

2 EM-DAT, The International Database (CRED) found that since the year 2000, deaths related to natural hazards have exceeded 1.1 million; over 2.7 billion people have been affected. Another concern is the economic impact of disasters. Over the last 12 years USD 1.3 trillion has been lost to disasters. The trend is rising and now exceeds, on average, USD 100 million per year over the last decade. Available at http://www.emdat.be/

Disability is a human experience that has attracted the attention of human beings across nations and cultures through all ages. While there are varying types and severities of disabilities, most persons with disabilities (PWDs) share the experience of discrimination in common. PWDs are often left out in the process of planning and implementations of social welfare projects and programmes, including health, education, security, transportation, housing and more. Disaster management did not start with the protection of PWDs as a central focus as objects of protection, rescue or care. Over time however, it has become apparent that PWDs have been left out of the scheme of government protection and provision in the event of disaster, whether man made or caused by natural events like storm, flooding, earthquake, Tsunamis or landslides.

The Yokohama Strategy and Plan of Action avers that the impact of natural disasters in terms of human and economic losses has risen in recent years, and society in general has become more vulnerable to natural disasters.[3] The Plan of Action further states that those usually most affected by natural and other disasters are the poor and socially disadvantaged groups in developing countries as they are least equipped to cope with them.[4] In the light of the rate at which natural and man-made disasters occur and consequently, the need for disaster management that will keep losses of lives and properties at the barest minimum (both for persons without and those with disabilities), it is imperative to discuss disaster management and disability in the context of the law. The law is an instrument of social engineering and therefore sets standards and regulates human conduct, including when the management of disaster becomes necessary. This will afford more vulnerable persons like those with disabilities, the needed and sometimes specialised attention they require.

This paper is divided into six parts comprising of the introduction which gives a preamble to the paper, the second part clarifies relevant concepts in the paper, the legal framework on

3 Yokohama Strategy and Plan of Action for a Safer World Guidelines for Natural Disaster Prevention, Preparedness and Mitigation World Conference on Natural Disaster Reduction Yokohama, Japan, 23-27 May 1994.

4 *Ibid.*

disability is the third part, while the fourth part focuses on the legal framework on disaster, and the fifth part of the paper examines the interconnection between disability and disaster management in terms of how disaster and disability connect around the law and lastly, the sixth part is the conclusion with recommendations.

Theoretical and Conceptual Clarifications

Disaster: In its Disasters & Emergencies Definitions Training Package, according to the WHO/EHA Panafrican Emergency, a disaster is "an occurrence disrupting the normal conditions of existence and causing a level of suffering that exceeds the capacity of adjustment of the affected community. It is the people who matter most, and without the people we have no disaster."[5] The United Nations Office for Disaster Risk Reduction (UNISDR) has also defined the term disaster as a

> serious disruption of the functioning of a community or a society at any scale due to hazardous events interacting with conditions of exposure, vulnerability and capacity, leading to one or more of the following: human, material, economic and environmental losses and impacts.[6]

The term disaster is sometimes interchangeably used with other terms like tragedy, adversity, catastrophe devastation, calamity and particularly emergency.[7] The UNISDR explained further, that in terms of the effect of the disaster, it can be immediate and localised, but is often widespread and could last for a long

5 WHO/EHA Panafrican Emergency Training Centre, Addis Ababa (Updated March 2002 by EHA) Disasters & Emergencies Definitions Training Package. Retrieved on 09 September 2017 at http://apps.who.int/disasters/repo/7656.pdf

6 United Nations Office for Disaster Risk Reduction (UNISDR), Terminology, 2009 version.

7 Emergency is sometimes used interchangeably with the term disaster, as, for example, in the context of biological and technological hazards or health emergencies, which, however, can also relate to hazardous events that do not result in the serious disruption of the functioning of a community or society.

period of time.[8] The effect may test or surpass the ability of a community or people to cope by relying on its own resources, and consequently may require assistance from external sources, which could include neighbouring jurisdictions, or those at the national or international levels.[9] Disasters occur in different scales and the scope of their impacts also differs. The following terms describe the scopes of disaster:

a) *Small-scale disaster:* A type of disaster only affecting local communities which require assistance beyond the affected community.

b) *Large-scale disaster:* A type of disaster affecting a society which requires national or international assistance.

c) *Frequent and infrequent disasters:* This depends on the probability of occurrence and the return period of a given hazard and its impacts. The impact of frequent disasters could be cumulative, or become chronic for a community or a society.

d) *A slow-onset disaster* is defined as one that emerges gradually over time. Slow-onset disasters could be associated with drought, desertification, sea-level rise, epidemic disease, etc.

e) *A sudden-onset disaster* is one triggered by a hazardous event that emerges quickly or unexpectedly. Sudden-onset disasters could be associated with earthquake, volcanic eruption, flash flood, chemical explosion, critical infrastructure failure, transport accident, etc.[10]

---

8 UNISDR, 2009, expounds that disaster damage occurs during and immediately after the disaster and it is usually measured in physical units (e.g., square meters of housing, kilometres of roads, etc.), and describes the total or partial destruction of physical assets, the disruption of basic services and damages to sources of livelihood in the affected area. It clarifies further that disaster impact is 'the total effect, including negative effects (e.g., economic losses) and positive effects (e.g., economic gains), of a hazardous event or a disaster. The term includes economic, human and environmental impacts, and may include death, injuries, disease and other negative effects on human physical, mental and social well-being'.

9 UNISDR 2009, *Ibid.*

10 The Sendai Framework for Disaster Risk Reduction 2015-2030 (para. 15). Available at:

Disaster Management: Disaster Management refers to the coordination and integration of all activities necessary to build, sustain and improve the capability to prepare for, protect against, respond to and recover from threatening or actual natural or human-induced disasters.[11] Disaster Management is a multijurisdictional, multi-sectoral, multi-disciplinary and multi-resource initiative.[12] Therefore, it is vital that the federal, state and local governments, civil society organisations (CSOs) and the private sector discharge their respective roles and responsibilities and complement each other in achieving shared goals of disaster management[13]. For achieving efficient disaster management, flow charts or plans such as the flow chart for disaster management of a National Preparedness Cycle – Arrow ring chart are often used. Below is a typical flow chart, illustrating an arrow ring chart sample which was created on the base of the figure illustrating the web page "National Preparedness Cycle Plan" from the Federal Emergency Management Agency (FEMA) website.[14]

> Planning makes it possible to manage the entire life cycle of a potential crisis. Strategic and operational planning establishes priorities, identifies expected levels of performance and capability requirements, provides the standard for assessing capabilities and helps stakeholders learn their roles.[15] The planning elements identify what an organisation's Standard Operating Procedures (SOPs) or Emergency Operations Plans (EOPs) should include for ensuring that contingencies are in place for delivering the capability during a large-scale disaster. [fema.gov/plan].[16]

11 National Disaster Framework Nigeria, Understanding the Framework. Retrieved on 14th October 2017 at http://www.preventionweb.net/files/21708_nigherianationaldisastermanagementf.pdf

12 National Disaster Framework Nigeria, *Ibid.*

13 National Disaster Framework Nigeria, *Ibid.*

14 The arrow ring chart example "National Preparedness Cycle" was created using the Concept Draw PRO diagramming and vector drawing software extended with the Target and Circular Diagrams solution from the Marketing area of Concept Draw Solution Park. www.conceptdraw.com/solution-park/marketing-target-and circular-diagrams Retrieved on 17 November 2017 from https://conceptdraw.com a2018c3/preview

15 *Ibid.*

16 *Ibid.*

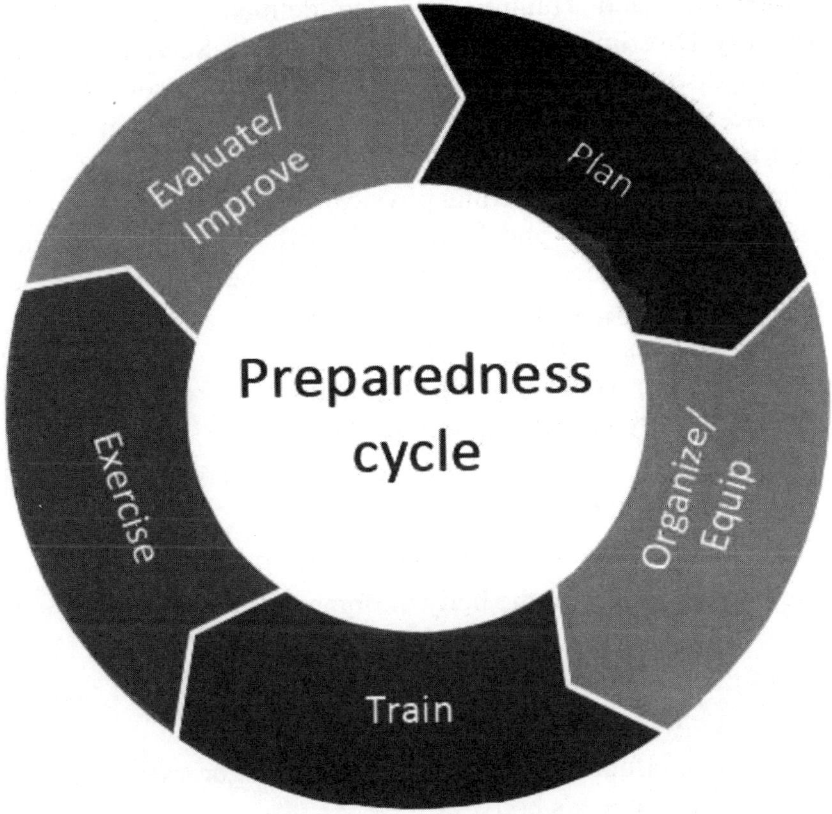

Disability: The Convention on the Rights of Persons with Disability (CRPD)[17] defines persons with disabilities to "include those who have long-term physical, mental, intellectual or sensory impairments, which in interaction with various barriers may hinder their full and effective participation in society on an equal

17 Convention on the Rights of Persons with Disabilities, G.A. Res. 61/106, 76th plen. mtg., U.N. Doc AIRES/61/106 (Dec. 13, 2006). The CRPD entered into force on May 2008 according to the UN News Centre, http://www.un.org/apps/news/story.asp?NewsID=26554#.VGzGqW1jJKl. See Loius O. Oyaro, 2015. Africa at Crossroads: The United Nations Convention on the Rights of Persons with Disabilities, American University International Law Review, Vol. 30, Issue 2, pp.347-377 at 347.

basis with others."[18] The Australian Government Department of Education and Training has also defined disability in its Disability Discrimination Act (DDA) 1992 in Section 4 of the Act as: a disorder, illness or disease that affects a person's thought processes, perception of reality, emotions or judgment or that result in disturbed behaviour. The definition of "disability" used in the DDA is broad, including physical, intellectual, psychiatric, sensory, neurological and learning disabilities. It also includes physical disfigurement and the presence in the body of disease-causing organisms, such as the HIV virus.

Other sources prefer to describe disability. For instance, the Commonwealth Disability Discrimination Act (DDA) 1992[19] defines disability, in relation to a person as:

(a) total or partial loss of the person's bodily or mental functions; or

(b) total or partial loss of a part of the body; or

(c) the presence in the body of organisms causing disease or illness; or

(d) the presence in the body of organisms capable of causing disease or illness; or

(e) the malfunction, malformation or disfigurement of a part of the person's body; or

(f) a disorder or malfunction that results in the person learning differently from a person without the disorder or malfunction; or

(g) a disorder, illness or disease that affects a person's thought processes, perception of reality, emotions or judgment or that results in disturbed behaviour; and includes a disability that:

(h) presently exists; or

---

18 Article 1, Convention on the Rights of Persons with Disabilities, 2006. The Convention was adopted by the United Nations on December 13, 2006, and opened for ratification on March 30, 2007.

19 Australian Government Department of Education and Training Factsheet Disability Discrimination Act, 1992. Retrieved on 04 December 2017 from: https://docs.education.gov.au/system/files/doc/other/dse-fact-sheet-1-dda_0.pdf

(i) previously existed but no longer exists; or

(j) may exist in the future (including because of a genetic predisposition to that disability); or

(k) is imputed to a person.[20]

The DDA definition of disability has been described as broad and including temporary and permanent disability that a person has now, may have in the future or is believed to have. The DDA definition encompasses impairments or diagnoses regardless of the impact an impairment has on a person's life functioning.[21]

The DDA covers present experience of disabilities, past experiences of disabilities, possible future experience of disabilities or even perceived disabilities.[22]

There are several other terms that are relevant or appropriate in discussing the connection between disability and disaster management including discrimination on the basis of disability, reasonable accommodation and universal design. They have been defined in the CRPD as hereunder:

Discrimination on the basis of disability means any distinction, exclusion or restriction on the basis of disability which has the purpose or effect of impairing or nullifying the recognition, enjoyment or exercise, on an equal basis with others, of all human rights and fundamental freedoms in the political, economic, social, cultural, civil or any other field. It includes all forms of discrimination, including denial of reasonable accommodation.[23]

Reasonable accommodation means necessary and appropriate modification and adjustments not imposing a disproportionate or undue burden, where needed in a particular case, to ensure to

20 Departmental Disability Definitions Fact Sheet of the Australian Government Department of Education and Training

21 The Australian Government Department of Education and Training has defined disability in its Disability Discrimination Act (DDA) 1992 in Section 4 of the Act.

22 Australian Government Department of Education and Training Factsheet Disability Discrimination Act, 1992. Retrieved on 04 December 2017 from: https://docs.education.gov.au/system/files/doc/other/dse-fact-sheet-1-dda_0.pdf

23 Article 2 of the Convention on the Rights of Persons with Disabilities (CRPD), 2006.

persons with disabilities the enjoyment or exercise on an equal basis with others of all human rights and fundamental freedoms.[24]

Universal design means the design of products, environments, programmes and services to be usable by all people, to the greatest extent possible, without the need for adaptation or specialised design. "Universal design" shall not exclude assistive devices for particular groups of persons with disabilities where this is needed.[25]

Communication includes languages, display of text, Braille, tactile communication, large print, accessible multimedia as well as written, audio, plain-language, human-reader and augmentative and alternative modes, means and formats of communication, including accessible information and communication technology.[26]

INTERNATIONAL LEGAL FRAMEWORK ON DISABILITY

The 2011 World Report on Disability by the World Health Organisation and the World Bank, avers that about 15% of the world's population will experience a disability at some time in their lives . In other words, about one billion persons around the world are already, or will become, persons with disabilities. Therefore, by protecting persons with disabilities, the convention plays a role in protecting some of the most vulnerable people in the world by ensuring their economic, social and political rights are secured.

The protection of the rights of PWDs did not start with the adoption of the CRPD. There had been a buildup of regulatory instruments, programmes and principles for promoting and protecting the rights of PWDs at the global level, particularly since after the second world war. The global community agreed on the undesirability of war and its destructive consequences on the human race, and found it necessary to protect and promote the rights of humans to life and dignity of the human person, among

24 *Ibid.*
25 *Ibid.*
26 *Ibid.*

other things. Thus, the Universal Declaration of Human Rights (UDHR) of 1945 was adopted and one of its key features is that it laid the foundation for modern human rights and also set the stage for the provisions in the Bill of Rights and standard for other human rights instruments. In its Article 2, the UDHR provided for all human rights "without distinction of any kind, such as race, colour, sex, language, religion, political or other opinion, national or social origin, property, birth or other status."

The UDHR however, did not mention impairment or disabilities specifically in the grounds of discrimination, nor did the International Covenant on Civil and Political Rights (ICCPR)[27] and the International Covenant on Economic, Social and Cultural Rights (ICESCR), which together form the International Bill of Rights.[28] However, the Bill of Rights forbid discrimination. The ICCPR prohibits discrimination as follows:

> Each State Party to the present Covenant undertakes to respect and to ensure to all individuals within its territory and subject to its jurisdiction the rights recognised in the present Covenant, without distinction of any kind, such as race, colour, sex, language, religion, political or other opinion, national or social origin, property, birth or other status.[29]

On its part, article 2 (2) of the ICESCR[30] similarly disallows discrimination in the implementation of human rights thus:

> The States Parties to the present Covenant undertake to guarantee that the rights enunciated in the present Covenant will be exercised without discrimination of any kind as to race, colour, sex, language, religion, political or other opinion, national or social origin, property, birth or other status.

27 International Covenant on Civil and Political Rights Adopted and opened for signature, ratification and accession by General Assembly resolution 2200A (XXI) of 16th December 1966 entry into force 23 March 1976, in accordance with Article 49.

28 Article 2 of the UDHR 1945, ICCPR and ICESCR 1966.

29 Article 2 (1) of the ICCPR. 1966.

30 International Covenant on Economic, Social and Cultural Rights Adopted and opened for signature, ratification and accession by General Assembly resolution 2200A (XXI) of 16 December 1966 entry into force 3 January 1976, in accordance with article 27.

The three instruments that comprises the Bill of Rights did not mention disability as a ground for non-discrimination, but all three contain a common clause: "other status", which has been applied to achieve non-discrimination on grounds of disability before the CRPD was adopted. While it was found useful, that provision was regarded as inadequate protection for PWDs. [31]

International Soft Law for the Protection of PWDs

As a result of the inadequate protection against discrimination for PWDs under the limited human rights regime, they continued to be viewed as objects of pity and consequently treated as objects of charity and welfare rather than subjects of rights under the law.[32] Such a perception minimised the role of cultural and physical barriers and rather magnified the impairment in the PWDs, a situation which necessitated that focus should shift to obvious physical barriers, and the manifold social, behavioral, stereotype-based barriers that lead to and potentially sustain the exclusion of persons with disabilities.[33] Human rights, and particularly of PWDs continued to emerge and from the 1970s, gained more prominence. The human rights regime of the UN in 1971 and 1975 when documents known as the Declaration on the Rights of Mentally Retarded Persons and the Declaration on the Rights of Disabled Persons were adopted respectively. The 1975 Declaration on the Rights of Disabled Persons stated, inter alia, that "the term 'disabled person' means any person unable to ensure himself or herself, wholly or partly, the necessities of a normal individual and/or social life, as a result of deficiency, either

31 Marianne Schulze. A Handbook on the Human Rights of Persons with Disabilities Understanding The UN Convention On The Rights Of Persons with Disabilities Edited by: Handicap International, 2009, p. 11.

32 Quinn/Degener, *Human Rights and Disabilities* – Chapter 1 "The moral authority for change: human rights values and the worldwide process of disability reform," in Marianne Schulze, *Ibid.*

33 Quinn/Degener, *Human Rights and Disabilities* – Chapter 1 "The moral authority for change: human rights values and the worldwide process of disability reform," in Marianne Schulze, *Ibid.*

congenital or not, in his or her physical or mental capabilities."[34] Although a "soft law", which means it had no binding force of law, the Declaration on the Rights of the Disabled proclaimed further that the rights, which persons with disabilities shall enjoy,

> shall be granted to all disabled persons without any exception whatsoever and without distinction or discrimination on the basis of race, colour, sex, language, religion, political or other opinions, national or social origin, state of wealth, birth or and other situation applying either to the disabled person himself or herself or to his or her family.

This definition obviously extended the scope of prohibition of disability discrimination by including other grounds on which PWDs are not to be discriminated against, compared with what was contained in the Bill of Rights. It also extends beyond PWDs to provide for those who care for them, such as their family members and makes all other human rights applicable to disabled persons thus:

> Disabled persons have the inherent right to respect for their human dignity. Disabled persons, whatever the origin, nature and seriousness of their handicaps and disabilities, have the same fundamental rights as their fellow-citizens of the same age, which implies first and foremost the right to enjoy a decent life, as normal and as full as possible.

The first indicators of reasonable accommodation may be discerned from another Article of the Declaration: "disabled persons are entitled to the measures designed to enable them to become as self-reliant as possible."

Although the declaration did not have binding force as an outright treaty document, it was an evidence of remarkable progress in the enhancement of the protection and promotion of the rights of PWDs.

Again, the UN declared 1981 as the International Year of Persons with Disabilities, while in 1982 the General Assembly adopted the World Programme of Action Concerning Disabled Persons

34 General Assembly Resolution 2856 (XXVI), 20th December 1971 and General Assembly Resolution 3447 (XXX) 9th December 1975.

which was followed by the proclamation of a Decade on Persons with Disabilities, which spanned from 1983 to 1992.[35] Still in 1982, PWDs were recognised in the authoritative interpretations, which the expert panels created under the UN Human Rights Treaties, may issue, for the first time.[36] For instance, the Human Rights Committee,[37] issued a general comment, which included a reference to persons with mental disabilities. On the right to liberty and security of the person, the committee held that this applies to "all deprivations of liberty, whether in criminal cases or in other cases, such as, for example, mental illness, vagrancy, drug addiction, educational purposes."[38]

By 1989, the UN adopted yet another specialised human rights treaty known as the Convention on the Rights of the Child (CRC), and it was the first treaty to include the first stand-alone article referring to the rights of persons (children) with disabilities unequivocally. The CRC also included "disability" as a potential ground of discrimination.[39]

Upon the accomplishment of the decade on PWDs in 1992, the UN Standard Rules on the Equalisation of Opportunities for Persons with Disabilities were adopted in 1993.[40] The standards spell out "a strong moral and political commitment on behalf of states to take action for the equalisation of opportunities for persons with disabilities." This was also a "soft law" and therefore more of a persuasive than binding instrument in terms of its legal status.

In keeping with the authority given by the UN to issue general comments on disability by the expert panels created under the

35 Marianne Schulze, A Handbook on the Human Rights of Persons with Disabilities Understanding The UN Convention On The Rights Of Persons with Disabilities Edited by: Handicap International September 2009, p. 12.

36 ICCPR, CESCR, CERD, CEDAW, CAT, CRC, CRMW.

37 The body constituted under the ICCPR to monitor the implementation of civil and political rights.

38 ICCPR General Comment 8, Liberty and security of the person.

39 See article 2 of the CRC.

40 The UN Standard Rules on the Equalization of Opportunities for Persons with Disabilities were adopted in 1993 by the UN General Assembly on 20th December 1993 by resolution 48/96 annex. A/RES/48/96.

UN Human Rights Treaties, in 1994, the committee under the ICESCR adopted a general comment on Persons with Disabilities which in its opening paragraph states:

> disability is closely linked to economic and social factors — conditions of living in large parts of the world as so desperate that the provision of basic needs for all – food, water, shelter, health protection and education must form the cornerstone of national programmes. Even in countries which have a relatively high standard of living, persons with disabilities are very often denied the opportunity to enjoy the full range of economic, social and cultural rights recognised.[41]

Similarly, in 1991, the committee under the Convention on the Elimination of All Forms of Discrimination Against Women (CEDAW) also mentioned the rights of women with disabilities and restated the need to specifically include them in State Party reports.[42] As noted earlier, the existing core treaties and soft laws were major strides in the crystallisation of the protection of rights of PWDs. The efforts were however grossly inadequate when compared with the level of disregard and violation of the rights of PWDs, notwithstanding the fact that all human rights provisions contained in the various instruments were also applicable to PWDs. Thus, the Convention on the Rights of Persons with Disabilities became necessary for specific protection and promotion of the rights of PWDs.

Convention on the Rights of Persons with Disabilities (CRPD)

The purpose of the CRPD as disclosed in article 1, is to "to promote, protect and ensure the full and equal enjoyment of all human rights and fundamental freedoms by all persons with disabilities, and to promote respect for their inherent dignity."

It covers various key areas such as self-determination, physical and programmatic access, personal mobility, health, education, employment, habilitation and rehabilitation, participation in political life, and equality and non-discrimination. The convention

41 ICESCR, General Comment 5, Persons with Disabilities.

42 CEDAW, General Comment 18, *Women with Disabilities.*

marks a shift in philosophy about disability from a social welfare concern/charity, to a human rights centred issue. It acknowledges that societal barriers and prejudices are themselves disabling and must be dismantled. There are eight underpinning principles of the CRPD (set out within it), namely respect for inherent dignity and individual autonomy; non-discrimination; full and effective participation and inclusion in society; respect for difference; equality of opportunity; accessibility; equality between men and women; and respect for the evolving capacities of children.[43] The intention of the principles is to ensure that PWDs are treated with respect, their human dignity is respected and all their human rights are accorded to them. Of great interest to this paper, are all its provisions, especially equality and non-discrimination as they are critical in the context of disaster management.

The Committee on the Rights of Persons with Disabilities opened an investigation into a signatory state for breaching their convention obligations in 2015. The investigation was triggered by article 6 of the optional protocol, which provides that an investigation will be carried out once the committee receives reliable information indicating grave and systematic violation of the human rights of persons with disabilities.[44]

African Regional Protection of the Rights of PWDs

African Charter[45]

Regional human rights documents were not left out in the recognition of PWDs in the various regions but this paper limits itself to the African region. The African human rights treaty known as the African Charter on Human and Peoples' Rights prohibits discrimination and makes all human rights equally applicable to all persons[46] thus:

43 Article 3 of the CRPD 2006, *op.cit.*

44 The government of the United Kingdom is being investigated, with the final report currently due to be released in 2017. "UN inquiry considers alleged UK disability rights violations". The Guardian. 20th October 2015.

45 Adopted 27th June 1981, OAU Doc. CAB/LEG/67/3 rev. 5, 21 I.L.M. 58 (1982), entered into force 21st October 1986.

46 Article 2 the African Charter on Human and Peoples' Rights (ACHPR), also known as the Banjul Charter, was adopted on 27th June 1981 and came into force on 21st October 1986, 21 LLM.

> Every individual shall be entitled to the enjoyment of the rights and freedoms recognised and guaranteed in the present Charter without distinction of any kind such as race, ethnic group, color, sex, language, religion, political or any other opinion, national and social origin, fortune, birth or other status.

The ACHPR makes provision for the protection of the rights of PWDs, calling for the right to special measures thus: "The aged and the disabled shall also have the right to special measures of protection in keeping with their physical or moral needs."[47]

African Women's Protocol

The African human rights regime also has protection in the African region specifically for women with disabilities. The African Women's Protocol explicitly provides for the rights of women with disabilities as follows:
State Parties undertake to:

> (a) ensure the protection of women with disabilities and take specific measures commensurate with their physical, economic and social needs to facilitate their access to employment, professional and vocational training as well as their participation in decision-making; [and]
>
> (b) ensure the right of women with disabilities to freedom from violence, including sexual abuse, discrimination based on disability and the right to be treated with dignity.[48]

African Charter on the Rights and Welfare of the Child

It is notable that the African Charter on the Rights and Welfare of the Child also, as a regional instrument, protects the rights of persons with disabilities. It provides for the protection of handicapped children by placing obligations on State Parties to ensure that:

47 Article 18 (4) of the African Charter on Human and Peoples' Rights (ACHPR), also known as the Banjul Charter, was adopted on 27th June 1981 and came into force on 21st October 1986, 21 LLM.

48 Article 23(a) and (b) of the African Women Protocol. (Emphasis mine).

1. Every child who is mentally or physically disabled shall have the right to special measures of protection in keeping with his physical and moral needs and under conditions which ensure his dignity, promote his self-reliance and active participation in the community.
2. State Parties to the present Charter shall ensure, subject to available resources, to a disabled child and to those responsible for his care, assistance for which application is made and which is appropriate to the child's condition and in particular shall ensure that the disabled child has effective access to training, preparation for employment and recreation opportunities in a manner conducive to the child achieving the fullest possible social integration, individual development and his/her cultural and moral development.
3. The State Parties to the present Charter shall use their available resources with a view to achieving progressively the full convenience of the mentally and physically disabled person to movement and access to public highway buildings and other places to which the disabled may legitimately want to have access to.[49]

The African human right regime is replete with avalanche of laws which can be found in various African instruments, but they are scattered in different documents and there is no single law that has all the African provisions together for harmony and easy access. The laws are also to be realised within domestic jurisdictions and that means that if any African country is not party to an instrument, its citizens cannot enjoy the benefits thereof. It is expected that countries that are parties to the African Charter and other instruments should include such provisions that protect and promote the rights of PWDs in their national laws and enforce them in their national laws. Although Nigeria is a party to the CRPD and all the African instruments highlighted, the effects are not too far reaching to date. Part of the reasons for

49 Article 13, African Charter on the Rights and Welfare of the Child African Comm'n On Human and Peoples' Rights, African Youth Charter 18 (2009), available at: http://www.au.int/en/sites/default/files/

this is that enforcement level is low, some of the laws are not yet domesticated, including the CRPD, and there is lack of political will to commit resources to the promotion and protection of the rights of PWDs.

International Disaster Management Framework

The law on disaster management is an emerging area of law. The focus of disaster management has evolved and the changing paradigm has been moving from supply of relief materials to victims of disaster, to disaster risk reduction as well as from state-centred intervention to involving governments, civil societies, groups and individuals.

United Nations Disaster Risk Reduction is the concept and practice of reducing disaster risks through systematic efforts to analyse and manage the causal factors of disasters, including through reduced exposure to hazards, lessened vulnerability of people and property, wise management of land and the environment, and improved preparedness for adverse events. Resilience is the ability of a system, community or society exposed to hazards to resist, absorb, accommodate to and recover from the effects of a hazard in a timely and efficient manner, including through the preservation and restoration of its essential basic structures and functions. There have been various initiatives towards disaster management and control at the international and national levels. Some of them will be considered here.

International Framework for Action for the International Decade for Natural Disaster Reduction of 1989

The International Strategy for Disaster Reduction (ISDR) was adopted during the 1999 IDNDR Programme Forum, held in July 1999 in Geneva, and ratified by the United Nations' Economic and Social Council (ECOSOC) and General Assembly during the second meeting of its commission on sustainable development. More than 20 speakers at the CSD spoke in favour of the strategy and the proposed institutional arrangements in November 1999, and a resolution to that effect was adopted.[50] The main thrust of the ISDR resolution was to reaffirm the need for strengthening an international framework for the improvement of early warning systems and disaster preparedness by developing an effective international mechanism for early warning, including the transfer of technology related to early warning to developing countries, under the auspices of the United Nations, as an integral part of future strategies and frameworks or any arrangements for natural disaster reduction.[51]

The Yokohama Strategy for a Safer World: Guidelines for Natural Disaster Prevention, Preparedness and Mitigation and its Plan of Action, 1994

Disaster prevention, mitigation, preparedness and relief are four elements which contribute to and gain from the implementation of sustainable development policies. These elements, along with environmental protection and sustainable development, are closely interrelated. Therefore, nations should incorporate them in their development plans and ensure efficient follow-up measures at the community, national, subregional and international levels.[52]

The International Strategy for Disaster Reduction of 1999

The United Nations (UN) International Strategy for Disaster Reduction Secretariat (UNISDR)[53] was established in 1999 as the successor to the Secretariat of the International Decade for Natural Disaster Reduction. UNISDR leads the organisation of the Global

50 The International Strategy for Disaster Reduction (ISDR) was adopted during the 1999 http://www.disaster.info.desastres.net/idndr/idndr.htm

51 General Assembly Resolution Adopted by the General Assembly [on the report of the Second Committee (A/54/588/Add.2)] 54/219. A/RES/54/219 3rd February 2000 Fifty-fourth session Agenda item 100 (b): International Decade for Natural Disaster Reduction: successor arrangements. Retrieved from on 17th Nov. 2017 from: http://undocs.org/A/RES/54/219

52 Yokohama Strategy and Plan of Action for a Safer World Guidelines for Natural Disaster Prevention, Preparedness and Mitigation World Conference on Natural Disaster Reduction Yokohama, Japan, 23-27th May 1994.

53 United Nations (UN) International Strategy for Disaster Reduction Secretariat (UNISDR retrieved on 21st October 2017 at: http://www.preventionweb.net/files/26097_unisdrassessment1.pdf

Platform for Disaster Risk Reduction, which meets every two years and has become the main global forum for guidance on the implementation of the Hyogo Framework for Action (HFA) and share experience among stakeholders.

The formal mandate of UNISDR is given by the UN General Assembly, and is to serve as the focal point in the United Nations system for the coordination of disaster reduction and to ensure synergies among the disaster reduction activities of the United Nations system and regional organisations and activities in socio-economic and humanitarian fields.[54]

Hyogo Framework for Action (HFA) 2005-2015: Building the Resilience of Nations and Communities to Disasters

The World Conference on Disaster Reduction was convened by the decision of the General Assembly,[55] and had five specific objectives:

(a) To conclude and report on the review of the Yokohama Strategy and its plan of action, with a view to updating the guiding framework on disaster reduction for the twenty-first century;

(b) To identify specific activities aimed at ensuring the implementation of relevant provisions of the Johannesburg Plan of Implementation of the World Summit on Sustainable Development on vulnerability, risk assessment and disaster management;

(c) To share good practices and lessons learned to further disaster reduction within the context of attaining sustainable development, and to identify gaps and challenges;

(d) To increase awareness of the importance of disaster reduction policies, thereby facilitating and promoting the implementation of those policies;

54 United Nations (UN) International Strategy for Disaster Reduction Secretariat (UNISDR retrieved on 21st October 2017 at: http://www.preventionweb.net/files/26097_unisdrassessment1.pdf

55 General Assembly resolution 58/214 of 23rd December 2003.

(e) To increase the reliability and availability of appropriate disaster-related information to the public and disaster management agencies in all regions, as set out in relevant provisions of the Johannesburg Plan of Implementation.

In terms of the expected outcome, while taking the above objectives into account, and drawing on the conclusions of the review of the Yokohama Strategy, states and other actors participating at the World Conference on Disaster Reduction ("the Conference") resolved to pursue the following expected outcomes for the next 10 years: The substantial reduction of disaster losses, in lives and in the social, economic and environmental assets of communities and countries.

The realisation of this outcome will require the full commitment and involvement of all actors concerned, including governments, regional and international organisations, civil society including volunteers, the private sector and the scientific community. The following strategic goals were to be adopted with the resolve of the conference with a view to attaining the expected outcome:

(a) The more effective integration of disaster risk considerations into sustainable development policies, planning and programming at all levels, with a special emphasis on disaster prevention, mitigation, preparedness and vulnerability reduction;

(b) The development and strengthening of institutions, mechanisms and capacities at all levels, in particular at the community level, that can systematically contribute to building resilience to hazards;

(c) The systematic incorporation of risk reduction approaches into the design and implementation of emergency preparedness, response and recovery programmes in the reconstruction of affected communities.

Sendai Framework for Disaster Risk Reduction 2015-2030

The Sendai Framework is the most recent initiative and sets out to consolidate the gains of previous moves with a view to improving on them. The present framework applies to the risk of small-scale and large-scale, frequent and infrequent, sudden and slow-onset disasters, caused by natural or man-made hazards as well as related environmental, technological and biological hazards and risks. It aims to guide the multi-hazard management of disaster risk in development at all levels as well as within and across all sectors. In terms of its expected outcomes, the substantial reduction of disaster risk and losses in lives, livelihoods and health and in the economic, physical, social, cultural and environmental assets of persons, businesses, communities and countries.

In terms of goals, it is intended to prevent new and reduce existing disaster risk through the implementation of integrated and inclusive economic, structural, legal, social, health, cultural, educational, environmental, technological, political and institutional measures that prevent and reduce hazard exposure and vulnerability to disaster, increase preparedness for response and recovery, and thus strengthen resilience.

The Sendai Framework is built on elements which ensure stability with the work done by states and other stakeholders under the HFA and introduces a number of innovations as called for during the consultations and negotiations. Commentators have identified the most significant shifts as a strong emphasis on disaster risk management as opposed to disaster management, the definition of seven global targets, the reduction of disaster risk as an expected outcome, a goal focused on preventing new risk, reducing existing risk and strengthening resilience, as well as a set of guiding principles, including primary responsibility of states to prevent and reduce disaster risk, all-of-society and all-of-state institutions engagement. In addition, the scope of disaster risk reduction has been broadened significantly to focus on both natural and man-made hazards and related environmental, technological and biological hazards and risks. Health resilience is strongly promoted throughout.

Also, to achieve its goal within national and local levels, article 27(a) of the Sendai Framework for disaster risk control advocates the need to

> mainstream and integrate disaster risk reduction within and across all sectors and review and promote the coherence and further development, as appropriate, of national and local frameworks of laws, regulations and public policies, which, by defining roles and responsibilities, guide the public and private sectors in: (i) addressing disaster risk in publicly owned, managed or regulated services and infrastructures; (ii) promoting and providing incentives, as relevant, for actions by persons, households, communities and businesses; (iii) enhancing relevant mechanisms and initiatives for disaster risk transparency, which may include financial incentives, public awareness-raising and training initiatives, reporting requirements and legal and administrative measures; and (iv) putting in place coordination and organisational structures.

The increase in the scope of stakeholders beyond governments, to include even businesses and households and the quest to mainstream and integrate disaster risk reduction into various sectors as well as create financial and other incentives, raising public awareness and ensuring coherence, are some of the innovations in the Sendai Framework, making it an improvement on earlier actions.

National Disaster Management Framework in Nigeria

As a country, Nigeria has had its fair share of disasters, including drought, desertification, flooding, epidemics, coastal erosion, dam failure, building collapse, oil spillage, maritime collision or accident, bomb explosion, communal clash, fire, air crashes and boat mishap.[56]

56 National Disaster Framework (NDMF) Understanding the Framework. Retrieved on 18th October 2017 from: http://www.preventionweb.net/files/21708_nigherianationaldisastermanagementf.pdf

Historically, salient disasters in Nigeria include, the yearly epidemics such as cholera, measles and cerebro-spinal meningitis. There were drought in the Sudan-Sahel zone of the country in 1972-74 and 1982-84. There were flood disasters in Ogunpa (Oyo State) in 1982, Lagos Bar Beach in 2001, Shiroro and Kaduna in 2003, Gombe in 2004, Jalingo in 2005, etc. Others include, the 2008 landslide in Agwu LGA (Enugu State); and 2005, 2006 and 2009 landslide in Agwagune (Cross River State). Also, Nigeria experienced numerous ethnic, political and religious violent conflicts in Kaduna, Lagos, Kano, Plateau, Bauchi, Borno, Taraba, Benue, Delta, Anambra, Ondo, and Osun.[57]

The National Disaster Management Framework (NDMF) is the instrument that functions as a regulatory guideline for effective and efficient disaster management in Nigeria. The framework defines measurable, flexible and adaptable coordinating structures, and aligns key roles and responsibilities of disaster management stakeholders across the nation. It describes specific authorities and best practices for managing disasters, and explains a paradigm shift in disaster management beyond mere response and recovery. The NDMF offers a holistic approach to disaster management.[58] It serves as a legal instrument to address the need for consistency among multiple stakeholders. It is a coherent, transparent and inclusive policy for disaster management in Nigeria.

The NDMF has stated its objectives as (i) Establishment of functional disaster management institutions at all levels of governance to prepare for, prevent, mitigate, respond to and recover from disaster events in Nigeria, and (ii) Development of capacity of relevant institutions and stakeholders for effective and efficient disaster management in Nigeria.

The NDMF clearly states the objectives of Disaster Response under thematic areas as follow:

i. ensure maximum survival of affected persons;

57 *Ibid.*

58 National Disaster Framework (NDMF) Understanding the Framework. Retrieved on 18 October 2017 from: http://www.preventionweb.net/files/21708_nigherianationaldisastermanagementf.pdf

ii. encourage self help;
iii. ensure preservation of property and the environment;
iv. establish mechanisms for integrated response;
v. bring order to a chaotic situation;
vi. keep the public informed through appropriate systems;
vii. ensure psychosocial support to affected.

It also identified the critical components of effective disaster risk reductions shall be the various regulations, standards, by-laws and other legal enforcement instruments. The federal, state and local governments shall ensure adherence to the legal enforcement instruments, some of which include:

i. land use regulations;
ii. urban planning and development standards;
iii. standard for environmental impact assessments;
iv. building codes;
v. fire codes;
vi. enabling acts/laws/bye-laws of the legislature on health and environmental issues; and
vii. relevant international conventions and treaties signed by the government of Nigeria.[59]

In terms of mainstreaming disaster risk reduction, the NDMF provides that disaster risk reduction shall be mainstreamed into developmental efforts at all levels of governance and it shall be integrated, amongst other things, into: spatial development planning including National Emergency Management Agency (NEMA) at the federal level, State Emergency Management Agency (SEMA) at the state level, and Local Emergency Management Authority (LEMA) at the local government level

59 National Disaster Framework (NDMF) Understanding the Framework. Retrieved on 18 October 2017 from: http://www.preventionweb.net/files/21708_nigherianationaldisastermanagementf.pdf

in order to strengthen the capabilities of federal, state and local governments to reduce the likelihood and severity of disasters.[60]

Nexus Between Disaster Management and Disability

It has been estimated that 10% of any population is likely to be disabled, and up to one in five of the world's poorest have a disability.[61] With a population of over 140 million, Nigeria has approximately 14 million people with disabilities.[62] The World report on disability, published in 2011 on its part, estimates that about 25 million Nigerians had at least one disability, while 3.6 million of these had very significant difficulties in functioning, while Natalie observed that there are varied estimates of persons with disabilities in Nigeria.[63]

The effects of disasters include destruction, usually leaving survivors with physical, sensory and psychosocial impairments. Participation of PWDs and accessibility experts in the disaster risk assessment and planning at the community level has been identified as the first essential phase. The second step to take, is the specific solutions on accessible environments, information and services should be identified. Through advance planning, organisations of PWDs and local governments can effectively

60 *Ibid.* The NEMA, SEMA, LEMA and other stakeholders, are to ensure the inclusion of spatial planners, Geographic Information System (GIS) experts in all phases of disaster risk reduction plans and programmes. Also, Disaster risk reduction strategies shall be integrated into governmental medium and long term goals of the National Development Plan and Educational Curriculum. Disaster risk reduction strategies shall be incorporated into education curriculum at all levels (Primary, Secondary and Tertiary) to inculcate the required knowledge, attitude, and skills in the youth.

61 See the UNESCO. Guide to community-based rehabilitation services. United National Educational, Scientific and Cultural Organisation, Paris. 1991; and . Mont D. Measuring Disability Prevalence. SP Discussion Paper No. 0706. Washington DC, USA: The World Bank. 2007 in Natalie Smith, (2011). The face of disability in Nigeria: A disability Survey in Kogi and Niger States, Disability, CBR and Inclusive Development, Vol 22, No.1, DOI 10.5463/DCID.v 22 i1.11, 2011, p. 35.

62 UNPD World Population Prospects: the 2004 revision. New York: UN Population Division. 2005.

63 Natalie Smith. The face of disability in Nigeria: A disability Survey in Kogi and Niger States, Disability, CBR and Inclusive Development, Vol 22, No.1, 2011. DOI 10.5463/DCID.v 22 i1.11,2011, p. 35.

address critical issues and coordinate with partners outside of the locality in the wake of disasters.[64]

Compared to the general population, PWDs face higher risks and are disproportionately affected by disasters. From available data, the mortality rate of the disabled population is two to four times higher than that of the non-disabled population in many disaster situations.[65] Disasters create a new generation of survivors with physical, sensory and psychosocial impairments. For every one person killed in a disaster, another three are injured or left with a long-lasting impairment.[66] Constructively however, PWDs have been found to make unique contributions, although often disregarded, to help reduce the risk of disasters and build resilient societies and communities.[67] The number of PWDs is expected to rise in the next decades due to range of factors including the increase in natural and human–induced disasters.[68]

Apart from the risk of acquiring a disability, persons with existing disabilities are more vulnerable and exposed to the destructive impact that disasters bear in terms of mortality, injury and adverse long-term health effects. Evidence gathered after the Great East Japan Earthquake in March 2011 indicates that mortality rates of persons with disabilities were two to four times higher than those without disabilities.[69]

64 United Nations ESCA Economic and Social Commission for Asia and the Pacific Committee on Disaster Risk Reduction Fourth session Bangkok, 27-29th October 2015. Item 6 of the provisional agenda Risk-sensitive development, E/ESCAP/CDR(4)/INF/4

65 Rehabilitation International (RI), the Nippon Foundation and UNESCAP, "Sendai Statement to Promote Disability-inclusive Disaster Risk Reduction for Resilient, Inclusive and equitable Societies in Asia and the Pacific" (24th April 2014). Available at http://www.riglobal.org/sendai-statement-to-promote-disability-inclusive-disaster-risk-reduction-for-resilient-inclusive-and-equitable-societies-in-asia-and-the-pacific/

66 CBM International, Technical brief for the post-2015 consultation process. Available from www.who.int/violence_injury_prevention/other_injury/en/disaster_disability.pdf.

67 *Ibid.*

68 United Nations, Department of Economic and Social Affairs, Population Division, World Population Prospects: The 2015 Revision (New York, 2015). Available from http://esa.un.org/unpd/wpp/Publications/Files/Key_Findings_WPP_2015.pdf.

69 Japan Disability Forum, What is behind the higher death rate of persons with disabilities? (Tokyo, 2013). Available from www.dinf.ne.jp/doc/english/twg/escap_121031/fujii.html.

The stark and often life-threatening impact of disasters on persons with disabilities is attributed, in part, to the absence of disability perspectives in all phases of disaster risk reduction (DRR). While "risk-informed decisions"[70] are often promoted in improving DRR, the experiences of PWDs leave much to be desired.[71]

How PWDs are affected by non-inclusion in Disaster Risk Reduction Plans

The deficiency of accessible environments, information and services has been identified as a critical factor. Deaf persons are often unable to receive early warning signals and any information that influences a life or death situation, as they are often transmitted only through audible means. Sign language interpreters are often not provided at emergency shelters. Wheelchair users are often unable to access evacuation routes, emergency shelters, temporarily housing units and bathrooms. Persons with intellectual disabilities and psychosocial disabilities are often not provided with effective communication and emotional support and are therefore often left in isolation.

Furthermore, although persons with disabilities are among the most at-risk groups during natural disasters, actors in the Asia-Pacific region depend on scarce evidence for their interventions due to the dearth of disaster data disaggregated by the status of disability.

Protection for PWDs in the context of Disaster

The CRPD has specifically provided for equality treatment of PWDs in situations of risk and humanitarian emergencies State Parties shall take, in accordance with their obligations under

70 United Nations. Sendai Framework for Disaster Risk Reduction (Sendai, 2015). Available from www.un.org/disabilities/documents/desa/3WCDRR/Sendai_2015 - 2030.pdf.

71 UNISDR. Key Findings: UNISDR 2013 Survey on Living with Disabilities and Disasters. Executive Summary. (Geneva, 2014). Available from www.unisdr.org/2014/iddr/documents/2013Disability SurveryReport_030714.pdf.

international law, including international humanitarian law and international human rights law, all necessary measures to ensure the protection and safety of persons with disabilities in situations of risk, including situations of armed conflict, humanitarian emergencies and the occurrence of natural disasters.[72]

In terms of international cooperation,

> States Parties recognise the importance of international cooperation and its promotion, in support of national efforts for the realisation of the purpose and objectives of the present convention, and will undertake appropriate and effective measures in this regard, between and among states and, as appropriate, in partnership with relevant international and regional organisations and civil society, in particular organisations of persons with disabilities.[73]

Conclusion

Disaster management has not been inclusive of persons with disabilities who often suffer discrimination and exclusion in many other spheres of life due to their disabilities. Recently however, the perception of disability as well as the position of law on the rights of PWDs is changing and their rights are becoming more crystalised, especially with the adoption of the CRPD, 2006 as well as other international laws which have also enhanced the status of PWDs at regional and national levels. While the law has also emerged for disaster management at all levels, it has not always factored the needs and possible contribution of PWDs to building resilience and disaster risk reduction in the ever increasing frequency of the occurrence of disaster as a result of natural and man-made causes.

The paper recommends that governments should consider the needs of PWDs, mainstream disability in disaster management plans, implementation and post disaster recovery activities. The principles of equality and non-discrimination needs to be reflected

72 Article 11 CRPD, 2006. Retrieved on 19th October 2017. http://www.un.org/esa/socdev/enable/rights/convtexte.htm

73 Article 32, CRPD 2006. Retrieved on 19th October 2017. http://www.un.org/esa/socdev/enable/rights/convtexte.htm

in all disaster management programmes, so that PWDs can be carried along not just as objects to be helped, but planning with them before and in the context of disaster. Disaster management laws should be made and observed at international, regional, national and local levels and awareness and preparedness drills should be given to everyone, and PWDs in the language that they understand, like Braille for example and constructions should adopt "universal design" that will serve everyone for accessibility to facilities. The CRPD has set down a standard for disaster management laws as well as disability laws at national and local levels and it should be so adopted and applied for the benefit of society at large.

6

DISASTER AND ITS MANAGEMENT IN NIGERIA: THE PLACE OF TORT LAW

Nwudego N. CHINWUBA

ABSTRACT

Disaster is a subject of much concern locally, regionally and globally. In Nigeria, the National Emergency Management Act provides a framework for dealing with disaster. This Act aims at dealing with disasters by providing for necessary actions. The Economic Community for West African States (ECOWAS) policies deal with natural disasters, their propensities, impact and actions driven to contain such disasters. The Sendal Framework-efforts by the United Nations Office for Disaster Risk Reduction directs attention and actions at natural and man-made disasters. These efforts underscore the attention which disaster has gained from the human community. In this chapter, a relationship is drawn between nature, humanity and law – a relationship which the discourse maintains all through, must be understood and underscored in any framework for dealing with disaster. In the light of this foundational hypothesis, the paper discusses law as a science springing from the basic order of the universe upon which all workable actions for order must be based. The relevance of tort law in disaster situations and disaster management is taken from the same basic hypothesis – that order is a prerequisite for human action and functionality and a fortiori, disasters should be acts of God and not man-made.

Keywords: Nature, Disaster, Law, Tort Law, Fault and Strict Liability

INTRODUCTION

Disaster is a state of affairs that requires attention from many perspectives. Disaster deserves attention, first from its application and spheres of impact and then its consequences and nature. Its impact cuts across the earth, humans and then impact on other species of creation. This overall impact and nature of disaster necessitates attention from all spheres of exposure, education and profession extending to the interaction of humans with creation, philosophy, sociology, to subject and professional areas such as economics, geography, law, religion, physics, medicine, etc. In this chapter, disaster and its management is viewed from the perspective of law and the discussion in law proceeds from the impact and relevance of tort as a subject of law.

Understanding law and its role in society is crucial to any discourse on disaster. Law draws its meaning from theoretical and substantive principles and application. The theory of law helps in appreciating the content of law while substantive principles are the practical application of law in its composition and procedure for application. Law springs from the complex inter-web of human life and experience and the nature and lessons learnt from experience. It is possible to deny that the human life is made-up of essence and existence but such denial does not change the reality that the human life in fact, comprises of an essence and existence. In essence, humanity is an expression of an essence conceived as an engine of the human vehicle. The human form itself can only be understood from the perspective of the idea of being human. If one takes a more critical look at the wonder of the human form and the orderliness of anatomy and its expression, this again makes it clearer that it is more rational than not, to conceive of life of being in essence and existence. The existence of the earth can only derive an explanation from the same perception as no one has been able to proffer an explanation of the root of the earth independently of the idea of the earth as a place of existence, that is, the earth as a place where all of creation simply found expression. There is though, one certainty and this is, order permeates creation and disorder is an exception to the order that permeates the earth and life within it. Order is seen in times and seasons. Order is seen in the anatomy and formation of life in all its forms.

Why for instance does the human being have nails since the fingers are fully functional without the nails? The only plausible explanation is to be found in the expression of the conception of order which predicates the human's claims to the earth in superior physical and mental evolution. Drawing from reasoning, it is logical and rational to state that rules of law are workable when they spring from a sociological environment requiring carefully drawn normative or intuitive solutions and/or resolutions brought to bear on practical expressions of law. It is only in this sense that law is so, and therefore, a science– a science of order.

This position implicates order and empathy or concern for good for the constituent occupants of each society, the earth and the earth space itself. It implicates an inclusive framework that reduces the vulnerable class to the barest minimum. This framework must be directed in any individual society to provide a conducive environment for the maximum actualisation of the mental and physical components of each person.

This perspective in fact forms the preamble of the Nigerian National Policy on Disaster Management which is stated as:

> In order to guarantee the adherence to the provisions of the 1999 Constitution of the Federal Republic of Nigeria, the Federal Government initiates, formulates and implements policies and programmes of activities to make the Nigerian environment peaceful, orderly and conducive for all its citizens. The policies of government are also targeted at preventing, reducing or managing disasters in all their ramifications whenever they occur.[1]

The question posed through this discourse is whether a steep inequitably stratified society can ever be positioned to achieve the stated ideals. Can Nigeria with such a tilted statistics of the citizenry in human and financial empowerment ever engage in real terms its objectives of disaster management? The next question is drawn in the light of the response from the first and stated as: can a state of disaster precede the actual occurrence of disaster as is generally known?

This chapter draws its thesis from this statement of objective of the National Policy on Disaster Management, 2013. It develops the proposition that the word 'disaster' is stratified and disaster may be a noun as well as a verb. Thus, the appropriate usage depends on the conception, composition and the state of evolution of each society. The thesis of this chapter is that the use of the word, 'disaster' itself requires proper clarification. The paper proceeds from this stage in two main sections. Section one is an analysis of disaster within the meaning ascribed to it by the Nigerian National Policy on Disaster Management on a comparative basis with that by the Government of India National Policy on Disaster Management, 2009. The second section deals with the place of tort law in disaster situations in three subheads; general comments on tort law; the pre-disaster situation and tort law in a disaster and post-disaster situation.

Disaster

The term 'disaster' envisages a sudden catastrophic event or phenomena. It is quite distinguishable from a hazard. The United

1 The Nigerian National Policy on Disaster Management available at www.preventionweb.net/files/21708_nigherianationaldisastermanagementf(last accessed 21/11/2017).

Nations Sendai Framework for Disaster Risk Reduction[2] defines hazard as "a potentially damaging physical event, phenomenon or human activity that may cause the loss of life or injury, property damage, social and economic disruption or environmental degradation. Hazards can include latent conditions that may represent future threats and can have different origins: natural (geological, hydrometeorological and biological) or induced by human processes (environmental degradation and technological hazards).[3]

Almost all definitions of disaster by major dictionaries use the word as a verb and adopt the word, 'sudden' in defining disaster. Thus, the Cambridge Dictionary defines disaster as an event that results in great harm, damage, or death, or serious difficulty.[4] The Oxford Dictionary defines it as a sudden accident or a natural catastrophe that causes great damage or loss of life.[5] Merriam Webster's Dictionary defines the word as a sudden calamitous event bringing great damage, loss, or destruction, natural disasters; and broadly: a sudden or great misfortune or failure.[6] However, at least one such definition uses the word as noun. According to that usage, 'the party was a disaster.'[7] Another definition excludes the description of sudden and defines 'disaster' "as a crisis situation causing widespread damage which far exceeds… ability to recover."[8]

The root of the word 'disaster' comes from an astrological sense of a calamity blamed on the position of planets and etymologically,

2 The United Nations Sendai Framework for Disaster Risk Reduction 2015-2030 Framework avaialable at www.preventionweb.net/files/43291_sendaiframeworkfordrren.pdf(last accessed 04/12/2017).

3 Sendai Framework for Disaster Risk Reduction 2015-2030 fn. 3

4 'Disaster' available at https://dictionary.cambridge.org/dictionary/english/disaster (2011-2013).

5 'Disaster' available athttps://en.oxforddictionaries.com/definition/disaster

6 'Disaster' available at https://www.merriam-webster.com/dictionary/disaster (last accessed 21/11/2017).

7 'Disaster' available at https://www.merriam-webster.com/dictionary/disaster (last accessed 21/11/2017).

8 'Disaster' available at www.cyen.org/innovaeditor/assets/Disaster_Management_Notes_and_Questions. (last accessed 21/11/2017).

the word is derived from Middle French *désastre.* Disaster is often seen from its moment of, or moments after, its occurrence. This has been the plague of disasters and its aftermath as far as Nigeria is concerned and also perhaps so, with other countries categorised as developing ones.

Disaster is in its most ordinary sense, a sudden accident or a natural catastrophe that causes great damage or loss of life. In a more technical sense, a disaster is a serious disruption, occurring over a relatively short time, of the functioning of a community or a society involving widespread human, material, economic or environmental loss and impacts, which exceeds the ability of the affected community or society to cope using its own resources.[9] However, as has been noted,

> disasters are not totally discrete events. Their possibility of occurrence, time, place and severity of the strike can be reasonably and in some cases accurately predicted by technological and scientific advances. It has been established there is a definite pattern in their occurrences and hence we can to some extent reduce the impact of damage though we cannot reduce the extent of damage itself.[10]

The thesis of this discourse is that a disorderly society where the reality and rationale of law are absent is a disaster. In this case, it is difficult to apply the normal definition of disaster to such a society, as disaster is a norm in such a society. When a society is however governed by law, the workings of the law bring about order and orderliness and a resultant tranquility. Law in this sense, denotes law operated as a science — a science of human condition and the rules that advance the strategy that brings about civilisation. In this sense, civilization denotes that state of evolution that is epitomised by empathy which characterises governance since government is of and by the people.

Where law is absent and a society is governed by arbitrary rules, epitomised by arbitrary institutions and its consequent

9 Technical sense of disaster readily available online.

10 'Disaster' available atwww.cyen.org/innovaeditor/assets/Disaster_Management_Notes_and_Questions (last accessed 21/11/2017).

disorder, anarchical institutions and power brokers as pursuers, a state of anarchy is in fact, in place and such a society cannot be described as pre-disaster situation but in fact is a disaster situation. In this case, it is therefore clear that a pre-disaster situation is not engaged and what is implicated is a disaster situation. This position implicates disaster induced through two different actors – human and nature-induced disaster. This distinction is clearer with the analysis above. In anarchical nations, disaster is almost altogether human-induced as a result of recklessness of conduct and habits ranging from pollution, flood, erosion to explosions, etc. Nature is practically ousted by the ferocity of recklessness from such a society. However, in a pre-disaster situation governed by law, disaster is more often nature-induced. In this case, disaster management apparatus and machineries focusing on response become very relevant while in the disaster cases what is required is prevention apparatus and mechanisms. Such a disaster situation requires a regime of tort law that safeguards the interests of anyone who is able to extricate him or herself from the majority. This intervention has the potential, if policy driven, to predicate a societal paradigm shift in the short or long term.

In normal evolving societies, disasters are now viewed academically from the perspectives of inappropriately managed risk and attendant consequences. These risks are the product of a combination of both hazards and vulnerability. Implicated in every disaster are action, reaction and response. Action is the state of affairs in a society before a disaster occurs. A reaction would normally follow when a society prior to disaster is in a disastrous condition such that the disaster occurring is but a continuing of a disordered pattern of an anarchical society. A response implicates well-defined laws driven to target disaster prior to its occurring.

A BRIEF OVERVIEW OF THE NIGERIAN AND INDIAN POLICY ON DISASTER AND ITS MANAGEMENT

The Nigerian National Policy on disaster management recognises the place of prevention in the question of disaster but fails to articulate the manner by which prevention can be achieved. The policy states the meaning of disaster:

> Nigeria like any other country is exposed to a wide range of social hazard which could be natural or man-made in which the lives of the citizens as well as infrastructures may be endangered or adversely affected... Furthermore, the event affects the social structure such as destruction or damage to governance, legal systems, buildings, communications and essential services. Disaster therefore, means a serious disruption of the functioning of the society, causing widespread human, material or environmental losses, which exceed the capability of the affected society to cope with using only its own resources.

The policy statement highlights types of disaster which disaster management organisations may be called upon to manage to include floods, destructive storms, fire, etc.

The Indian policy does not proffer the meaning of disaster. It provides a context of the policy:

> Disasters disrupt progress and destroy the hard-earned fruits of painstaking developmental efforts, often pushing nations, in quest for progress, back by several decades. Thus, efficient management of disasters, rather than mere response to their occurrences has in recent times received increased attention both within India and abroad. This is as much a result of the recognition of the increasing frequency and intensity of disasters, as it is an acknowledgement that good governance in a caring and civilised society, needs to deal effectively with the devastating impact of disaster.

On disaster risks in India, the Indian policy states:

> ...Vulnerability to disasters/emergencies of Chemical, Biological, Radiological and Nuclear (CBRN) origin also exists. Heightened vulnerabilities to disaster risks can be related to expanding population, urbanisation and industrialisation, development within high-risk zones, environmental degradation and climate change...

> In the context of human vulnerability to disasters, the economically and socially weaker segments of the population are the ones that are most seriously affected...

These two positions from the Indian policy underlie the earlier proposition made in terms of the relationship between society prior to disaster, and disaster, and post-disaster. Articulating the background to an issue presents a progressive perspective from which resolutions may spring up.

Happily, although the Nigerian policy does not capture the effect of inequity and disorder on the prognosis of disaster, it does recognise that most of the highlighted disaster types are preventable. The Indian policy identifies one way in which this can be achieved. It articulates as its objectives the promotion of a culture of prevention, preparedness and resilience at all levels through knowledge, innovation and education. To this end, it adopts a vital approach of community-based disaster management, including last mile integration of the policy, plans and execution; capacity development in all spheres; consolidation of past initiatives; cooperation with agencies at national and international levels and multi-sectorial synergy. The words 'culture' and 'paradigm shift' run through the policy.

The Nigerian policy fails to capture the *apriori* position of disaster as lying in culture and attitude. As a result of this failure in the Nigerian policy, the average Nigerian fails to see disaster in disorder and anarchy until anarchy is transmuted from habitual hazard to hellish catastrophes. Why will fire not erupt in a society where petrol is moved as if petrol and water are alike? In Nigeria, at the moment, people move petrol by hand, bicycles, motor cycles, cars, etc. Why will tankers not cause major tragedies when the simplest checks on the movement of such heavy duty vehicles are not in place? Why will trailers moving heavy containers not form disasters when the simplest measure of ensuring that such containers are double-hedged on the vehicles are not done? The long lines of heavy duty trucks on the Eko and Apapa flyover bridges in Lagos are overlooked until a catastrophe happens. Man-holes all over Lagos are unclosed. The dirt and blithe which occur due to carelessness and the culture of filth get very little attention in the face of destructive erosions. Pollution from vehicles and the higher forms from industries go on as a normal part of the life of the people and the atmospheric life. The failure of government: federal, state and local tiers to implement proper physical development and planning result in underground percolation of water and sewage. Builders habitually device means of channelling sewage underground to avoid frequent evacuation of sewage. In the absence of constant power, the indiscriminate use of generators compound the disastrous state of affairs in the nation for the majority of Nigerians. The blaring of horns to disorient and weaken the synergy of constructive thoughts are all aspects of habitual disaster and hazardous living *apriori* to catastrophes.

In the light of the welcome approach of the Indian policy, tort law has a key tool which if engaged will result in a move from an *apriori* society in a state of disaster to a disaster management one through the use of education, knowledge and law.

General Comments on Tort Law

Tort law is essentially received in Nigeria and remains essentially undeveloped. Tort law is especially positioned to aid in dealing with disaster management as governmental efforts can only go so far and mainly target the immediate aftermath of disaster. Tort law not only aids in putting government on its toes to pay attention to possible sources of disaster but aids it in taking effective actions aimed at removing or reducing vulnerability and risk. On a private law level, tort law equips individuals with the power through law to maintain boundaries and enforce them. Tort law works through the payment of damages usually un-liquidated and the use of injunctions. In England, tort law has developed in their historical setting from the socio-economic and philosophical trends of the day necessitating legislative reforms from time to time to shore up aging common law positions. Consequently it is imperative to understand the basic developments and evolution of tort law to be able to apply such developments to local circumstances.

Traditionally, it is said that tort law aims at compensating the victim. Tort law also aims at assisting not only contract but also constitutional laws and indeed human rights in their scope and objectives. The two key and mutually supporting functions of tort law are to compensate people when their rights are infringed and in providing a mechanism for redress, to thereby define and uphold those rights. The courts strive to strike a balance between promoting corrective justice and remedying wrongs and not contributing to the creation of an overly litigious society that believes there must be a remedy for every misfortune. In attempting to achieve this balance, judges consider the balance of risk and responsibility at both an individual and a social level.Tort law in essence protects not only individual interests in physical, emotional well-being and security of property but also economic interests.

It is therefore not surprising that questions surrounding economic activities have yielded debates on companies and production activities internalising the costs of potential claims in tort law to facilitate tortious claims as they arise. On the one hand, the claim is hereby made that tort law is validated by principles which aim at aiding the internalisation of costs of economic activities to ensure that victims of tortious actions are protected. Internalisation would usually promote insurance and deterrence in innovative and risky activities. Internalisation however, is viewed by economists as an inefficient market mechanism and a wrong approach to the theoretical basis for tort law. Internalisation presupposes governmental intervention which for the economist, is unnecessary for an optimal market efficiency. For this view, the right approach is for private bargains to be utilised in resolving questions of tortious harms where economic activities is involved. Thus, the proper question is not, which of two interacting activities caused the harm that one of them suffered but rather, how is the total value of the production of both activities to be maximised? On this view, "under certain conditions, the total value of production would be maximised by private bargains. The most important of these conditions, are first, that there be clearly defined and alienable legal entitlements, and second, that the costs of striking a bargain – the so-called transaction costs – be sufficiently low. The claim that value maximising bargains will be struck under these conditions has come to be known as Coase Theorem."[11]

Key considerations when pursuing or defending a claim in tort will include focus on the specific requirements for establishing a successful claim, the nature of the remedies that may be available, the defences that might apply and specific issues arising in relation to liability. While every case will be fact-specific, a body of case law has evolved, with some statute, which will help practitioners to consider where in the spectrum their case might fall and what matters might need to be considered.

There is a distinction between tort claims and contract claims, although there are occasions (such as professional negligence claims) where a claimant may be able to pursue a claim in tort and/or contract in the alternative. Whatever the tort (with the exception of trespass to land and defamation where a claimant does not need to prove loss), to establish tortious liability, the claimant must satisfy the court of three points: that the defendant had a legal duty to act in a particular fashion; that the defendant's conduct was in breach of that duty; and that the claimant suffered injury or loss as a direct result of the defendant's breach. There are some defences to a tort claim that are generally applicable and these defences are often run in the alternative or in addition to the general defences available.

---

11 For more detailed discourse see, https://pdfs.semanticscholar.org/f23d/cbe4b4b3668096d4ece21c8dea89f6afe26a

Tort Law in 'Pre'-Disaster Situation

Generally tort law operates only when injury or damages have already occurred. Trespass and libel are actionable without proof of damage but the law sees the damage as an intricate aspect of the wrong, thus once there is a trespass the law proceeds from the position that the trespass in itself is a damage requiring no further proof of damage. It is the gravity of these wrongs that make them as such. Trespass and defamation fall into this category with some minor exceptions in respect of slander.

A pre-disaster society which targets prevention is one in which law prevails and is applied in the light of human experience to provide decent living experience for the populace to facilitate superior human evolution and self-actualisation. To this extent, tort law has evolved to fully utilise the remedy of injunction and in particular the *quia timet* injunction in dealing with imminent and potential injuries to rights and in particular rights in land and property. The necessity to set up prevention agencies and mechanisms cannot but be underscored with disaster situations and the *quia timet* injunction is a tool that can be used effectively for prevention of forseeable harm timeously.

The *quia timet* injunction

The *quia timet* injunction ordinarily means timely intervention by the court because of the fear that a particular harm may occur which could cause irreparable damage. Generally, tort law envisages that the action has taken place before injunction is granted. Two cases have been used recently to analyse the use and extent of the *quia timet* injunction. In *London Borough of Islington v Elliot and Morris,*[12] the English and Welsh Court of Appeal examined the use of *quia timet* injunction and although on the facts of the case, the injunction was not granted and thought to be unnecessary, the court reinforced the relevance of the injunction and approved its

12 [2012] EWCA Civ 56,

use in *Hooper v Rogers*[13] and *Lloyds v Symonds.*[14] In the latter two cases the court granted injunctions, *quia timet* to restrain something (which as the plaintiff alleged) they were threatening or intending to do. In *Hooper v Rogers,*[15] the defendant had cut a track across a steep slope which provided the foundation of the plaintiff's farmhouse. The evidence was that this had exposed the slope to a process of soil erosion which would eventually undermine the farmhouse and cause it to collapse.

Russel LJ expressed the view that the probability of future damage is not absolute standard and what was important is doing justice between the parties. In *Merck Sharp Dohme Corp. v Teva Pharma B.V.*[16] the plaintiffs sought a *quia timet* injunction to stop the defendants from infringing on a valid patent which it held. The plaintiffs claimed they would suffer potentially unquantifiable damage should a *quia timet* injunction not be granted against the defendants who appeared intent on surreptitiously launching a generic version of the patented pharmaceutical product of the claimants. The claimants based their fear on the conduct of the defendants in failing to respond promptly to enquiries and silence on the question whether they intended to infringe the claimant's patent by surreptitiously launching the generic version of the patented product. These fears combined with the previous record of the defendants in similar circumstances with another company formed the major pleadings of the claimants in this case. After a careful analysis of existing authorities, including those of other European jurisdictions, the Court of Appeal granted a *quia timet* injunction against the defendants.

This analysis demonstrates that by analogy, the *quia timet* injunction could be used in the face of many actions undertaken by individuals, organisations and government as a speedy way to forestall continuing potentially catastrophic consequences from already known dangerous and hazardous activities. This is a

13 [1975] 1 Ch 43
14 [1998] EWCA Civ 511
15 [1975] 1 Ch 43
16 [2012] EWHC 627

compelling tool in containing activities of these nature in Nigeria. It is thus imperative to educate citizens on the availability of such an injunction for such actions. The advantage of using principled legal tools as opposed to fiat executive and arbitrary actions by government and its agencies is that law impacts on the culture of the people through reason and provides therefore, a better platform for dealing with sociological issues assailing a society.

Tort Law in Disaster/Post Disaster Situation

Tort law becomes fully functional in disaster or post-disaster situations because of the very nature of tort law. Tort law aims at compensating loss or injury and therefore becomes fully engaged when injury or damage has occurred. In these circumstances, while other torts may be engaged indirectly, the main torts which are triggered in disaster situations are negligence and strict liability torts. Negligence encompasses occupier's liability as well as vicarious liability. In this discussion, these complimentary principles will not be in focus. Negligence is treated in this discourse in a broad manner with the aim of highlighting its main principles.

Strict Liability Torts

The identifiable distinctive feature of strict liability torts is in the natural placement of the burden of proof. Strict liability tends to shift the natural burden of proof which is traditionally on the claimant to the defendant. In addition, the two established strict liability torts deal with land and interests in property. Trespass to land is sometimes also placed as a strict liability tort despite its location with intentional torts. Strict liability torts are very helpful to victims of disaster who may deploy the torts in asserting their pre-disaster rights. The two main strict liability torts are nuisance and the rule in *Rylands v Fletcher.*[17] Nuisance is a tort divided into private and public. Private nuisance envisages an unreasonable continuous state of material interference with

17 (1868) LR 3 HL 330

the use and enjoyment of another person's interest in property which causes damage. The public perspective envisages an action which materially affects the reasonable comfort and convenience of life of a class of people in the society. In *Adediran v. Interland Transport Limited,*[18] public nuisance was described as one which inflicts damage, injury, or inconvenience to the generality of the population or upon all of a class who come within its ambit. The case also decided that a private person can sue for public nuisance.

The rule in *Rylands v Fletcher* traditionally protects a property or interests thereon (exceptionally a person) where there has been a damage as a result of escape from the defendants property of an accumulation amounting to a non-natural use of property.[19]

Since the decisions in *Cambridge Water Company v Eastern Counties Leather Plc,*[20] and *Transco Plc v Stockport Metropolitan Borough Council,*[21] from the English jurisdiction and *Burnie Port Authority v General Jones Pty Ltd.*[22] from Australia, and *Smith v Inco Limited*[23] from Canada, it does appear that both the rule in *Rylands v Fletcher*[24] and nuisance have been thoroughly streamlined and at least can be said to be very close to the requirements for negligence to apply or assimilated within those requirements except to the extent that the latter two torts of strict liability regime are now said to govern land or interests in it while the tort of negligence is more extensive and does cover personal injury arising without proof of intention. It can therefore not be in doubt that the strict liability torts of the rule in *Rylands v Fletcher*

18 (1991) LPELR-SC.119/1987; (1991) NWLR (PT.214) 155

19 N Chinwuba 'Mr. Fletcher as the Quintessential Nigerian: Can a more in-depth analysis of the rule in *Rylands v. Fletcher* reveal a uniquely Nigerian Tort'(2013) 1 (1) *Journal of Comparative Law in Africa* 25 – 46; N Chinwuba 'Strict Liability or Reasonable Foresight in Rylands v. Fletcher: Why the Nigerian Position should be contextual' (2013) 31 *Journal of Private & Property Law* 103-125; N Chinwuba & E P Amechi 'Taming the Strict Liability Regime in Tortious Acts in Nigeria: the Relevance of the Forseeability of Damage Test for *Rylands v Fletcher* Rule in Oil Pollution Litigation' (2017) 1 BL.J 1-25

20 [1994] 2 AC 264

21 [2003] UKHL 61; [2004] 2 AC 1

22 (1994) 179 CLR 520, (1994) 120 ALR 42

23 (2011) O.R. (3d) 321

24 (1868) LR 3 HL 330

and nuisance are very vital tools in dealing with the consequences of disaster. It must be noted however that with respect to the rule in *Rylands v Fletcher,* only catastrophic escape now qualifies for liability as insurance and internalisation are to be preferred with respect to the subjects falling under the rule in modern times. This is the overall import of the decisions in *Transco Plc v Stockport Metropolitan Borough Council*[25] and *Smith v Inco Limited.*[26]

In the light of the above analysis, it is clear that there is need to pay a special attention to negligence as a tort. This section thus deals with the elements that would establish the tort of negligence. While the nature of vicarious liability is outside the scope of this discourse, it is worth noting that vicarious liability caters to the efficiency of negligence in tort law by ensuring that the impecuniosity of a negligent employee is passed on to his principal while the justice of law creates an escape route from this for the principal who does not have control of the employee through the ancillary principles governing liability for independent contractors.

Negligence — The broad Principles applicable to the tort of negligence

It is clear that negligence in ordinary parlance is a key factor in pre-disaster societies. Pre-disaster societies have been identified in this discourse as mainly, developing nations and in particular, the habitual state of hazardous pattern of living in Nigeria has been highlighted. In spite of this unfortunate pattern of living, it remains important that the rules developed in the application of negligence as a tort are followed. This is in line with the thrust of this chapter that law is predicated on order and law is a prerequisite for an orderly environment that reduces the propensity for disaster. For this reason, this section deals with the rules applicable to a proper finding of negligence on the part of a defendant.

25 [2003] UKHL 61; [2004] 2 AC 1

26 (2011) O.R. (3d) 321

i. Duty of Care

Negligence is a broad category of tortious claim that could be termed the umbrella of all torts. Indeed, it can as well be the case that all of tort law is underpinned by the question of negligence.

As originally formulated, the elements of negligence were distilled as legal duty, breach and damages. Lord Atkin in *Donoghue v Stevenson*[27] referred to the Biblical neighbour and its transmutation to a legal neighbour. Since then, the courts have been faced with streamlining the neighbour to 'tort'. The evidence provided by case law and statutory interventions would appear to also have some moral background which is to be found in the relationship between the neighbour and his God, that is, that we are to forgive others even as our heavenly father forgives us. The court finds its intervention in the middle course. How far should we hold a fallible human being liable for his undesired actions which may cause damage? In this sense, the courts strive to do justice but in a wider framework not in ad-hoc sense that does justice between immediate litigants.

The questions then are formulated in a far-reaching sense to form both questions of law and fact. This is the contribution of the judiciary to social justice. It now appears accepted that the question –'duty or no duty' ultimately depends in most (and perhaps all) cases on the courts' assessment of what is fair, just and reasonable, but that question is asked not in the context of justice as between the particular parties to the case but in the context of all cases of that type.[28] This position is well put by Lord Browne-Wilkinson:

> In English law, questions of public policy and the question whether it is fair and reasonable to impose liability in negligence are decided as questions of law. Once the decision is taken

27 [1932] AC 562

28 See the views of *Winfield & Jolowicz Tort 18th ed.*, by W.V.H. Rogers (Sweet & Maxwell 2010)

> that, say, company auditors, though liable to shareholders for negligently auditing, are not liable to those proposing to invest in the company... that decision will apply to all future cases of the same kind.[29]

Thus, the first question is whether there is a recognisable duty. In *Donoghue v Stevenson*, the question was, whether a duty was owed by manufacturers to consumers. Once the answer was in the positive, then the next question is one of fact, whether the defendant owed the duty to the particular claimant before the court. In this way the approach of the court translates to a rule of law not arbitrariness. Nigerian courts have generally capped the application of negligence in application and scope to pre-1960 subjecting it to a question of fact. It has to be stated that the question of duty, breach and indeed damages, involve questions of facts and law quite in contradistinction to the positions taken in *UTB v Ozoemene*[30] and *Ngliari v Mothercat.*[31]

It would appear that the approach to be taken in establishing a duty of care can be restated to be that there must be sufficient proximity between the defendant and claimant. Proximity here, not being mere geographical measurement but the question whether the claimant is a person whom the defendant should have in contemplation in his action.[32] The other questions used in determining duty at this stage is the fairness, justice and reasonableness of imposing such a duty. The court would also inquire whether the defendant has assumed some responsibility to the claimant with regard to the matter from which the damage arose. A final point is that the court does not close the areas in respect of which duty may be imposed but does so by analogy to existing accepted duty relations.

---

29 Barret (A.P.) v. London Borough of Enfield [2001] 2 AC 550, [1999] UKHL 25, [1999] 3 All ER 193, [1993] 3 WLR 79

30 (2007) LPELR-SC.129/2001; (2007) 3 NWLR (Pt.1022) 448; (2007) 1-2 S.C (Part II) 211

31 [1999] 13 NWLR 426; (1999) LPELR-SC.223/1993

32 See Anns v Merton [1978] A.C. 728; Junior Brooks Ltd v. Veitchi Co. Ltd. [1983] 1 A.C. 520; Yuen Kun Yeu v Att-Gen. of Hong Kong [1988] A.C. 175

Where the court decides that a duty exists then it moves to the next question of whether such a duty has been breached. Again, at this point, it is not enough that the claimant has suffered damage as a result of the action of the defendant. It must be shown that the defendant breached a duty to the claimant.

ii. Breach of Duty

With respect to breach, the court proceeds on the principle of reasonableness and the standard is applicable to the ordinary reasonable man. Breach of duty, more than the other two elements of negligence, implies some measure of discretion. Thus, although the court must assess the standard based on the facts of the case before it, the intuitions of the court/judge cannot be ruled out. Nevertheless, in many common law jurisdictions, there are factors employed in determining reasonableness. These include, the magnitude of the risk involved; the desirability of avoiding the risk; the exigency of the defendant's action; and practicability of the precautions.

iii. Causation and Remoteness

In this area, the law through the judiciary and legislature, can be said to be engaged in negotiating a balance between compensating the individual who has suffered a wrong and the persons who should be making the compensation on the one hand, and on the other hand, how far the law should go in making the wrong-doer responsible for the suffering of the victim of a wrong. The reason for these continuing efforts at balancing is rooted in the original formulation of the rule – a person is to love his neighbour as himself. This rule of morality is restated as having a legal equivalent and basis. In a similar vein, the rule that God forgives us our sins and therefore we must forgive others is also given vent in limiting, with the aid of logical legal rules and guiding factors, potential claimants. In the context of an industrial society, economic questions are considered and justificatory basis for tort law are rooted through its effect on industrialisation and economic welfare of society.

It is now clear that establishing negligence is a progressive action. First, the plaintiff must establish that the defendant owes him a duty of care, which is both in fact and also in law. Secondly, the plaintiff must also establish that the duty has been breached by the defendant. Certain factors exist to help the courts arrive at this decision. The burden to prove the existence of a duty and breach of it is on the claimant/plaintiff. However, in a few cases, the burden may shift to the defendant because, the court draws such inference from the case that only the defendant can, in fact, proffer an explanation as to how the plaintiff sustained the injury in question.

In these first two parts, the work is for the plaintiff and in the limited cases where the burden of proof shifts, it is for the defendant to convince the court to progress to the third element of causation and remoteness of damage. The damage here is concerned with the damage suffered by the claimant and not the damages by way of remedy.

This part which aims at linking the defendant sufficiently with the injury suffered by the claimant entails two balancing acts of the court as noted above. Here we have the question of private determination of the case as it is before the court and long term impact of that decision. In this later question, two things are at play in the working of the courts: (1) while compensating the defendant is right and is the main objective of tort law; but (2) will compensating the defendant create a precedent which will profit the claimant and in the long run be good for the entire legal system in a given society, economically and socially? This is often described as policy effect.

Thus, in arriving at a decision in negligence, the approach of the court in a case between two individuals, an individual and a company and a company and another company, should be driven along this philosophy. Thus, when a case involves two individuals, the impact would normally be socially and historically driven while if corporations are involved, a lot more is at play: social, economic etc. This is because, corporations involve weighted issues of employment, insurance etc. which all impact directly on the economy and the social life of the people.

According to Winfield, distinguishing the second stage of enquiry from the first necessitates noticing a shift in nomenclature from the 'historical' or 'mechanical' to one of obvious fairness and legal policy, about setting limits to the responsibility which a wrong doer bears for his negligence.[33] According to Markesinis and Deakin:

> Notwithstanding, the importance of notions of personal responsibility, it is impossible to give an account of this area without returning over and over again to certain central issues of policy which are to be found throughout the tort of negligence and which affect other torts involving damage. These include the search for an economically efficient framework of liability and, in conjunction or sometimes in competition with this, for a system of loss shifting and loss spreading which reflects social values and concerns about the allocation of risks and responsibilities. In a doctrinal sense, it is impossible to write about causation without acknowledging that inherent in the law's account of causation is also a view of what constitutes *fault*.[34]

iv. Causation in Fact

This usually involves the 'but-for' test. Would the injury of the plaintiff have occurred in spite of the action of the defendant? If this is the case, then the defendant is absolved at this preliminary stage – usually. The claimant must prove the existence of a causal link on the balance of probabilities, which is taken to mean a likelihood of more than 50%. If the court finds that it was as likely as not, that the injury would have occurred without the defendant's negligence, the action will fail even if there is an admission of carelessness.[35] If, however, 'but-for', the action of the defendant, the claimant would not have sustained the injury, then, the defendant is liable and then the court would usually then be obliged to proceed to the next stage of the enquiry.

However, the test often runs into hitches, where for instance, the wrong of the defendant is followed by another wrong which

33 *Winfield & Jolowicz Tort* 18th ed., by W.V.H. Rogers (Sweet & Maxwell 2010) 308

34 *Markesinis and Deakin's Tort Law* 6th ed., by S. Deakin, A. Johnston & B. Markesinis (Clarendon Press – Oxford 2008) 248

35 See Barnett. See further *Wilsher v Essex Area Health Authority* (1988) AC 1074; see also *Bolitho v City & Hackney Health Authority* [1998] AC 232

does not substantially alter the initial scenario, but does add to it. Thus, in *Baker v Willoughby,*[36] as a result of the defendant's negligence, the claimant suffered an injury to his leg. Damages for this injury had been fully assessed. However, before trial was concluded, armed robbers shot the claimant several times on the same leg resulting in the amputation of the leg. Ordinary justice would have limited the liability of the defendant to the initial injury but legal justice demanded that suing the robbers even to judgment would have inadequately compensated the claimant since they would only be held accountable for injury to an already damaged leg. The original tort feasor had to bear full responsibility.[37] Again, where there is more than one wrong doer each, contributing fully or substantially to the cause of the claimant's injury, all the wrong doers will be held liable.

In *Fairchild v Glenhaven Funeral Services Ltd,*[38] the House of Lords had to qualify the 'but-for' test to allow recovery in a case where it was impossible on the state of scientific evidence to determine which one or more of several employers, all admittedly in breach of duty, caused the claimant to suffer a fatal illness. According to Markesinis and Deakin,

> These cases suggest that although the 'but-for' test is based on notions of the limits of individual responsibility, and in particular on the precept that the defendant should not be liable for a loss which he personally did not cause, such a principle may come into conflict with the aim of ensuring that the victim of tortuous conduct is fully compensated for losses caused by fault.[39]

In the same vein, Lord Nicholls notes, even the sophisticated versions of the 'but-for' test cannot be expected to set out a formula whose mechanical application will provide infallible threshold guidance on causal connection for every tort in every circumstance. In particular, the 'but-for' test can be over-exclusionary.[40] There could also be a probabilistic cause, as in a

36 [1970] AC 32
37 See however, *Jobling v Associated Dairies Ltd.* [1982] AC 794
38 [2003] 1 AC 32
39 *Markesinis & Deakin's Tort Law* (2nd Ed.) 247.
40 *Kuwait Airways Corp. v Iraqi Airways Co.* (Nos. 4 & 5) [2002] AC 883,1092

situation where the court is faced with several probable causes of the claimant's injury of which the defendant's action is one. What should be done? Clearly one of the probable causes is responsible for the claimant's damage but there is no certainty which one actually is the cause.

In *McGhee v National Coal Board,*[41] the plaintiff contracted dermatitis after working in a kiln where he was exposed to cement dust. The defendants were not faulted for this as this was incidental to his contract of employment. However, the employers were in breach of their common law duty of care in failing to provide washing facilities at the place of work. Medical evidence about the causes of dermatitis was such that it was not possible to say that, had washing facilities been provided, the pursuer would have escaped the disease. In this case, the defendant's failure had substantially contributed to the claimant's injury. The House of Lords held that the claimant was entitled to succeed. In Fairchild, the result of the case was that all contributors were made to bear responsibility equally rather than send the claimant away. This position was varied in *Barker v Corus (UK)*[42] and proportionate liability imposed in relation to the extent of the defendant's liability.

This case demonstrates the necessity for theoretical principled basis for practical application of the law. Damages could easily have been awarded without explanations but the courts have consciously provided legal reasoning that impacts on the individual and the society on both short and long term basis for the preservation of society. As Lord Nicholls puts it:

> Any other outcome would be deeply offensive to instinctive notions of what justice requires and fairness demands. The real difficulty lies in elucidating in sufficiently specific terms the principle being applied in reaching this conclusion. To be acceptable the law must be coherent. It must be principled. The basis on which one case, or one type of case, is distinguished from another should be transparent and capable of identification. When a decision departs from principles normally applied, the

41 [1973] 1 WLR 1

42 [2006] UKHL 20

> basis for doing so must be rational and justifiable if the decision is to avoid the reproach that hard cases make bad law.[43]

There is also the question of loss of a chance for recovery which appears not to be receiving favourable consideration. Loss of chances cases appear to be treated as a failure to prove causation.[44]

v. Causation in Law or Forseeability

As Winfield puts it,

> the law must draw a line somewhere, some consequences must be abstracted as relevant, not on grounds of pure logic, but simply for practical reasons....The dominant (though by no means conclusive or all-embracing) question in the determination of the scope of the defendant's liability is whether the harm in respect of which he is sued was a foreseeable consequence of his negligent act.

However, after the 'but-for' test has been established in favour of the plaintiff's injury, the next question is to determine whether there as a matter of law the defendant's action remains the reason for the plaintiff's injury. Certain factors can be distilled as aiding the court to arrive at a conclusion on this arm of the negligence enquiry. One such principle is for the court to ascertain whether there has been a legal break in the chain of causation. This could then lead to the question of foresight; that is, whether the plaintiff's damage and its kind could have been foreseen or the chain had been broken either by natural events or by the plaintiff's own action.

Where the chain has been broken by natural events, then the defendant is absolved but where the plaintiff's action is contributory, then the question of contributory negligence may set in and where none of this is the case, then the court must proceed to determining forseeability with respect to damage and

43 *Fair Child v Glenhaven Funeral Services Ltd.* [2002] UKHL 22,36; [2003] 1 A.C. 32

44 See *Hotson v East Berkshire Health Authority* [1987] AC 750; Gregg v Scott [2005] UKHL 2; [2005] 2 AC 176

its kind. The principle contained in the previous section may have led to the conclusion that the defendant's wrong was a cause of the claimant's loss in the sense that it was (or is to be treated as if it was) a necessary condition to produce that loss or made a material contribution to it. No defendant is responsible *ad infinitum* for all the consequences of his wrongful conduct, however remote in time and however indirect the process of causation, for otherwise human activity would be unreasonably hampered. The dominant question therefore from this point is "was the harm or damage suffered by the claimant a foreseeable consequence of the action of the defendant?"

Two cases dominate this aspect of requirements for establishing negligence — *The Wagon Mound*[45] and *Re Polemis.*[46] The test of forseeability involves the assessment of facts against a legal standard because all the facts of the particular case have to be brought into account. The foresight here need not be precise both in terms of duty, damage and manner of its occurrence. In *Stewart v West African Terminals Ltd,*[47] Denning MR stated:

> It is not necessary that the precise concatenation of circumstances should be envisaged. If the consequence was one which was within the range which any reasonable person might foresee (and was not an entirely different kind which no one would anticipate), then it is within the rule that a person who has been guilty of negligence is liable for the consequences.

This stage breaches the dire each of causation in law and is the full province of the court. It is a legal negotiation driven by the determination whether the defendant should be absolved or held accountable for his wrong. This aspect is now total legal reasoning and forms the actual exhibition of expertise or lack of it in legal reasoning. The parties and their actions or inactions have no more role to play, in the sense that this part of the proceedings now falls to the deduction of the judge. Two main variables in the work of the court are at play – one is the overall determination of

45 [1961] A.C. 388

46 [1921] 3 KB 560

47 [1964] 2 Lloyds Rep. 371, 375;

whether or not the defendant is liable and the second, is if indeed he is liable to what extent should he be liable?

After the establishment of the 'but-for' relationship between the claimant and the defendant, the court now assesses whether to move to the next stage involving the determination of whether there has been an act which could be regarded as breaking or affecting the causal chain. Such an act could be the act of a third party, an external event to the causal event or the act of the claimant himself. Where the act of the claimant is involved, it may absolve the defendant completely or make the claimant liable to the extent of his contribution to the damage. Aside from the reasons proffered below, because here there is an overlap between the English and Nigerian position as we have local legislations governing contributory negligence different from common law principles and the English 1945 Contributory Negligence Act.[48]

CONCLUSION

The human journey is one of aspiration, objective, hard work and consideration for others. The chapter identifies 'others' to include the earth and other occupiers of the earth aside from humans. In this sense, there is a responsibility of self-awareness, self-preservation and stewardship for others. The stewardship for others entails a policy directive that captures equality in the sense of equal opportunity for decent living free of apprehension and undue concern for one's safety. Stewardship falls on government to realise the aspirations of a people to express themselves maximally as humans in a conducive environment for self-actualisation and effective participation in nation-building. Prevention of disaster and management of disaster entails a multifaceted approach from multi-disciplinary dimensions but this discourse has focused on

48 On Natural occurrences, see *Carslogie Steamship Co. Ltd. v Royal Norwegian Government* [1952] AC 292; *Lamb v Camden London Borough Council* [1985] QB 625. On Natural Act of a Third Party, see *The Oropesa* (reported in Winfield); *Rouse v Squires* [1973] 3 QB 889; see however, *Wright v Lodge* [1993] 4 ALL ER 299

law with special emphasis on the role of tort law. Tort law has evolved complementarily with constitutional law, contractual and ancillary rights and criminal law in protecting the rights of individuals and societies in themselves and the provision of orderly tranquil society.

The chapter has demonstrated ways in which tort law can achieve this role through the utilisation of the payment of unliquidated damages and the use of injunctions, particularly the *quia timet* injunction. The cases of *Merck Sharp Dohme Corp. v Teva Pharma B.V.*[49]; *Hooper v Rogers*[50]; *Lloyds v Symonds*[51]; *and Smith v Inco Limited*[52] demonstrate the resolution of the courts to make justice available to all by facilitating the creation and preservation of boundaries. Negligence, strict liability torts of the rule in *Rylands v Fletcher* and nuisance and indeed, trespass to land are all torts which are relevant in dealing with disaster in pre-disaster situations as identified in this work as well as disaster and post-disaster societies in their context in this discourse. It remains incumbent on government at all levels to close the wide stratification of society that creates vulnerable groups who become incapable of caring for themselves or so battered that to these groups, it does not matter what happens one way or the other in a circle of hopelessness.

49 [2012] EWHC 627
50 [1975] 1 Ch 43
51 [1998] EWCA Civ 511
52 (2011) O.R. (3d) 321

7

EXAMINING THE LAW ON FAMILY RESETTLEMENT IN DISASTER AFFECTED AREAS FROM MATRIMONIAL STABILITY PERSPECTIVE

Roseline O. EHIEMUA

ABSTRACT

This paper adopts a descriptive and evaluative approach as it argues from the premise that there is no adequate law in Nigeria on family resettlement in disaster affected or prone areas, especially from the perspective of ameliorating the physical and psychological impact such disasters have on matrimonial stability. The paper's use of the concept of 'family resettlement' is limited to local and international support-rehabilitation process of local victims of disaster within a country, not in a 'refugee' situation, which this paper chooses to refer to as 'an expanded family law perspective.' From this family law angle, the paper asserts that just providing shelter, food and clothing is not adequate to resettle a family or guarantee matrimonial stability. The essential arguments of this paper, however, are that, the family being the smallest unit of society experiences emotional disorientation and physical loss when the wives and/or daughters are raped or captured, the male bread winner or the supportive wife is killed, family property or business is destroyed during a disaster occurrence arising from war, terrorism or ecological factors. The trauma is better imagined than experienced and could lead to chain effects such as prostitution, emotional detachment between affected spouses capable of leading to broken marriages, the development of criminal tendencies, and the destruction of the aspirations and potentials of the offspring of affected families. These family experiences on the short and long run come back to hunt and hurt society. The paper ultimately recommends an amendment to the existing legal frameworks that should address these problems likely to affect the stability of marriages and well-being of the children by suggesting some provisions that should constitute part of the legal instrument which, for instance, led to the formulation of NEMA.

Keywords: Disaster, Law, Family Resettlement, Matrimonial Stability

I. Introduction

The term 'disaster' implies a great misfortune causing enormous material damage and human suffering: it may refer to an event that causes widespread damage, physical injury and intense mental trauma; or, a state in which victims find themselves during and after the event.[1] It could also be defined as a calamitous event arising from war, ecological or natural cause that seriously disrupts the functioning of a community or society, and results in human, material, economic and environmental destruction which overwhelms the community's or society's ability to cope with its left-over resources.[2] Thus, disaster is a complex, multidimensional

1 Lourdes Ladrido – Ignacio and Antonio P. Perlas, 'From Victims to Survivors: Psychosocial Intervention in Disaster Management in the Philippines' (1995-1996) 24 International Mental Health 3.

2 International Federation of Red Cross and Red Crescent Societies (IFRC) 'What is a disaster?' <www.ifrc.org.en/what-we-do/disastermanagement/aboutdisaster/what-is-a-disaster> accessed 19th November 2017; Government of Kenya, Ministry of State for Special Programmes, Office of the President, 'Draft National Policy for Disaster Management in Kenya' March 2009, 3 <www.ifrc.org/docs/idrl/1058En.pdf> accessed 19th November 2017

phenomenon which affects the ecological, economic, material, psychological, social and spiritual dimensions of the human life.[3] The rehabilitation of the human life is paramount to humanity's response to ameliorating the impact of disaster. By 'matrimonial stability', the paper refers to a condition in which spouses in the family remain together and the unity, peace, moral obligations of the marriage and spousal companionship before a disaster are not threatened by divorce or separation after the disaster.

Nigeria, like the rest of the world, is vulnerable to a variety of disasters, whether related natural phenomena such as flooding and droughts, or man-made disaster, which results from genocide, armed conflict, industrial accident, oil spillages, or epidemic. All these factors have made Nigeria prone to disasters, as evidence of historical and contemporary experiences abound. However, beyond such events, the impact of disaster on most families which constitute the affected community is a negative one. The disaster becomes a condition that inevitably affects a family structure in various ways.[4]

Over the years, respect for the rights and needs of the family and awareness of the responsibilities of states have developed both at international and regional levels. It is pertinent to note that, states have a duty to protect and assist families in their time of need. For instance, Article 18(1) of the African Charter on Human and People's Rights provides that the family is the natural unit and basis of the society and it shall be protected by the state, which shall take care of its physical health and moral need.[5]

This paper therefore asserts that the duty of government to protect the family unit and its stability is 'all-inclusive' and ramifying. Government at all levels must realise that, while it is important to focus on providing for temporary accommodation, medical attention and relief materials for families affected by

3 Ladrido – Ignacio and Perlas (n 1) 4.

4 Jean Lipman-Blumen, 'A Crisis Framework Applied to Macrosociological Family Changes: Marriage, Divorce and Occupational Trends Associated with World War II' (1975) 37 Journal of Marriage and Family 889.

5 The African Charter on Human and People's Rights (Ratification and Enforcement) Act, 1983, Cap A9 LFN 2004.

disaster, the consequences of disaster on the mental, psychological, economic and the social well-being of families impact negatively on matrimonial stability. The protection of the family during disaster requires that appropriate measures be put in place, which will allow dispersed or disoriented family members to be re-integrated or rehabilitated both physically and mentally, as well as promote social and economic sustainability of the family as a unit.

Examining the Concept of the Family from Legal and Sociological Perspectives

The role of the family as the central unit of human society is recognised in all cultures and traditions. The integrity of the family is protected under the constitution of the Federal Republic of Nigeria, 1999 and other international instruments. There is a universal consensus that the family, as a fundamental unit of the society, is entitled to respect and protection. A right to family resettlement during disaster is inherent in recognising the family as a group or unit.

In defining the term 'family', there are however some problems. For instance, Douglas posits that one would think that the concept 'family' appears to have a global significance and meaning appearing in Article 16 of the Universal Declaration of Human Rights (UDHR) where it is proclaimed as "the natural and fundamental group unit of society." He contends that the meaning of the phrase as spelt out in this instrument and in other international instruments is not elucidated; thereby giving opportunities to different countries, with widely different cultures, to provide their understanding of what otherwise looks like a uniform concept.[6]

Burton agrees that the term, family, is difficult to define in modern times, since contemporary human rights law has accepted that two persons who have never met but are linked by blood

6 Gillian Douglas, *An Introduction to Family Law* (2nd edn, Oxford University Press 2004) 1

– such as the unmarried father and his child were found to be *de facto* family.[7] Douglas and Burton's views corroborate the position of Bromley, an older family law scholar, who posits that:

> ... the word 'family' is one which is difficult to define precisely. In one sense, it means all blood relations who are descended from a common ancestor; in another, it means all the members of a household including husband and wife, children, servants and even lodgers.[8]

The family, obviously, is an important construct (as a concept and segment) in every society. This may have prompted the introduction of section 17(3) (h) to the 1999 Constitution of the Federal Republic of Nigeria. The paragraph stipulates that "the evolution and promotion of family life is encouraged." As the nucleus of a community, what affects the family is of vital interest to all.[9] Evidently, forming a family through marriage is important in many societies. The concept of marriage as a social contract is pervasive. It is one in which the security of partners in the bargain is of paramount interest; which is why many legal systems have integrated into their laws all forms of security for the parties in a marriage, in order to forestall social conflicts that could challenge any modern constitution's claim to the promotion of justice and equality, and which could also become anarchic to society.

The family is seen as a social institution that performs essential functions for the society. What these functions are vary from one society to another. They include:

i. ***Raising Responsible Children and Providing Emotional Security:*** If a society is to persist beyond one generation, it is necessary that adults should not only bear children, but also provide them with the basic requirements of life. Family members can provide crucial emotional support and care for each other. Parents can furnish the love and

7 Frances Burton, *Family Law* (Cavendish Publishing 2003) 3.

8 Peter Mann Bromley, *Family Law* (6th edn, Butterworth 1981) 1.

9 Peter A. Anyebe , 'Family Law and Inheritance in Nigeria' in E. Azinge and N. Aduba (eds), *Law and Development in Nigeria: 50 Years of Nationhood* (Nigerian Institute of Advanced Legal Studies 2010) 1142.

security that children need as they are growing up. As Schaffer has argued:

> Families are ideally suited for the bringing up of children: they are small, intimate groups, making it easy for children to acquire consistent rules of behaviour; they are linked to various outside setting (other families, work, leisure, and so forth) to which children can gradually be introduced; and they are usually composed of individuals deeply committed to the child whose security and care can therefore be guaranteed. The family is thus the basic unit within which the child is introduced to social living.[10]

ii. ***Economic Support:*** The second function involves providing economic support for her members. Although the family may be a self-sufficient unit as virtually every member engages in activities geared towards providing for needs such as food, clothing, shelter, etc.

iii. Another basic function of the family is providing care and support for the elderly/aged and physically challenged especially when there is no social security for them by government. Keeping families intact after disaster promotes this social responsibility of the family and can help to sustain marital stability as well.

iv. Spousal companionship is a significant contribution of the family to the well-being of the society which disasters can threaten or destroy and for which the need to preserve in the aftermath of disaster should be of interest to all stakeholders. This is a vital element of the family unit.

The advantages of family life are not limited to the benefits received by the members themselves. Families benefit the state. Ronald Reagan captured a popular perception of the family in this regard, that:

> Strong families are the foundation of society. Through them we pass our traditions, rituals and values. From them we receive the love, encouragement and education

10 C. Schaffer, *Making Decision About Children: Psychological Questions and Answers* (Blackwell 1990), 204.

> needed to meet human challenges. Family life provides opportunity and time for the spiritual growth that foster generosity of spirit and responsible citizenship.[11]

THE IMPACT OF DISASTER ON FAMILY SECURITY AND EMOTIONAL STABILITY

When there is a disaster, the ability of the family to function as a unit is restricted, if not completely eroded; which ultimately affects the nature and structure of society. Studies on disasters tend to cut across multiple disciplines including: clinical, social and developmental psychology, psychiatry, epidemiology, sociology and law;[12] each of which views the subject from a different perspective. Over the years, a gradual change has occurred on the perception of the impact of disaster. Rather than emphasize the physical environmental effect of the destruction wreaked by disaster, an approach that dwells on structural and economic losses, the focus has shifted to the human person.[13] In recent times, much attention is now given to the emotional effects of disaster which results from the loss of a loved one, properties and the source of livelihood.[14] The fear and the helplessness experienced by all disaster victims create physical, emotional and cognitive reaction that may make recovery difficult.

The far-reaching impact of disaster can engender a range of psychological, social and economic consequences. The question is: how far can these consequences affect the stability of a marriage? Answering this question has proved surprisingly difficult. Studies on disaster encounter a number of formidable methodological obstacles. As a result, available literature on disaster tend to focus on the impact of disaster on individuals, children, women and

11 Quoted by White House Working Group on the Family (1985) citing Jonathan Herring, *Family Law* (2nd edn, Pearson Education 2004) 8.

12 George A. Bonanno, et al *Weighing the Costs of Disasters: Consequences, Risks* and *Resilience in Individuals, Families and Communities'* (2010) 11 Psychological Science in the Public Interest 3.

13 *Ibid*

14 Harry Estill Moore and H.J. Friedsam, *Reported Emotional Stress Following a Disaster* (1959) 38 Social Force 135.

communities, while failing to consider the potential impact of disaster on marriage stability. In some of the available literature on the long term effects of disaster, Erikson,[15] Hodgkinson and Stewart[16] reveal that there have been reports of rising incidents of divorce and marital conflicts in the wake of disaster. On the other side, Quarantelli[17] and Granot[18] take exception to this finding, claiming that the effect of disaster on marriage is minimal.

Certainly, there is no empirical research in Nigeria on the impact of disaster on matrimonial stability but the story of one of the recent flood victims brings to the fore the challenges faced by spouses and families at the occurrence of a disaster. The story in Nigeria of a flood victim identified as Mrs. Lember Vitalis is quite pathetic. It was reported that Mrs. Lember had been squatting in another woman's kitchen since she lost her home to the recent flood disaster in Benue State. She had no home to house her new born babies, and she lost every means and hope of survival.[19] It was further reported that her husband absconded since he could not provide for the basic needs of his family. Mr. Vitalis Kember who left his home shortly after his wife gave birth to a set of triplets in Benue State only returned to take full responsibility of his wife and children after they had received help from well

15 Kai T. Erikson, *Everything in Its Path: Destruction of Community in the Buffalo Creek Flood* (Simon and Schuster 1976) citing Hayim Granot 'Impact of the Gulf War on Marriage and Divorce in Israel' (1985) 25 *International Journal of the Sociology of the Family* 39.

16 Peter E. Hodgkinson and Michael Stewart, *Coping with Catastrophe: A Handbook of Post-disaster Psychosocial Aftercare* (Routledge 1991), citing Hayim Granot, 'Impact of the Gulf War on Marriage and Divorce in Israel' (1985) 25 *International Journal of the Sociology of the Family* 39.

17 E. L. Quarantelli, 'An Assessment of Conflicting Views on Mental Health: The Case of Traumatic Events' in Charles Figley (eds), *Trauma and Its Wake,* vol. 11 (Brunner/Mazel 1985) citing Hayim Granot, 'Impact of the Gulf War on Marriage and Divorce in Israel' (1985) 25 *International Journal of the Sociology of the Family* 39.

18 Hayim Granot, 'Impact of the Gulf War on Marriage and Divorce in Israel' (1985) 25 *International Journal of the Sociology of the Family* 45.

19 'Photos: Benue Flood Victim Squatting in Kitchen Delivers Set of Triplets' <https://www.lailasblog.com/benue-flood-victim-delivers -set-triplets-kitchen/> accessed 20th November 2017.

meaning Nigerians.[20] This is one of the ways disaster affects matrimonial stability. There are obviously many unreported stories of other negative effects such as female spouses hawking or prostituting her female endowments to survive, sustain oneself, family, and children; or do so to fulfil a need when the husband dies or absconds as a result of the disaster.[21] The loss of means of livelihood alone has wide-ranging negative impact on marriage in ordinary social situations, let alone during disasters. The loss of means of sustenance affects parents' physical and mental health; it can impact on marital relationships and increase the likelihood of divorce and desertion.

Another sad tale shared by another victim of disaster also provides a different angle of its effect on family. Fatima, one of the victims of the Boko Haram insurgency in the North-East Nigeria, living in Kabbar Malla host community in Maiduguri, recounts her experience. Her husband used to be a trader and the family owned their home. She fled in fear three years ago with her children and spent one month in the forest after Boko Haram attacked her village and killed her husband. To survive, they relied on leftover from communities they passed along on the way. She now lives at Kabbar Mallam host community with her children in a crowded room, with a leaking roof. She does not work as people do not want to give job to someone of her age. She has not been able to pay rents for three months. To feed her family, her eldest daughter buys sachets of water from a vendor and hawks them on the street. If she is unable to make money from the sales of water, the children are sent begging to survive.[22]

There are many challenges faced by families that threaten marital stability in the aftermath of disaster. The negative impact takes various dimensions which include: child trafficking, sexual

20 Naija xtreme, 'Man Who Ran Away after His Wife Gave Birth to Triplets in Benue Finally Returns' (18th October 2017) <https://naijaxtreme.com/man-ran-away-wife-gave-birth-triplets-benue-finally-returns-photos/> accessed 20th November 2017.

21 Some of the marital experiences or observations are unprintable from unofficial or oral narratives about family victims of disaster.

22 OXFAM, 'Displaced from Home: Stories of Survival and Solidarity in Nigeria' <https://www.oxfam.org.en/was-africa-crisis-nigeria/displaced-homes-stories-survival-and-solidarity> accessed 19th November 2017.

abuses and violence, prostitution, destruction of the aspirations and potentials of the offspring, etc.

Legal Frameworks on the Protection and Rehabilitation of Victims of Disaster in Nigeria: A Review

1. The Constitution of the Federal Republic of Nigeria, 1999

It is the responsibility of government to protect citizens from natural and man-made disasters. This responsibility, although not expressly stated in the Constitution, could be gleaned from section 14(1) (b) which provides that: "the security and welfare of the people shall be the primary purpose of government." The protection of the welfare of the people by extension includes the protection of the welfare of the family as a unit.[23] If the family is the basic unit of the society then, the protection of the family unit, particularly during disaster, should be the primary purpose of government. In order to fulfil this constitutional mandate, several agencies have been established by the Federal government.

2.) National Emergency Management Agency (Establishment) Act, 1999

This Act established the National Emergency Management Agency (NEMA)[24] and a State Emergency Management Committee for each state of the federation.[25] The functions of NEMA, amongst others, are:

a. To formulate policy on all activities relating to disaster management at the national level;

23 Winnifred O. Olanipekun, 'The Illusory Right to Family Life' (Guardian, 12th March 2016) <https://www.guardian.ng/opinion/the-illusory-right-to-family-life/> accessed 20th November 2017.

24 National Emergency Management Agency (Establishment) Act, 1999, s 1(1), CAP N 34, Laws of the Federation of Nigeria, 2004.

25 *Ibid,* s 8(1).

b. co-ordinate and promote research activities relating to disaster management at the national level;

c. monitor the state of preparedness of all organisations or agencies which may contribute to disaster management in Nigeria;

d. collate data from relevant agencies, so as to enhance forecasting, planning and field operation of disaster management;

e. educate and inform the public on disaster prevention and control measures; and

f. distribute emergency relief materials to victims of natural and other disasters and assist in the rehabilitation of the victims where necessary.[26]

In furtherance of its objectives, NEMA, in conjunction with other stakeholders in disaster management, has developed strategic plans for management of disaster in Nigeria. These strategic plans are:

i. National Disaster Response Plan (NDRP) 2001

This plan was approved by the Federal Executive Council in 2001 as a guideline for responding to disasters in Nigeria. NDRP establishes a process and structure for the systematic, coordinated and effective delivery of federal assistance to address the consequences of any major disaster or emergency declared by the President of the Federal Republic of Nigeria. The NDRP applies to a major disaster or emergency which includes a natural catastrophe, fire, flood or explosion, regardless of the cause; or any other occasion or instance for which the President determines that federal assistance is needed to supplement state and local government efforts and capabilities.[27] However, the

26 *Ibid* s 6

27 NEMA, 'Foreword to the National Disaster Management Framework'<http://nema.gov.ng/national-disaster-management-frame-work/2/>accessed 25th November 2017.

NDRP had some deficiencies, for instance, issues like risk reduction, prevention, preparedness, mitigation and recovery were not captured by the NDRP. This necessitated the development of the National Disaster Management Framework (NDMF) to correct implementation gaps and increase efficiency and effectiveness of disaster management in Nigeria.

ii. *The National Disaster Management Framework (NDMF), 2010*

The NDMF provides a mechanism that serves as a regular guideline for effective and efficient disaster management in Nigeria. It offers a holistic approach to disaster management. It serves as a legal instrument which addresses the need for consistency among multiple stakeholders. It is a multi-jurisdictional, multi-sectoral, multi-disciplinary and multi-resources initiative. It spells out the roles and responsibilities of the federal, state, local governments, and civil organisations in disaster management. The framework defines measurable, flexible and adaptable coordinating structures, and aligns key roles and responsibilities of disaster management stakeholders across the nation. It describes specific authorities and best practices for managing disasters and explains a paradigm shift in disaster management beyond mere response and recovery. It has eight sections which focus on various thematic areas. Section seven is most related to the focus of this paper, because it dwells on strategies required for bringing back disaster affected areas and its victims to normalcy through rehabilitation and reintegration. The disaster rehabilitation and reintegration strategy includes: re-establishment of adequate supply of housing to replace that which has been destroyed; restoration of public infrastructure and social services damaged by the accident, enhancing resilience and resuscitation of livelihood of affected communities. Thus, the disaster rehabilitation and reintegration planning includes:

i) The involvement of all relevant stakeholders in rehabilitation and reintegration;
ii) Define stakeholders' roles and responsibilities for disaster, rehabilitation and reintegration;
iii) Consultatively define appropriate vision and approaches for rehabilitation and reintegration of displaced persons, families and groups.

Although, the NDMF provides for the key processes required to achieve a coordinated, coherent and consistent approach to disasters management in Nigeria, it fails to identify and define specific strategies for family rehabilitation and reintegration. There is therefore the need for a more integrative response to disaster management in Nigeria, taking into consideration the needs of the family as a group from coordinated or combined economic, psychological and matrimonial perspectives. This would involve regular contacts with the victims, counselling from relevant experts and meeting with physical obligations by all stakeholders with the governments at different levels taking the lead.

International Conventions and Treaties on Resettling Displaced Victims

1.) Regional Responses to Disaster Management in Africa

Nigeria is a member of the African Union (AU) that was established under the Constitutive Act of the African Union, 2001. Although the AU Act does not specifically refer to disaster response, it provides for member states to "work with relevant international partners in the eradication of preventable disease and the promotion of good health on the continent."[28] The AU also adopted the Dares

28 The Constitutive Act of the African Union, 2001, Article 3(n) <http://www.achpr.org/instruments/au-constitutive-act/> accessed 25th November 2017.

Salaam Declaration on Feeding of Infants and Young Persons in Emergency Situations in Africa in 1999.

The African Regional Strategy for Disaster Risk Reduction was developed through the initiative of the AU, the New Partnership for Africa Development (NEPAD) and the United Nations International Strategy for Disaster Reduction (ISDR) and adopted by the highest decision-making organ of the AU in 2004.[29] The initiative is aimed at assisting AU member states in enhancing disaster risk reduction, in order to reduce the suffering of the communities and destruction of the environment.

Further, the African Union Convention for the Protection and Assistance of Internally Displaced Persons in Africa, 2009, known as the Kampala Convention was ratified by Nigeria on April 17th, 2012. The Kampala Convention provides that states shall bear the primary duty and responsibility for providing protection and assistance to IDPs, without discrimination of any kind.[30] Accordingly, the Convention imposes obligations on states to assess the needs and vulnerabilities of IDPs and host communities,[31] and to provide adequate humanitarian assistance to IDPs, in all phases of displacements.[32] Where appropriate, such assistance must be extended to local and host communities.[33] The Kampala Convention highlights the specific needs of separated and unaccompanied children, female heads of household, expectant mothers, mothers with young children, the elderly and the disabled.[34] It stresses the need to protect displaced people against sexual and gender-based violence, harmful practices, and recruitment of children for unwholesome practices as well as human trafficking and smuggling.[35]

---

29 UNISDR, 'African Strategy for Disaster Risk Reduction' <www.unisdr.org/files/4038-africanregionalstrategy/.pdf> accessed 28th November 2017

30 African Union Convention for the Protection and Assistance of Internally Displaced Persons in Africa, 2009, Article 5(1).

31 *Ibid,* Article 5(5).

32 *Ibid,* Articles 3, 4, 5, 7, 9, 10 and 11

33 *Ibid* Article 9(2)(b)

34 *Ibid* Article 9(2)(c)

35 *Ibid* Articles 9(2)(d)

The provisions of the Kampala Convention appear laudable. There are however no specific strategies toward achieving many of its objectives. It also does not stress the need to preserve the sanctity of marriage, as no matrimonial stability clauses are provided with regard to preserving spousal companionship, preserving the unity and abiding faith of a couple in their marriage or the prevention of the likelihood of marriage breakdown in the event of a disaster. In this paper, this is considered very important as part of a coordinated effort to reduce to the barest minimum the harrowing effects of disasters on a society.

2.) International Response on Disaster Management

Nigeria is a party to a number of treaties and conventions that facilitate international disaster response. These are:

i. The Convention on the Rights of the Child, 1989

The child rights approach to disaster management aims to ensure that risk management, especially disaster response, uses the child rights approach by recognising the following articles in the convention on the rights of a child in responding to any type of disaster. Article 6 provides that every child has an interest in right to life. Article 9 states that parties shall ensure that a child shall not be separated from his or her parents against his/her will. In Article 20, a child temporarily or permanently deprived of his/her family environment shall be entitled to special protection and assistance. Article 22 states:

> Parties shall provide … cooperation in any efforts by the United Nations … to protect and assist such a child to trace the parents or other members of the family…for reunification with his/ her family.

Article 34 asserts that parties should undertake to protect the child from all forms of sexual exploitation and sexual abuses. Article 39 directs all parties to take all appropriate measures to promote physical and psychological recovery and social integration of a child victim of any form of neglect, exploitation, or abuses.

ii. The Sendai Framework for Disaster Risk Reductions 2015-2030

The Sendai Declaration is an outcome and declaration of the heads of states and governments, ministers and delegates who participated in the Third United Nations World Conference on Disaster Risk Reduction. It recognises the increasing impact of disasters and their complexity in many parts of the world, and calls all stakeholders to action, recognising that the realisation of the new framework depends on unceasing and tireless collective efforts.[36]

The Sendai Framework is a 15 year, voluntary, non-binding agreement which recognises that the state has the primary role to reduce disaster risk, but that responsibility should be shared with other stakeholders including local governments, the private sector and other stakeholders. It aims to achieve the following outcomes: the substantial reduction of disaster risk and losses in lives, livelihoods and health and in the economic, physical, social, cultural and environmental assets of persons, businesses, communities and countries.

This framework builds on achievements and elements established under the predecessor agreement — the Hyogo Framework for Action: Building the Residence of Nations and Communities 2005-2015 — and introduces a number of important innovations, including a strong emphasis on disaster risk management, as opposed to disaster management. The Sendai Framework underscores the fact that disaster risk reduction is essential to achieving sustainable development.

36 United Nations Office for Disaster Risk Reduction (UNISDR) 'Sendai Declaration' <www.unisdr.org/we/inform/publications/43300> accessed on the 24th November 2017.

RECOMMENDING INCLUSION OF SPECIFIC LEGAL CLAUSES ON FAMILY RESETTLEMENT, MATRIMONIAL STABILITY AND OTHER RELATED ISSUES FOR DISASTER AREAS IN NIGERIA IN THE NATIONAL DISASTER MANAGEMENT FRAMEWORK (NDMF) (2010)

Clause 1: Pre-disaster Measures for Prone Areas

In order for families to effectively respond to disaster, there is need to put in place adequate preparatory measures at both federal and state levels, particularly for disaster-prone areas. These shall include:

a. Community training and capacity-building of parents, teachers, students and staff on regular basis, comprising life saving skills during disaster.

b. Integrating disaster management skills into school curricular.

c. Building the resilience of school facilities to disaster impact in order to protect parents and children in the event of a natural disaster.

d. Establish a unit to be known as National Emergency Family Registry in all local government areas of the country. It will provide a platform to help family reunification when individuals are separated from their families during disaster. It will provide an avenue for survivors and their loved ones or spouses to locate themselves and reunite.

Clause 2: Post-disaster Measures

a. Prepare for evacuation and relocation: This should include, keeping children together with their parents, and providing accommodations for the victims from temporary to permanent settlements.

b. Emotional well-being: Regardless of age, all individuals who experience disaster are affected in so many ways and it may take time to return to normal conditions. In this regard, there should be regular visitation and counselling of both children and spouses by trained psychologists, social workers and lawyers. Disaster distress or help contacts should be provided to family members to connect with trained professionals. The services of such helpline staff will be free and referrals to any support agencies will be free, and health facilities will also be at no financial cost.

c. Economic support: Programmes aimed at mitigating the economic loss will be rewarding for families. These should include;

   - Direct financial assistance to families to assist them to start off a business, to cultivate a new farm, or provide instant employment.
   - Provision of training facilities to acquire formal and informal skills. This program would help families to cope with emotional and financial stress, as they are kept busy physically and mentally.

d. Other provisions to be based on the results from discoveries or findings of family disorientation and divorces as a result of disaster occurrences.

CONCLUSION

The fact that focus has shifted or is shifting to the human factor in scholarly, legal, social, economic and political responses to disaster is a welcome development. Much attention, however, need to be given to marital stability because it is essential to human rehabilitation in such a calamitous event as disasters. Marital stability is vital to family reunion and happiness and the control of crimes and depressions that are likely seamy consequences of disasters. The first place to look into in this consideration is the available legal instruments put in place to enforce coordinated responses to disaster. Although in Nigeria, enacting a law is one thing, but effective or committed enforcement or implementation is another; but an appropriate, comprehensive and well thought-out law is a starting point to arresting or reducing the harrowing effect of such a common menace on the family.

In Nigeria, the existing legal framework on disaster management is silent on the need to guarantee marital stability of victims of disaster. There are stories of family disorientations and hopelessness which are unreported from one disaster occurrence to another in Nigeria whether from flooding incident, natural disaster or war or militancy, all of whom have become part of Nigeria's historical and contemporary experiences. It is an area which this paper intends to stimulate further scholarly interests and researches and hopefully instigate enough studies to influence public and government's attention toward seeing marital stability as vital to human rehabilitation in the aftermath of disaster.

8

RIGHTS OF CHILDREN IN DISASTER MANAGEMENT

Omolade O. Olomola

Abstract

Children are vulnerable and by their nature need protection from vagaries of life. The situation with children is precarious and requires urgent attention as the productive population. This chapter considers the factors which are germane for the protection of children from risks associated with disaster. Disasters are either natural or man-made, whichever form it takes, children ought to be protected. In reality it may be difficult to exempt children completely from natural disasters, however, preventive measures should be put in place to alleviate the associated challenges. Natural disasters are violent events that are outside the control of humans and they are caused by the forces of nature and may result in loss of life, injury, and damage to property. On the other hand, man-made disasters are due to carelessness of human. The destructive nature of disaster is a matter of consideration in protecting children and for sustainable development.

Even the juvenile justice system protects children by its philosophy of concern, care and reformation. Also, the Child

Rights Convention (CRC) and other International instruments set standards and principles for survival, development, protection and participation of children.

In times of war and armed conflicts, most suffering inflicted on children during armed conflicts could be prevented or alleviated if there was greater respect for existing international law. In the event of armed conflicts, children are entitled to the general protection and special protection.

Owing to the nature of disaster and attendant risks, it is germane to adopt a holistic approach for protection of the rights of children. Disasters are inevitable, even the fiercest hurricane and volcanic eruption send warning signals before its eventuality. It is important for governments and stakeholders to come together to prepare a soft landing in time of disaster for the best interests of children.

Keywords: Vulnerable, Prevention, Child, Disaster

Introduction

There is no universally accepted legal definition of a child and the word 'child' may depend on the context in which it appears. For instance, there are many statutes which contain the description of who is regarded as a child or children as the case may be. Section 307 of the Criminal Code merely describes a child for the purpose of liability for murder. In addition, Under the Children and Young Persons Law, a child means a person under the age of fourteen (14) years. However the debate surrounding the definition of a child is gradually being settled in the light of the passage of Child Rights Act in 2003 and subsequent adoption in some states of the federation.

However, for the purpose of our present discussion, it is safe to adopt the definition of the word 'child' as contained in two international Instruments as follows:

a. The United Nations Convention on the Rights of the Child 1989; and

b. The African Charter on the Rights and Welfare of the Child 1991.

It is important to state that this definition was adopted in the Child's Right Act of Nigeria.[1]

Nigeria is signatory to both instruments and has gone ahead to domesticate the United Nations Convention on the Rights of the Child as the Child Rights Act of 2003(CRA). Under Article II of the African Charter on the Rights and Welfare of the Child, a child means every human being below the age of 18 years. Also under Article I of the U.N. Convention on the Rights of the Child 1989, a child is every human being below the age of 18 years unless, under the law applicable to the child, majority is attained earlier. As expected, Section 277 of the Child's Rights Act 2003 states that "child" means, a person under the age of eighteen (18) years; "child of the family" in relation to the parties to a marriage, means:

(a) a child of both of those parties;

(b) any other child, not being a child who is placed with those parties as foster parents by a local authority or voluntary organisation, who has been treated by both of those parties as a child of their family.

The main purport of this chapter is to emphasise on the rights of children, and these rights must be recognised and enforced in situations of disaster. The abuse of children's rights would have negative ripple effects against sustainable development.

1 Child's Rights Act 2003 (LFN 2004 c 50).

Vulnerability of Children

It is common knowledge that women, children and young adults[2] are considered vulnerable. What are the reasons behind their vulnerability? The general reasons are commonly linked to a number of factors which includes low self-esteem on the part of the young adult and a poor self-image.[3] Runaway children are at more risk of being victims of sexual exploitation. An abuser isolates and targets the vulnerable for exploitation. Children by nature, generally, can easily become victims of sexual exploitation. Sexual exploitation is not the only challenge that a child is faced with; there are other factors which exposes their vulnerability.

World Health Organization (WHO) and United Nations Children's Fund (UNICEF) categorises "adolescents" as individuals in the 10-19 years age bracket and "youth" as the individuals between 15-24 years of age bracket.[4]

Vulnerability as a concept is used in several areas in order to refer to the potential for poor outcomes, risk or danger. The main challenge now is how to effectively address overall reduction in vulnerability, in line with disaster management.

The dictionary[5] definition of the word 'vulnerable' states: "capable of being physically or emotionally wounded; open to attack or damage." This means, it is a state of being susceptible to danger.

According to Shilpa Khanna Arora *et al*, the literal meaning of the term "vulnerability" is the state or condition of being weak

2 A person who is in his or her late teenage years or early twenties: Definition of "young adult" Cambridge Advanced Learner's Dictionary and Thesaurus (CUP Cambridge 2017) <https://dictionary.cambridge.org/dictionary/english/young-adult> accessed 15th November 2017.

3 'Why are children and young people vulnerable?' <https://www.sefton.gov.uk/social-care/children-and-family-care/child-sexual-exploitation-(cse)/why-are-children-and-young-people-vulnerable.aspx >accessed 10th October 2017.

4 UNDESA,'Definition of Youth'<http://www.un.org/esa/socdev/documents/youth/fact-sheets/youthdefinition pdf> accessed 19th October 2017.

5 'Vulnerable' Merriam-Webster's Collegiate Dictionary<https://www.merriam-webster.com> accessed 16th October 2017.

or poorly defended.[6] The concept of vulnerability in relation to adults denotes the ones who are more exposed to risks because of their status. Vulnerability can be in the form of deprivation (food, education, and parental care), exploitation, abuse, neglect, violence, and infection with HIV and other deadly viruses.[7] Thus, vulnerability is a position which may graduate from a level of resilience to total helplessness.[8]

The summary of the provisions of Article 24 and 27 of the CRC stipulate that there must be recognition of the rights of the child to enjoy the best quality of physical and mental health that is possible. Countries have the duty to ensure that no child is prevented from enjoying the best health care services and in addition, there must be safe homes, food, clothing, schooling and protection.[9] It is a recognised fact that children in developing countries who live in poor socioeconomic conditions are more vulnerable to health and other challenges. Factors which promote children's vulnerability can be identified as environmental disruptions like disasters, poverty and health issues. These factors can either positively or negatively affect the total well-being of a child.[10]

Vulnerability is thus, a set of conditions that make individuals and communities more susceptible to diseases or disabilities, as a result of individual, social and programmatic elements.[11] Situation of being vulnerable can be minimised through a supportive environment.

6 S Khanna Arora and others, 'Defining and Measuring Vulnerability in Young People' [2015] 40 (3) IJCM 193 <https://www.ncbi.nlm.nih.gov/pmc/articles/PMC4478662/> accessed 16th October 2017.

7 S Khanna Arora and others (n561).

8 *Ibid.*

9 The United Nations Convention on the Rights of the Child (CRC)1989 art 24 and 27.

10 Daniel Ignacio da Silva and others, 'Vulnerability in child development: influence of weak family bonds, substance abuse and domestic violence' (2014) 23 Florianopolis <http://www.scielo.br/scielo.php?script=sci_arttext&pid=S0104-07072014000401087>accessed 16 October 2017.

11 Ayres JRCM and others, 'From Natural History of Disease to Vulnerability' in Parker R and Sommer M, Routledge Handbook in Global Public Health (Routledge, New York 2011).

It is necessary to recognise and understand vulnerability in order to provide appropriate interventions.[12] The present challenges facing children now are offshoots of changing times. These include upsurge in urbanisation, political violence, natural and man-made disasters, changing family forms, inadequate food supplies and poverty. These challenges create new levels and dimensions of threats for the children victims much more than problems of health and disability.[13]

World Bank's "Orphans and Vulnerable children (OVC)" toolkit defines vulnerability as "the group of children that experience negative outcomes, such as the loss of their education, morbidity and malnutrition, at higher rates than do their peers."[14]

The World Bank's toolkit identified the following categories as vulnerable children:

1. Street children
2. Children in the worst forms of child labour
3. Children affected by armed conflict
4. Children affected by HIV/AIDS
5. Children living with disability, and
6. Local OVC groups.

It should be noted that disasters and the aftermath is also a very important indication of vulnerability in children. It is quite easy to associate vulnerability with sexual exploitation but it goes beyond that. Many vulnerable children fall into multiple categories and the list is increasing with the new experiences of conflicts, violence and disasters in developing countries in Africa.

A comprehensive research carried out in Africa by Skinner *et al,*[15] to generate the definition of 'vulnerability' defined

12 *Ibid*

13 Engle PL and others, 'Child Development: Vulnerability and Resilience'[1996] 43 SSM(5)<http://www.ncbi.nlm.nih.gov/pubmed/engle> accessed 16th October 2017.

14 World Bank'OVC Core Definitions'<http://info.worldbank.org/etools/docs/library/164047/howknow/definitions.html>accessed 15th October 2017.

15 Skinner D and others, 'Towards a Definition of Orphaned and Vulnerable Children' [2006]10 AIDS Behav (6) <https://www.ncbi.nlm.nih.gov/pubmed/16639543> accessed 14th October 2017.

vulnerable children as "the ones not having certain of their basic rights fulfilled." This definition is very appropriate and holistic in terms of the rights of children in other situations of disaster and violence. Apart from the general susceptibility of a child, being an orphan is also a major determinant of vulnerability. The definition is also centered on the three fundamental aspects of dependence:

1. Material aspects — money, food, clothing, shelter, health care and education;
2. Emotional aspects — care, love, support, space to grieve and containment of emotions;
3. Social aspects — absence of a supportive peer group, of role models to follow, or of guidance in difficult situations, and risks in the immediate environment.[16]

Since vulnerability is not yet a clearly defined terminology, assessment of the extent of such a problem is in bits and not total. The concept of vulnerability generally refers to sets of exposed individuals who suffer undue risks. Vulnerability is a relative term with varying degrees and changes with time and between countries, thus it is highly contextual.[17] Whilst this fact is not in doubt, the status of a child as being vulnerable because of the age, size and capability can be regarded as universal. Being a child is a constant factor all over the world, because naturally, it is time bound and soon as a child attains the age of majority, the society prescribes a new status for him or her. When the child becomes mature and an adult, he/she can start taking decisions on his/her own welfare. It is therefore right to state that children and young people that are separated from their parents are clearly vulnerable to disaster situations.[18] Children are specifically vulnerable in situations of armed conflict and/or natural disasters, and they require special attention to ensure their protection and well-being.

Children need special support and attention during disaster crises to meet their basic needs and ensure that their rights are

16 *Ibid.*

17 S Khanna Arora and others, *op cit* (n561)

18 *Ibid.*

protected. During emergencies, school attendance is disrupted, there may be separation from families, they could be recruited as child soldiers or physically assaulted.

There is a robust operational framework within which humanitarian agencies work to help protect and assist children in damaging events.

DIMENSIONS OF DISASTER

Disasters occur when the human, physical, economic or environmental damage from an event, or series of events, devastates a community's capacity to handle such. Disasters can happen through natural forces or it could be man-made. It may take a process of time gradually like in volcanic eruptions,[19] it could also come swiftly.[20]

Natural disasters include the following:

a. Geophysical – earthquakes, volcanoes, landslides, avalanches, tsunamis
b. Meteorological – storms, cyclones, hurricanes, typhoons, blizzards
c. Hydrological – floods, storm surges
d. Climatic events – droughts, extreme temperatures, wildfires
e. Biological – epidemics, infestations.[21]

19 Volcanoes usually have a life of many thousands of years. Once a volcano has begun to erupt, it usually takes about ten years before that particular eruption comes to an end. Sometimes the eruption lasts for hundreds of years. 'Volcanic Eruptions' Scholastic <https://www.scholastic.com/teachers/articles/teaching-con tent /volcanic-eruptions/>accessed on 15th November 2017.

20 The greatest variables of volcanic eruptions are when and how the first volcanic tremors are detected. If they are not detected, then there is no warning signal for measurement of the interval. Not all volcanoes have active measuring going on. Some occur where measurement is not considered necessary. Where measurement is being conducted, the answer will vary with the type of volcano and what is going on in the earth below it. <https://www.quora.com/How-long-does-it-take-for-a-volcano-to-erupt-after-detection-of-the-first-volcanic-tremors>accessed 15th November 2017.

21 International Federation of Red Cross and Red Crescent Societies,'Types of disasters: Definition of hazard' <http://www.ifrc.org/en/what-we-do/disaster-management/about-disasters/definition-of-hazard /> accessed 15th October 2017.

Types of Disaster

Disasters can take many different forms, and the duration can range from an hourly disruption to days or weeks of ongoing destruction. Below is a list of the various types of disasters – both natural and man-made or technological in nature – that can impact a community. All these can have serious consequences on a community.[22] The following is another classification of the types of disasters.[23]

Natural Types of Disasters

Agricultural diseases and pests[24]
Damaging winds[25]
Drought and water shortage[26]
Earthquakes[27]
Emergency diseases (pandemic influenza)[28]
Extreme heat[29]
Floods and flash floods[30]
Hail[31]
Hurricanes and tropical storms[32]
Landslides and debris flow[33]
Thunderstorms and lightning[34]
Tornadoes[35]

22 <http://restoreyoureconomy.org/disaster-overview/types-of-disasters/> accessed 20th October 2017.
23 *Ibid.*
24 <http://www.usda.gov/documents/agricultural_pests_and_diseases.pdf> accessed 20th October 2017.
25 <http://www.nssl.noaa.gov/education/svrwx101/wind/>accessed 20th October 2017
26 <http://www.ready.gov/drought> accessed 20thOctober 2017.
27 <http://www.ready.gov/earthquakes> accessed 20th October 2017.
28 <http://www.ready.gov/pandemic> accessed 20th October 2017.
29 <http://www.ready.gov/heat> accessed 20th October 2017.
30 <http://www.noaa.gov/education/svrwx101/floods/>accessed 20th October 2017.
31 <http://www.noaa.gov/education/svrwx101 hail/ > accessed 20th October 2017.
32 <http://www.ready.gov/hurricanes> accessed 20th October 2017.
33 <http://www.ready.gov/landslides-debris-flow> accessed 20th October 2017.
34 <http://www.nssl.noaa.gov/education/svrwx101/thunderstorms/>accessed 20th October 2017.
35 <http://www.nssl.noaa.gov/education/svrwx101/tornadoes/>accessed 20th October 2017.

Tsunamis[36]
Wildfire[37]
Winter and ice storms [38]
Sinkholes[39]

Man-made disasters include conflict, environmental degradation, pollution and industrial accidents. Complicated crises otherwise referred to as complex emergencies happen as a consequence of a combination of man-made and natural disaster. These generally involve violence and loss of life, the displacement of populations and extensive damage to societies and economies.

Hurricanes and tropical storms are among the most powerful natural disasters because of their magnitude and destructive capability. Tornadoes are short-lived but violent, and capable of triggering winds in excess of 200 mph. Both earthquakes and tornadoes strike unexpectedly without warning. Flooding is the most common of natural hazards, and requires the knowledge of the natural systems of our environment, including floodplains and the frequency of flooding events. Wildfires are more rampant during drought. Man-made and technological types of disasters include the following:

a. Hazardous materials
b. Power service disruption and blackout
c. Nuclear power plant and nuclear blast
d. Radiological emergencies
e. Chemical threat and biological weapons
f. Cyber attacks
g. Explosion
h. Civil unrest

Disasters can be caused by humans. Hazardous materials tragedies include chemical spills and groundwater contamination. Workplace fires are more common and can cause significant

36 <http://www.ready.gov/tsunamis> accessed 20th October 2017.
37 <http://www.ready.gov/wildfires> accessed 20th October 2017.
38 <http://www.nssl.noaa.gov/education/svrwx101/winter/>accessed 20th October 2017.
39 <htttp://ga.water.usgs.gov/edu/sinkholes.html> accessed 20th October 2017.

property damage and loss of life. High-risk targets include military and civilian government facilities, international airports, large cities and high-profile landmarks. Cyber-terrorism involves attacks against computers and networks done to intimidate or coerce a government or its people for political or social objectives.[40] Communities become vulnerable to threats posed by extremist groups who use violence against both people and property. The number and scale of these complex emergencies had increased in the past decade with the proliferation and escalation of internal conflicts.

Disaster Related Risks (DRR) and Children Liabilities

The devastation caused by disaster such as hurricane, flood, fire and the likes is usually terrifying to everybody generally and to children especially. Children are to be involved in interventions in order to reduce their fear. It is imperative to enlighten the children in the simplest manner with details which will not be scary to them. The ages of children and maturity will determine levels of understanding and contributions in making decisions which will affect their lives, thus, a child's age will affect the response during and post-disaster. For example, the younger ones below ten years old may refuse school attendance out of fear and worries. The teenagers become more argumentative and show a decline in school performance.[41]

There is a rising indication of the ability of children to act as protagonists for action to reduce climate and disaster risks in their communities. Children have great intuitions and are able to combine their personal experiences with external information.[42] They are able to transfer their perceptions of risk to others to bring about changes in behaviour that will reduce risks and vulnerabilities. This

40 <http:www.//restoreyoureconomy.org/disaster-overview/types-of-disasters/>accessed 20th October 2017.

41 American Academy of Child and Adolescent Psychiatry (AACAP), 'Helping Children after a Disaster'Facts for Families information sheets 2008 (36).

42 Tanner T,'Children Communicating Climate and Disaster Risks' IDS in Focus Policy Briefing [2009]13(3)<http://www.ids.ac.uk> accessed 20th October 2017.

understanding is necessary to formulate policies and programmes which are well-suited to meet the needs of children and create an enabling environment for their participation and agency.[43]

A small deed can have tremendous impact where there is access and opportunity. Education is indeed a game changer in DRR and development work in the community. A Charter was put together by some children through the help of UNISDR and Plan International in 2012. The Charter elaborates on how children could play participatory roles in disaster risk reduction.

Children's Charter on Disaster Risk Reduction

This is an action plan for disaster risk reduction for children by children. The five point Children's Charter for Disaster Risk Reduction was developed through consultations with over 600 children from 21 countries across Africa, Asia and Latin America and identifies children's priorities for Disaster Risk Reduction (DRR). It was launched at the Global Platform for DRR in Geneva in 2011.[44]

a) **Schools must be safe and education must not be interrupted.** "I felt unhappy when I saw the school destroyed by the storm. I did not go to school because the school was destroyed. It happened at night and in the morning, my friend and I went to school and found it had been destroyed." (A child from Lao PDR).[45]

b) **Child protection must be a priority before, during and after a disaster.** "We do not feel protected by anyone in our community" (A child from Mozambique). In India, children suggested

43 Tanner T, *op cit*, (n597).

44 Children's Charter for disaster risk reduction <http://plan-international.org/about-plan/resources/publications/emergencies/childrens-charter-for-disaster-risk-reduction> accessed 20th October 2017.

45 UNISDR and Plan International 2012, 'Children's Action for Disaster Risk Reduction: Views from Children in Asia'.

training in life-skills which they feel would "enable them to protect themselves from risks and troubles" and that they be provided special care when they are traumatised by disasters. "We should be treated with dignity during emergencies and should take protective measures to mitigate/stop physical and sexual abuse."[46]

c) **Children have the right to participate and to access the information they need.**
"I am part of the Emergency Committee that has been created in the community. Our goal is to reduce the disaster risk by implementing the principles of the Community Emergency Plan." (A child from Dominican Republic).[47]

d) **Community infrastructure must be safe, and relief and reconstruction must help reduce future risk**
"Build bridges because every year children miss school in the rainy season when they have to cross gullies, rivers and water channels huge enough to drown them." (A child from Lesotho).[48]

e) **Disaster Risk Reduction must reach the most vulnerable**
"In my area, there were three children about 4-5 years old. Once they went on the river dyke to avoid the flood water which could make them wet on the street, but they slipped and fell in the river and were drowned because they could not swim." (A child from Philippines) "To me, there is a big difference, because many disabled children are not treated properly by providing them the necessary information since they are not being taken into account due to the fact of being disabled." (A child from the Philippines).[49]

---

46 *Ibid.*

47 UNISDR and Plan International, *op cit,* (n600).

48 *Ibid.*

49 See generally Child-Centered Disaster Risk Reduction: Building resilience through participation<http://plan-international.org/about-plan/resources/publications/emergencies/child-centred-disaster-risk-reductionbuilding-resilience-through-participation> accessed on 13th October 2017.

Child Protection and Disaster Management

According to international standards, a child is any person under the age of 18 years. The inter-agency minimum standards for Child Protection in Humanitarian Action defines child protection in emergencies as "the prevention and response to abuse, neglect, exploitation, and violence against children in times of emergency caused by natural or man-made disasters, conflicts, or other crises." These include all forms of physical and psychological abuse, sexual and gender-based violence, and deprivation of basic needs. The minimum standards emphasise children's specific rights as defined in the UN Convention on the rights of the child and support the UN Sustainable Development Goals.

It is important to note that the majority of people affected by emergencies are often children.[50]

Disaster management is a complex process involving international, national and local organisations, each with a distinct role to play. To respond to disaster situations, a coordinated effort is required.

The risks to children defined in the minimum standards for Child Protection in Emergencies include:[51]

a. physical, sexual, and psychological violence
b. neglect
c. Injuries
d. harmful practices including child's early and forced marriage
e. psychological distress and mental disorders of children; associated with armed conflicts and armed groups
f. child labour
g. unaccompanied and separated children
h. justice for children

50 NHCR Global Trends Report 2013.

51 International Federation of Red Cross and Red Crescent Societies, 'Briefing Child Protection in Emergencies' Geneva 2016<http://www.ifrc.org> accessed on 21st October 2017.

Children are among the weakest, most dependent and vulnerable populations in emergencies.

Statistics from jurisdictions from around the world have revealed that the risk of violence against children during and after disasters can escalate radically.[52] Among the most neglected forms of child protection in emergencies include child-focused gender-based violence, trafficking and migration, child labour and violence against children with disabilities.[53] Child protection in emergencies is considered a "life-saving intervention."[54] Children have great individual personalities; they are pliable creatures with great adaptability.[55] It is trite to assume that early childhood trauma will eventually affect a child's personality or result in mental illness.[56] It has however been discovered that children exposed to disaster are twice as likely to develop psychological disorders as their peers, either in childhood or later.[57] Their active participation as agents of change should be encouraged and ensured, whether they are exposed to disaster risk or being affected by an emergency situation. In other words, their voices should be heard before, during and after the impact of a damage-generating event. The basic principles of right-based approach on the children are provision, protection and participation.[58] There are other factors to be considered along with the three principles. Issues such as life-cycle, gender equality, multiculturalism, equity and inclusion, and the participation and principles of humanitarian aid should also be considered.[59]

52 *Ibid* See also PS Pinheiro, World Report on Violence Against Children. (United Nations, New York 2006).

53 A Jabry (ed), After the Cameras Have Gone – Children in Disasters(2nd edn Plan International, London 2005).

54 Gurvinder Singh and others, 'Predictable, Preventable: Best Practices in Addressing Interpersonal and Self-directed Violence During and After Disasters'(IFRC and Canadian Red Cross 2012) <http://www.ifrc.org/PageFiles/94522/ViolenceIn Disasters -English-1up.pdf>accessed 17th October 2017.

55 AM Clarke and others, Early Experience, Myth and Evidence (Open Books London, 1976).

56 D Black, 'Children and Disaster'[1982] 285 BMJ 989.

57 *Ibid.*

58 UNICEF/TACRO 'Actions for children and youth resilience: Guide For Governments'(Panama2013)

59 *Ibid*

Disasters are intricate happenings with multiple causes, thus, its management requires all-embracing, preparation to deal with the complexities. There is no nation without a disaster management plan, no matter its simplicity. Disaster is not a concept or happening which provides for a choice. The best way to deal with disaster is 'by saving for the rainy day.' This means that there must be adequate preparation with deep foresight before the happening of a damaging event. National disaster management plans are associated to the most commonly experienced disasters in that country or region and the funds available to tackle such. Government funded aid programmes coordinate national responses to disasters in some countries like Nigeria. They may also receive interventions from international and foreign aids organisations.

Different regimes in Nigeria tried to document policies to establish organisations that will affect the lives of the people positively in the time of emergencies. The objective of such organisations is mainly to alleviate sufferings in the face of disaster and also remedy the situation within their capabilities. It is also part of the responsibilities of the government to take into consideration the welfare of the people. In order to survive and be in dominion, human beings must conquer and control his environment. The sad aspect is that the efforts are always limited by force majeure because of the nature of disaster which is always sudden and unannounced.

Children and Rights

The Constitution of the Federal Republic of Nigeria, 1999 which is the grundnorm guarantees certain fundamental rights to every person including children. These rights are contained in Chapter IV of the Constitution, (Sections 33-46).
These include:

The right to life (S.33)

The right to dignity of the human person (S.34)

The right to personal liberty (S.35)

The right to fair-hearing (S.36)

The right to privacy and family life (S.37)

The right to freedom of thought, conscience and religion (S.38)

The right to freedom of expression and the press (S.39)

The right to peaceful assembly and association (S. 40)

The right to freedom of movement (S.41)

The right to freedom from discrimination (S.42)

The right to acquire and own immovable property anywhere in Nigeria (S.43)

The right against compulsory acquisition of one's property without compensation (S.43)

There is provision in the constitution for special procedure for enforcement of these rights by the courts through the Fundamental Rights (Enforcement Procedure) Rules.

These rights are sacrosanct and must be respected irrespective of whose rights are in issue; once the individual is of the homo sapien race which is human beings.

The foundation of rights-based approach is on the fact that all children are "rights holders." All children, irrespective of their economic status, race, colour, sex, language, religion, national, ethnic or social origin, property, disability, birth or any other status have the same rights and are entitled to the same protection by the State.[60] Children living in pre and post-disaster communities are human beings with full potential to contribute to society and be agents for positive change. Children must be able to participate in matters affecting them and be empowered to speak up for the fulfilment of their rights.

60 UNHR, Protection and promotion of the rights of children working and/or living on the street – United Nations Human Rights, Office of the High Commissioner. Office of the High Commissioner for Human Rights United Nations Children's Fund Consortium for Street Children Aviva.

The international legal framework establishing the obligations of states in relation to children is very comprehensive. The Convention on the Rights of the Child 1989 is the main international document for the promotion and protection of the rights of the child, and it applies to all children in all circumstances. This instrument has wide acceptance and ratification which shows the importance that states accord to the protection and promotion of the rights of children. The convention is exclusive as the first legally binding instrument to take a holistic approach of the rights of the child and it takes into consideration a whole range of rights like civil, political, economic, social and cultural.

The CRC establishes a framework of obligations for different state actors and it is indeed a groundbreaking instrument in recognising all children as rights holders and reaffirms the general principles of best interests of the child, non-discrimination, participation, survival and development as the framework for all actions concerning children. All children, irrespective of their economic status, race, colour, sex, language, religion, national, ethnic or social origin, property, disability, birth or any other status have the same rights and are entitled to the same protection by the State.

To every right, there are always corresponding obligations and in this instance child should participate in case of exposure to risks associated with disaster or emergency situations. Their opinions should be heard before, during and after the impact of a disaster in order to have a fair assessment of disaster.[61] Research shows that the voices of children are often not heard and do not influence humanitarian decision-making.[62] Under human rights law, the states are accountable for respecting, protecting and fulfilling children's rights within their territories. While states play the role of the principal duty bearer for all children, other non-

61 Their active participation as agents of change should be encouraged and ensured. See generally Actions for children and youth resilience Guide For Governments 2013 UNICEF/TACRO Panama.

62 A Jabry (ed) op cit, (n608)

state entities, professionals and individuals are also recognised by the convention on the rights of the child as duty bearers in the fulfilment of children's rights. Protecting children and preventing unwholesome experiences implies taking a holistic approach that understands children's relationships as interdependent and interconnected, and therefore recognises that rights can be violated and also protected. This can only work when there is accountability which is to ensure that when children's rights are violated; the actors are isolated and dealt with.[63]

In terms of public policy, specific actions must be taken in the four priority areas relating to the welfare of children and youth, including: protection, education, water, sanitation and hygiene, health and nutrition. These sectors must work jointly to establish working mechanisms and applications which guarantee effective execution of plans, projects and programmes, either for the prevention and mitigation of disaster risk, or for preparation and response to emergencies or disasters, or for the early recovery and reconstruction. To accomplish this, it is important to identify the state authorities who have the jurisdiction to create such a coordinated and efficient management view.[64]

The implementation of disaster risk reduction actions which build or strengthen resilience, and which focus on children and youth, should take into consideration the following basic components:[65]

1. The organisational and institutional capacity of the state in its various fields to undertake such actions.
2. Knowledge of disaster risk and the underlying factors, and its relationship and impact on children and youth.

---

63 UNHR, Save the Children, 'Keys to Successful National Child Protection Systems'E/ICEF/2008/5/Rev.1, para. 12. 3 http://www.ohchr.org/Documents/Issues/Children/Study/OHCHR Brochure Street Children.pdf accessed 17th October 2017.

64 UNICEF and RET 2013, Action for Children and Youth Resilience: Guide for Governments Panama.

65 *Ibid*

3. The role of information and awareness-raising amongst the general population regarding the well-being of children and youth in relation to disaster risk reduction. This should highlight the role of the family, community and the state as guarantors of the rights of children, adolescents and young people.

Preventive Measures and Post-Disaster Protection of Children

The term resilience is always used in associating children with disaster situations. In order to build resilience in children and youth, there is a need to assess and understand the issues which affect them in terms of disaster risk and its management.[66] A thorough assessment of the disaster itself is necessary whether natural or man-made in order to isolate and calculate the vulnerabilities to which children and adolescents are exposed.

It is also necessary to ascertain the short, medium and long term effects of disasters on all round development of the local community, in addition to the national impact with respect to equitable, egalitarian and sustainable progress.

It is pertinent to ascertain that children have well-defined human rights and obligations of states to achieve these rights. Achievement of these rights is based on some condition precedents which would help the children to grow up into well-rounded adults.[67] These conditions must be taken into consideration at pre and post-disaster periods. They must be protected against circumstances which threaten their self-esteem or mental integrity.

With the recent upsurge in the occurrence of disasters all over the world, the new dimension of approach now is on pre-disaster action and reaction. Governments are now diverting attention from provision and distribution of relief materials to prevention and mitigation.

66 UNICEF/TACRO *op cit* (n619)
67 *Ibid*

This is in response to the United Nations (UN) Declaration in 1989 which designed 1989 to 2000 as International Decade for Natural Disaster Reduction (IDNDR).[68] The intention of the UN is that member countries within the space of ten years should have put in place an assessment of risk posed by environmental hazards (such as flooding), long term preparedness and prevention plans and warning systems.[69]

In 2016, UNICEF Child Protection team established and operationalised Child Protection Information Management System (CPIMS) for the emergency states. Together with the Nigerian military, UNICEF developed a protocol for handover of detained children to civilian authorities. UNICEF and the Civilian Joint Task Force (CJTF) developed an action plan for the CJTF to stop recruitment and use of children. In addition, 423 stakeholders were trained at the federal and state levels on monitoring and reporting mechanism on grave child rights violations.[70] In 2016, the UNICEF Education team in partnership with SUBEB (State Universal Basic Education Board) has facilitated access to temporary safe learning environment to 106,882 children affected by conflict; 200,302 children received learning materials.[71] Traditional disaster management is increasingly being complemented with Disaster Risk Reduction (DRR) and Climate Change Adaptation (CCA). By and large, a novel policy is created, focusing on prevention and mitigation of disaster and climate risks before they have become apparent emergencies. Emergency response planning is gradually becoming less shock-driven and is guided more by vulnerabilities before, during and

68 A Emeribeole, 'Managing Flood Disaster in Nigerian Cities: Issues and Strategies Toward Meeting the Challenges in the Modern World–a Case Study of Owerri Metropolis Imo State, Nigeria' 2015 <https://www.fig.net/resources/proceedings/fig_proceedings/fig2015/papers/ts05a/TS05A_emeribeole_7587.pdf>accessed 14th October 2017.

69 *Ibid.*

70 <https://www.unicef.org/nigeria/NG_wwd_sitreps17.pdf>accessed 14th October 2017.

71 *Ibid.*

after emergencies like early recovery.[72] Development planning is gradually becoming more risk-informed and climate-sensitive.

In this evaluation, risk is a disaster that has not occurred but is likely to occur if not countered by risk management and controlling efforts. While conventional disaster management concentrates on emergency preparedness and disaster response, DRR focuses on preventing and mitigating disaster risks before they manifest as emergencies. Examples of DRR interventions include 'hardware' such as safe schools and elevated hand pumps designed to withstand extreme climate events as well as 'software' such as development programmes targeting people affected by past disasters and exposed to current or future hazards and climate risks.[73] The economic evidence available supports comprehensive and preventive risk management approaches.[74] Community-based DRR, early warning and emergency preparedness have consistently proven to be more cost-effective than humanitarian action, often by a factor of 1:4 to 1:24.[75]

History has proven to us that planning for the unforeseen helps to mitigate the harshness of disaster in Nigeria. It may not be completely possible to meet force with force with the challenges of disaster management.[76]

Organised disaster management in Nigeria dates back to 1906, when the fire brigade was established with functions that meant beyond firefighting to the services during emergencies.[77]

There was later to be an adhoc arrangement in the office of the head of state because emergency response to disaster was considered a security issue. This arrangement continued until 1972/1973, when the country experienced a devastating drought. The drought had negative socio-economic consequences and cost

72 UNICEF/EAPRO 'Protecting Children From Poverty, Disaster and Climate Risks Linking Social Protection with Disaster Risk and Climate Change Adaptation in East Asia and the Pacific' Reflections from a Symposium –UNICEF East Asia and Pacific Regional Office (EAPRO) 2014 <http://www.unicef.org/eapro> accessed 14th October 2017.

73 *Ibid.*

74 United Nations and World Bank 2014.

75 IFRC 2010.

76 Emeribeole A *op cit* (n 68).

77 *Ibid.*

the nation the loss of many lives and properties, a development that led to the establishment of the National Emergency Relief Agency (NERA) in 1976. NERA was charged with the responsibility of collecting and distributing relief materials to disaster victims. Considering the limited scope of the Agency, the government in 1993 decided to expand its scope to include all areas of disasters. This was backed up with Decree 199 of 1993, which raised the status of the Agency to an independent body under the presidency as unit in the secretary to the government of the federation's office.

NERA later became National Emergency Management Agency (NEMA), with a new vision to build a culture of preparedness, prevention, response and community resilience to disaster in Nigeria.[78]

In recent times, disasters have become a regular 'visitor' in many countries and jurisdictions. As conflicts are bound to be in any society, so also disaster and its attendant challenges is not constrained or location-bound. Disasters have been without borders; with tsunamis in Asia, recent hurricanes Harvey and Irma in the States of Texas and Florida, United States of America and the Caribbean disasters. Not leaving Africa out, there have been a number of flood disasters, earthquakes, land and mud slides and a lot of man-made disasters.

Disaster management involves many diverse activities. These activities can be grouped into five main stages viz: assessment, mitigation, preparedness, response, and recovery.

The first three activities are performed before the occurrence of disaster, while the fourth and fifth take place during and after the occurrence of disasters respectively.[79]

i. Assessment

This involves inventorying (identification and recording) the sensitivity and vulnerability of a region to certain types of

78 Emeribeole A *op cit* (n623).

79 'Preparedness, Response, and Recovery Considerations for Children and Families' (Workshop Summary Forum on Medical and Public Health Preparedness for Catastrophic Events, Board on Health Sciences Policy) Institute of Medicine National Academies Press Washington (DC)(Mar 21st 2014) <https://www.ncbi.nlm.nih.gov/books/NBK195865/> accessed 17th October 2017.

hazards. At this stage, the levels of risks, the danger to human life, environment and structures are considered and determined. The assessment will provide identification of development that increases them, thus establishing the culture of prevention.

ii. Mitigation

This entails making necessary provisions to ensure that the region is less vulnerable to known risks and danger. Mitigation activities may include: land use and planning; moving settlement away from areas susceptible to such risks and dangers such as flood and storm areas; and the establishment and enforcement of building code etc.

iii. Preparedness

This involves planning of emergency aid, development of scenarios and monitoring systems, and establishment of early warning system, public information and awareness of likely hazards, community involvement in disaster management programmes, establishment of disaster management and reduction at local, state and national levels and establishment of proper communication channels.

iv. Response

This happens after the occurrence of the disaster which would have caused untold human suffering and damages to the environment. At this stage, rescue teams will attempt to save lives, injured people will be cured and nursed and relief will be supplied to traumatised survivors. This is the most sensational stage of disaster reduction and management system.

v. Recovery

This stage involves assessment of damages, rehabilitation, cleaning of the environment and social and economic reconstruction. It also entails the first three stages of disaster management process viz; assessment, mitigation and preparedness, all of which are central to strategic development aimed at preventing or minimising the effect of future disasters.

Convoy of Hope, an NGO dealing with disasters and children also suggests the inclusion of monitoring assessment, response, and recovery as a model for taking care of children in emergency.[80] The 2010 National Commission on Children and Disasters made the following recommendations in taking care of children.

1. Mental Health
2. Child Physical Health and Trauma
3. Emergency Medical Services and Pediatric Transport
4. Disaster Case Management
5. Child Care and Early Education
6. Elementary and Secondary Education
7. Child Welfare and Juvenile Justice
8. Sheltering, Housing, and Evacuation[81]

Holistic Approach to the enforcement of child rights

It is indeed true that children are rarely mentioned with respect to humanitarian accountability standards, they are also not overly exempted. It is essential to create an optimistic environment for children as this would make it easier to restore a sense of normalcy to their lives. It is important to adopt a child-friendly approach to programming in the aftermath of a disaster that promotes children's rights and also takes into account their views and concerns. Thus, children's and adolescents' active participation in decisions that affect their lives and in planning and implementing of relevant programmes and activities is to be ensured. Not only will participation have a positive effect on their mental well-being and help them to regain control over their lives, it is also an effective protection strategy.[82]

---

80 <https://www.convoyofhope.org/what-we-do/disaster-services/?gclid=EAIaIQobChMIyu2Y0fPE1gIVwbftCh0pFwpaEAMYAiAAEgKpJ_D_BwE>accessed 14th October 2017.

81 National Academies Press, *Op cit* (n634)

82 Brookings-Bern Project on Internal Displacement,Human Rights and Natural Disasters(Operational Guidelines and Field Manual on Human Rights Protection in Situations of Natural Disaster Pilot version 2008)<https://www.brookings.edu/idp>accessed 14th October 2017.

The CRC is the first human rights treaty which unambiguously grants children certain civil and political rights. Certain articles in the CRC expressly grant the children opportunity to air their views and participate in issues affecting their lives. Some of such Articles are replicated below:

Article 8: Preservation of identity: The State has an obligation to protect, and if necessary, re-establish basic aspects of the child's identity. These include name, nationality and family ties.

Article 12: The child's opinion: The child has the right to express his or her opinion freely and to have that opinion taken into account in any matter or procedure affecting the child.

Article 13: Freedom of expression: The child has the right to express his or her views, obtain information and make ideas or information known, regardless of frontiers.

Article 14: Freedom of thought, conscience and religion: The State shall respect the child's right to freedom of thought, conscience and religion, subject to appropriate paternal guidance.

Article 15: Freedom of association: Children have a right to meet with others and to join or form associations.

Article 16: Protection of privacy: Children have the right to protection from interference with privacy, family, home and correspondence, and from libel or slander.

Article 17: Access to appropriate information: The State shall ensure the accessibility to children of information and material from a diversity of sources, and it shall encourage the mass media to disseminate information that is of social or cultural benefit to the child and take steps to protect him or her from harmful materials.[83]

Conclusion

The philosophy behind children's rights is well founded and directed towards the achievement of the continuity of the human race in an atmosphere of peace, harmony, development and happiness. The various laws and instruments on the rights of the child discussed so far emphasise summarily that "in all actions concerning the child undertaken by any person or authority, the best interests of the child shall be the primary consideration."[84]

However, despite copious provisions in our municipal laws and international instruments on the rights of the child, there appears to be a gap between law and practice resulting in gross inability of the child to realise these rights at present.

In accordance with international human rights law, children are entitled to special measures of protection above and beyond those provided to adults until they reach the age of 18. The best interests of children should serve as the guide for humanitarian decisions concerning them. At the same time, children and adolescents should not be considered as helpless victims, but rather as active participants. As stated earlier in this chapter, children have started making important contributions in all stages of emergency situations. In addition, children and young people have demonstrated proficiency, and their participation improves the quality and the spread of emergency outreach. Like any individual, children have a right to be supported in making these contributions.

Children have a right to information in humanitarian crises. Information for children has to be age-appropriate. Children and young people can be effective communicators in emergency situations. Children have their own peculiar needs and concerns.

83 UNICEF/EAPRO, The Participation of Children and Young People in Emergencies: A guide for relief agencies, based largely on experiences in the Asian tsunami response (UNICEF EAPRO 2007).

84 African Charter on Human and Peoples' Rights 1981 art 4(1), Cap A9 LFN 2004 as amended.

Boys and girls of different ages have to be included in consultations to ensure that humanitarian agencies address their priorities.

Participation brings benefits to children, families and communities. It contributes to children's education and development. It helps children to protect themselves from abuse and exploitation.

Lack of involvement of children disregards what they are capable of doing. It also undermines them by sending the wrong message to the community and decision makers that it is acceptable to exclude children from decision-making, information, consultations and contributions and also that children have no role in the public sphere. Children's participation in emergencies is not about evidence, efficiency and effectiveness. It is all about rights and its enforcement and these same rights apply in emergencies as in other times like during the disaster response.[85]

Finally, children are particularly vulnerable in situations of armed conflict or natural disasters, and they require special attention to ensure their protection and well-being. Thus, recognition of the rights of the child is paramount. Hence, the core concerns of the UN Convention on the Rights of the Child clearly state that children have the right to protection from abuse, neglect and exploitation. Where children's bodily integrity is at risk in crises with widespread and/or systematic violence, providing safe spaces in which children can play, learn and have adequate shelter is critical.

Nutrition and health is also critical especially in developing countries where child malnutrition may already be a significant risk factor, even small-scale natural disasters have the potential to greatly worsen the health of affected children. Access to education restores a sense of normalcy and provides a safe and protected environment in which children can grow physically and intellectually.

During a period of crisis and when fleeing, disaster families can easily become separated. Deprived of parental care, separated children are often forced to head their own households, having

85 *Brookings-Bern Project on Internal Displacement Op cit* (n637)

to work to gain a livelihood in order to support themselves and their younger siblings. Reuniting children with their parents or caregivers, and providing services to support them during this process is a key component of humanitarian response. It is also critical to the protection of minors from exploitation or abduction in the midst of the chaos caused by a humanitarian crisis.[86]

86 UNICEF, Children and Disaster Relief UNICEF in Emergencies<http://www.unicef.org/emergencies/Protecting Children in Emergencies> (Save the Children) <http://www.savethechildren.org/site/c.8rKLIXMGIpI4E/b.6192515/k.319F/Protecting_Children_in_Emergencies.html>; Minimum Standards for Child Protection in Humanitarian Action < http://cpwg.net/minimumstandards/ >Inter-Agency Network for Education in Emergencies: <http://www.ineesite.org/en/>

9

ENVIRONMENTAL LAW AND DISASTER MANAGEMENT: AN EXAMINATION

Amari C. OMAKA &
Caroline U. AGOM

ABSTRACT

The UN defined disaster as a serious disruption of the functioning of a society, causing widespread human, material or environmental losses, which exceed the ability of the affected society to cope using its own resources. Similarly, section 6(2) of National Emergency Management Agency Act 1999 defined natural disasters to include any hazards arising from any crisis, epidemic, drought, flood, earthquake, storm, train, road, aircraft, oil spillage or other accidents; including mass-deportation or repatriation of citizens from any other country. The work examined the "Environmental Law and Disaster Management: An Examination," drawing inspiration from Nigerian legal system, Indian legal system and international law. We adopted conceptual method of data collection, via books, journals, statutes, case law and internet materials. It was found, among other things, that outside the use of law and statutes to control natural ecological and other hazards, there are also social and scientific methods of checkmating the menace. These include: preventing hazards or minimising their effects, predicting hazards and issuing warning signals, identifying and avoiding sites where hazards are likely to occur, restricting land use, developing emergency preparedness measures, spreading economic losses over a large population through insurance and taxation, restructuring a community so that it is less vulnerable to hazards, etc. We recommended inter alia that there should be pre-disaster preparatory and preventive measures for disaster risk reduction; putting measures in place for emergency responses including search, evacuation, rescue and salvage in event of any natural disaster. Finally, we concluded by adopting the recommendations made by the 1994/95 Constitutional Conference as cited in the Supreme Court case of Attorney General of Lagos State vs. Attorney General of the Federation of Nigeria and 35 others, for hazard reduction in Nigeria.

Keywords: Environment, Law & Policy, Ecological Disaster, Management

INTRODUCTION

Environmental laws are those areas of law, national,[1] regional,[2] and international[3] which aims at the protection of the environment.[4] In the words of Justine Thornton and Silas Beckwith,[5] "like many legal terms, environmental law may be seen as having a central 'core of meaning' surrounded by what may be called a penumbra of uncertainty." It is often difficult to draw the boundary of environmental law. While laws that relate to protection of natural environment and resources can be said to be central to

1 Such as the law of any nation.
2 Such as laws of the African Union, European Union etc.
3 Like laws and conventions made under the aegis of the United Nations.
4 Richard Macrory, Environmental Law, Microsoft Encarta Encyclopaedia, Standard 2005.
5 *Ibid.*

environmental law, the scope has now been widened to include laws that are designed to protect the quality of life of particular groups of people. These include laws like sanitation laws, health and safety of employee laws, traffic laws, consumer laws, wildlife and conservation law, etc.[6] Thus, environmental law includes the legislation and the received English common law of general application, which can be used to protect the environment and address environmental concerns. Such laws include the legislation passed at federal, state, and local government levels to address the problems caused by air, noise or water pollution, toxic chemical use and waste management.[7]

Environmental law is the field of law dealing with the maintenance and protection of the environment, including preventive measures such as the requirement of environmental impact statements to measure, assign responsibility and provide clean up for accidents that harm the environment.[8] It also embodies all the natural and man-made rules that ensure sanity in our relationship with our environment, housing and development, of which non-compliance will elicit some sort of sanction against the defaulter. Environmental law, therefore, can simply be said to be any rule or law, which is made or used for the protection, maintenance, conservation and improvement of the environment, including those for the redress for environmental harm.[9]

Thus, environmental law developed from various principles of the law of torts, criminal law, law of insurance, land law, property law, industrial law, energy law, petroleum law, public health law and mineral laws. Environmental law utilises the common law principles of nuisance, strict liability, trespass, negligence, damages, and riparian rights and among others in the advancement

6 *Ibid.*

7 Okonkwo T.R, *The Law of Environmental Liability,* Lagos. AEDE publications 20003.p12

8 Garner, A. B. (ed.) (2004). *Black's Law Dictionary 7th Edition*. St. Paul US: West Publishing Co. P. 215

9 Omaka, C.A. *The Nigerian Conservation Law,* Lagos: Lion's Unique Concept Publications, 2004. See also Black's Laws Dictionary 7th ed. West Group, St. Paul, Minn 1999. P. 555.

of its goals.[10] These legislations and common law principles put in place to safeguard man's surroundings, flora and fauna are the very gamut of environmental law. Environmental law, therefore, can simply be said to be any rule or law, which is made or used for the protection, maintenance, conservation and improvement of the environment, including those for the redress for environmental harm.

By natural disasters we mean unintended destruction of lives and properties, the environment, infrastructure, means of livelihood, domestic and public live-lines by man-made and natural incidents/accidents, negligence and other un-envisaged vicissitudes. They include wildfires, windstorms, drought, volcanic eruptions, earthquakes, tsunamis, avalanches, etc. Under Myanmar law, incidents such as damage to crops by pests, plant disease and outbreak of epidemics, violent insurgencies and industrial accidents are classified as natural disasters.[11] According to the United Nations, disaster is defined as "a serious disruption of the functioning of a society, causing widespread human, material or environmental losses, which exceed the ability of the affected society to cope using its own resources."[12] Similarly, section 6(2) National Emergency Management Agency (Establishment, etc.) Act [13] defined "natural or other disasters" to include any disaster arising from any crisis, epidemic, drought, flood, earthquake, storm, train, roads, aircraft, oil spillage or other accidents and mass deportation or repatriation of citizens from any other country.

Ecological problems have dealt a blow on human environment over the years. Professor Eugene Arene, an erudite university don, has conducted a comprehensive research on these problems. In his

10 Okonkwo, *Op cit.*

11 Natural Disaster Management Law. (The Pyidaungsu Hluttaw Law No. 21,2013)

12 C.E. Pratap (2012). Disaster Management and Law - A Human Rights Perspective. http://www.lawyersclubindia.com/articles/Disaster-Management-and-Law-A-Human-Rights-Perspective-4497.asp#.VlYmTF4R6nd. Downloaded January 12, 2016.

13 This Act established the National Emergency Management Agency (NEMA) via Act 12 as amended by Act 50 of 1999, to manage disasters in Nigeria. It has been tackling disaster related issues through the establishment of concrete structures.

treatise, the erudite professor pointed out that[14] throughout history, natural disasters have exacted a heavy toll of death and human suffering. According to him, natural hazards such as earthquakes, landslides, tidal waves, hurricane, tornadoes, floods, volcanic eruptions and wildfires have claimed more than 2.8 million lives worldwide in the past 20 years, adversely affecting 820 million people. Since 1949, at least 17 individual disasters have killed more than 10,000 people each; on two occasions – in Bangladesh and China – single disasters took more than a quarter million lives each. In *Nichols v. Marshland*,[15] the court considered tornadoes, earthquakes, windstorms, lightening, floods, etc. as ecological natural disasters scarcely outside man's control, and which ordinarily should be anticipated. In fact, progress in scientific and technical understanding of natural hazards, as well as in techniques to mitigate their effect, led to the establishment of an International Decade for Natural Hazard Reduction (IDNHR), which was 1990-2000.

Nigeria has had a number of emergency situations arising from disasters — natural and man-made. The natural phenomena include tropical storms, land erosion, windstorms, floods, drought, desertification, human diseases, coastal erosion, livestock diseases, crop pests and diseases, wildfire, harmattan haze and landslides. Other potential hazards include earthquakes and volcanoes. The major man-made hazards include civil strife; road, water and air accidents; and technological episodes such as oil spills, hazardous wastes dumping and industrial accidents.[16] All of these call for urgent action for the strengthening of emergency preparedness to reduce peoples' vulnerability and cushion the impact of disasters on our settlements, economy and environment. In Nigeria, this fact has been recognised. That is why the Constitution of the Federal Republic of Nigeria makes environmental protection a duty under section 20. The said section provides as follows:

14 E Arene: An Assessment of Policies for A Healthier Environment. Kuru. November 1991

15 (1875) L.R. 10 Ex.255.260, per dictum of Bramwell B.

16 See Nigerian National Policy on Environment 2016.

> The state shall protect and improve the environment and safeguard the water, air and land, forest and wildlife of Nigeria.

By virtue of a provision like section 20 of the 1999 Constitution, it can be seen that the said laws are essentially about how to protect and improve the environment, and to safeguard the water, air and land, forest and wildlife of Nigeria. It has been shown in the judgment given in *Attorney General of Lagos State v. Attorney General of the Federation of Nigeria and 35 others*[17] that the discussions and considerations that led to the formulation of the text of section 20, which for the first time in the constitutional history of Nigeria came to be inserted in the Constitution under the Fundamental Objectives and Directive Principles of State Policy. The Constitutional Conference Committee's discussion, recommendation and resolution on the subject make it clear beyond argument that it was after deliberating on such purely environmental matters as "afforestation and reforestation, devastations through mining activities, desertification, disasters and emergencies commission and the federal environmental protection etc."[18] that the conference came up with that provision. According to the Supreme Court:

> Land can be safeguarded by relevant environmental laws which may prohibit any activities on land that could lead to, say landslide or erosion or pollution with toxic waste with its catastrophic effect. And by similar laws, for the purpose of the reclamation and rehabilitation of devastated land and environment; and to safeguard land against desertification and effect of indiscriminate burning activities. It is in the light of all these that the scope of section 20 of the Constitution should be ascertained.[19]

Nigerian National Policy on Disaster Reduction

Some of the notable natural disasters include: earthquakes, landslides, tsunamis, windstorms, volcanoes, wildfires, sandstorm and dusts-spells, drought, desertification, erosion, expansive soil,

17 (2003) FWLR PT 168 at 909, ratio 11, particularly at p.1003

18 *Ibid.*

19 *Ibid*, ratio 15 particularly at p.1006, pars. H-A.

flood, etc. Within the ecosystem, these ecological hazards distort the interrelationship existing between man and nature. As a matter of national policy, the following strategies are required to mitigate the negative impacts of natural and man-made disasters on the lives of the people.

A) Prepare comprehensive hazard maps and vulnerability analysis for the country by:

i. compiling historical data of disaster occurrence;
ii. analysis of meteorological, seismological, agricultural and environmental records;
iii. employing satellite imagery and the GIS system to plot the hazard maps.

B) Establish very effective early warning systems for meteorological, geophysical, biological, social and industrial hazards by:

i. enhancing the meteorological services;
ii. effective monitoring of pests and disease epidemics;
iii. resuscitation of seismographic stations and the existing seismological centres;
iv. development of reliable biological indicators;
v. building of a viable network for early warning information dissemination.

C) Develop and maintain prompt emergency response mechanisms and contingency plans by:

i. making an inventory of all existing resources for emergency response for easy marshalling at times of disasters;
ii. establishing a body to coordinate emergency response to reduce duplication of efforts and enhance accountability;
iii. formulating a national emergency policy and an emergency plan.

D) Mount a sustained public awareness and education programme on hazard preparedness by:

i. engaging military and para military forces as well as voluntary organisations in drills on emergency response including search and rescue;
ii. preparing curriculum and integrating emergency preparedness into school.[20]

National Policy on Drought and Desertification

Drought and desertification are the most important environmental problems affecting the northern states. Government, and at some instances in collaboration with donor countries, international organisations and institutions, has done a lot to combat desertification, and mitigate the effects of drought since early 1970s. Efforts have been made through more efficient utilisation and alternative sources of fuel wood, promulgation of state edicts, afforestation and reforestation programmes. Despite all these efforts, desertification has continued its down-south movement which currently renders the areas north of Latitude 15° either desertified or prone to desertification. This process coupled with the effects of drought have continued to cripple the socio-economic lives of the people living in the affected areas. The negative impacts on the lives of the people call for urgent actions to check the desert encroachment, rehabilitate affected areas and institutionalise drought ameliorating measures. The strategies needed to achieve this include:

a. develop a National Action Programme to Combat Desertification and Mitigate the effects of drought towards the implementation of the Convention to Combat Desertification (CCD) in Nigeria.

b. Integrating public awareness and education on causes and dangers associated with drought and desertification, as well as the constraints of the CCD.

---

20 See Nigerian National Policy on the Environment 2016.

c. strengthening of national and state institutions involved in drought and desertification control programmes;

d. promoting of sustainable agricultural practices and management of water resources;

e. encouraging of viable afforestation and reforestation programmes using tested drought resistant and/or economic tree species;

f. encouraging the development and adoption of efficient wood stoves and alternative sources of fuel wood;

g. establishing drought early-warning systems;

h) involvement of the local people in the designing, implementation and management of natural resources conservation programmes inimical to combating desertification and ameliorating the effects of drought;

i. intensifying international cooperation and partnership arrangements in the areas of training, research, development and transfer of affordable and acceptable environmentally sound technology and provision of new and additional technical and financial resources;

l. inventorising degraded lands, and implement preventive measures for lands that are not yet degraded or which are slightly degraded;

m. adopting an integrated approach to address physical, biological and socio-economic aspects of desertification and drought;

n. intensifying co-operation with relevant Inter-and Non-Governmental Organisations in combating desertification and mitigating the effects of drought;

o. strengthening of the nation's food security system;

p. establishing, reviewing and enforcing cattle routes and grazing reserves.[21]

---

21 See Nigerian National Policy on the Environment 2016.

National Policy on Flood and Erosion

Flooding in one form or the other affects at least 20% of the nation's population. It cuts across the society from the urban residents to the rural dwellers. Flooding is a threat to physical infrastructures, including residential accommodation, commercial, and industrial properties, roads, rail lines, bridges, port installations, etc. It also destroys farmlands, including standing crops. Losses due to flooding run into many billions of naira per year.

On the other hand, 85% of the total land area of Nigeria is under severe sheet, rill and gully erosion with over 2000 active gully erosion sites spread and around the country. Erosion leads to impoverishment of the soil as nutrients are washed away, loss of livelihood as farmlands become wasteland thus increasing the menace of rural urban migration, and pollution and siltation of available sources of drinking water. Human lives and properties, especially buildings are endangered as they collapse into gullies.

Coastal erosion is widespread along the nation's 800km long coastline with estimated mean shoreline retreats of 2 - 30 metres per year. The worst affected areas include Victoria Beach in Lagos, Awoye/Molume in Ondo State, Ogborodo/Escravos and Forcadọs in Delta State, Brass in Bayelsa State and Ibeno - Eket in Akwa Ibom State.

To achieve effective management of urban, river and coastal flooding and stabilise all gully and coastal erosion sites and enforce management practices aimed at preventing/minimising the incidence of erosion, the following strategies are required:

For flooding:

a. enforce compliance with planning/urban laws/edicts;
b. build embankments and levees along rivers and coastline prone to flood;
c. establish rainstorm early warning system;
d. establish and monitor weather stations, river and tidal gauges;
e. ensure appropriate management of dams;

f. ensure proper maintenance of existing urban drainage channels; and
g. enforce environmental sanitation laws in towns and cities.

For soil and coastal erosion, it will be necessary to:

a. prepare and implement a comprehensive national policy on soil and coastal erosion and flood control;
b. formulate and enforce regulations for soil and water conservation, especially in erosion-prone areas;
c. carry out national watershed delineation and characterisation for use as a basis for development of an aggressive management and enforcement programme to protect and maintain the quality of the nation's lands water and coastal resources and implement the programme;
d. prepare periodic masterplan on the management of soil and coastline erosion and flood, and advise the federal government on the financial requirements for the implementation of such plans;
e. carry out feasibility and scientific studies on soil erosion and related flood problems for the design of appropriate integrated remedial control measures;
f. carry out public enlightenment campaigns on environmental degradation arising from poor land and water management practices;
g. provide and promote training on environmental issues as they relate to flood, erosion, land degradation and water conservation;
h. promote integrated ecosystem management with other agencies connected with agriculture, land use, soil and water conservation, rural development and coastal resources management including environmentally sound recreational use;

i. strengthen national capacity through personnel development, provision of training facilities and research on combating climate-related ecological problems;

j. strengthen capacity of the Environmental Management Support System (EMSS) for Remote Sensing data gathering, GIS facilities and development of disaster/ environmental data bank;

k. support agro-forestry and integrated Coastal Zone Management;

l. encourage planted fallow in abandoned farmland using soil enriching species;

m. promote conservation farming and use of organic fertiliser and soil conditioners; and

n. establish viable contingency plans for tackling socio-economic and other problems resulting from coastal and other erosional disasters.[22]

Among the natural ecological disasters briefly discussed, some are peculiar to Nigeria. These are windstorms or line squalls, dust-spells, droughts, erosion, flood, wildfires and desertification.

Other Specific Relevant Laws and Procedure

National Emergency Management Agency (Establishment, etc.) Act[23]

In Nigeria, the main enabling law for disaster management and control is the National Emergency Management Agency (Establishment, etc.) Act, 1999, which is an Act to establish a National Emergency Management Agency (NEMA) and a State Emergency Management Committee for each state in the federation; and to make provision for other related matters.

22 See Nigerian National Policy on the Environment 2016.

23 National Emergency Management Agency (Establishment, etc.) Act, is an Act of the National Assembly of the Federal Republic of Nigeria,1999.

The Agency (NEMA)

The mission and vision of NEMA is to coordinate resource towards efficient and effective disaster prevention, preparation, mitigation and response in Nigeria; and to build a culture of preparedness, prevention, response and community resilience to disaster in Nigeria.

Functions of NEMA

The Act vested the authority of managing disasters in Nigeria on NEMA. According to the enabling law; the Agency shall among the other things;

(a) formulate policy on all activities relating to the disaster management in Nigeria and co-ordinate the plans and programmes for efficient and effective response to disasters at national level;

(b) monitor the state of preparedness of all.

Specifically, section 6 of the Act bestows NEMA with the following responsibilities. The Agency shall:

(a) formulate policy on all activities relating to disaster management in Nigeria and co-ordinate the plans and programmes for efficient and effective response to disasters at national level;

(b) co-ordinate and promote research activities relating to disaster management at the national level;

(c) monitor the state of preparedness of all organisations or agencies which may contribute to disaster management in Nigeria;

(d) collate data from relevant agencies so as to enhance forecasting, planning and field operation of disaster management;

(e) educate and inform the public on disaster prevention and control measures;

(f) co-ordinate and facilitate the provision of necessary resources for search and rescue and other types of disaster curtailment activities in response to distress call;

(g) co-ordinate the activities of all voluntary organisations engaged in emergency relief operations in any part of the federation;

(h) receive financial and technical aid from international organisations and non-governmental agencies for the purpose of disaster management in Nigeria;

(i) collect emergency relief supply from local and foreign sources and from international and non-governmental agencies;

(j) distribute emergency relief materials to victims of natural or other disasters and assist in the rehabilitation of the victims where necessary;

(k) liaise with State Emergency Management Committees established under section 8 of this Act to assess and monitor, where necessary, the distribution of relief materials to disaster victims;

(l) process relief assistance to such countries as may be determined from time to time;

(m) liaise with the United Nations Disaster Reduction Organisation or such other international bodies for the reduction of natural and other disaster;

(n) prepare the annual budget for disaster management in Nigeria; and

(o) perform such other functions which in the opinion of the Agency are required for the purpose of achieving its objectives under the Act.

Powers of the Council[24]

The Council shall have power to:

(a) manage and superintend the affairs of the Agency;

(b) subject to the provisions of this Act, make, alter and revoke rules and regulations for carrying on the functions of the Agency;

(c) fix the terms and conditions of service, including remuneration of the employees of the Agency after consultation with the Federal Civil Service Commission; and

(d) do such other things which in the opinion of the Agency are necessary to ensure the efficient performance of the functions of the Agency.

The Act[25] established for each state of the Federation of Nigeria, a State Emergency Management Committee[26] consisting of:

(a) the deputy governor of the state who shall be the chairman;

(b) the secretary to the state government;

(c) one representative each from:

i. the State Ministry of Women and Social Welfare;
ii. the State Ministry of Health;
iii. the State Ministry of Works;
iv. the State Fire Service;
v. the Federal Airport Authority of Nigeria;
vi. the State Environmental Protection Agency;
vii. the Agency;
viii. the Commander of the Disaster Reaction Units;
ix. the Nigeria Police Force;
x. the Federal Road Safety Commission;
xi. the Nigeria Security and Civil Defence Corps;
xii. the Nigerian Red Cross Society;
xiii. the National Maritime Authority in coastal States.

24 See section 7

25 See Part III of the Act

26 Section 8(1)(2)(3)

Section 3 provides that the members of the State Committee shall:

(a) be persons of proven integrity;

(b) other than ex-officio members, hold office on such term as may be specified by the governor of the state;

(c) be appointed by the governor of the state; and

(d) be paid such allowances as the governor of the state may, from time to time, determine.

Section 4 provides that the State Committee may make rules regulating its own proceedings.

Section 5 provides that the provisions of section 4 of this Act shall apply mutatis mutandis to the State Committee.

Functions of State Committee[27]

The State Committee has the duty to:

(a) notify the Agency of any natural or other disasters occurring in the state;

(b) respond to any disaster within the state and may seek assistance from the Agency if it deems fit in each circumstance;

(c) carry out disaster management activities in the state as may, from time to time, be recommended by the Agency; and

(d) be accountable to the Agency for all funds accruing to it for purposes of discharging its functions under this Act.

27 Section 9

Facilities and Infrastructures of NEMA for Coordination of Disaster Management

Mission Control Centre (MCC)

The Mission Control Centre located in the headquarters of NEMA is a computer-based satellite technology that uses the COSPAS-SARSAT hi-tech system/facility. The system is designed to pick distress alerts and location data to assist in Search and Rescue Operation, using spacecraft and ground facilities to detect and locate the signals of distress beacons operating on 406 MHz. When there is a distress alert from the beacon located on a ship or aircraft, the satellite system transmits the signal to the ground segment facilities from where the data is processed and transmitted to the appropriate Mission Control Centre (MCC). The MCC now alerts the nearest Rescue Coordinating Centres (RCC) or Disaster Reaction Units (DRU) of the Army, Air Force, and the Navy for appropriate and immediate action.

Geographic Information System (GIS)

The agency has established a functional GIS lab for early warning and precision in response to the management. The lab collects spatial data, analyses and prepares useful information that helps to aid responses to disaster. It is one of the significant facilities of the Agency for its disaster risk reduction programme.

Mobile clinics

Time is a crucial element in disaster management. In recognition of this, the Agency obtained approval and procured facilities for mobile clinics that have been strategically stationed in Abuja, Kaduna, Lagos and Port Harcourt to be deployed in the event of any major disaster. While arrangements are on to acquire more of the clinics, the Agency has also procured a good number of ambulances stationed in Abuja and all the six zonal offices.

Helicopter for Search and Rescue

The Agency has in place a helicopter for search and rescue. The helicopter is stationed in Abuja and manned by officers of the Nigerian Air force. Necessary arrangements are being concluded to acquire additional helicopter which we hope that with time there would be one for each of the six geo-political zones.

Policies and Strategies of NEMA

i. Contingency Stockpiling

To ensure prompt and efficient response to disasters, the Agency established warehouse in its six zonal offices and Abuja and continous to stock them from time to time with relief and rehabilitation items. This is to ensure timely intervention in response to disasters.

ii. Search and Rescue (SAR)/Epidemic Evacuation Plan

The plan which was approved in 2008 allows for a system where all stakeholders can draw on collective strength of one another and build a formidable group of highly mobile, motivated, dedicated and trained workforce of disaster managers.

iii. Collaboration with Security Agencies

Security agencies constitute the major stakeholders of the Agency. Over the years, NEMA has ensured a successful collaboration with the security agencies through regular meetings, workshops, simulation exercises and training, utilising every opportunity to advance the cause of disaster management in the country.

iv. National Disaster Response Plan

The plan is a document approved by the Federal Executive Council which states the roles of stakeholders and the call out plan in response to disasters. The document is presently undergoing a review to update and made more functional.

v. Training and Capacity Building

NEMA believes in the preparation of the human resources for optimal performance. The department of training was established which develops the curriculum and coordinate the human resources development of the agency in addition to catering for the training needs of the stakeholders. The staff are continuously trained locally and abroad on modern skills which qualify them to be good disaster managers as obtain in other part of the world.

vi. Advocacy/Stakeholder Meeting

The agency engages in advocacy to ensure the preparedness of the stakeholders. Through this, it assesses available equipment and facilities at the disposal of the stakeholders. In the absence of such or where seriously inadequate in supply, it encourages the provision. The Agency also ensures a close tap on the stakeholders through regular meeting and activities.

vii. Disaster Risk Reduction

Hitherto, disaster management in the country was mainly humanitarian relief supplies with huge funds expended annually on perennial emergencies. The paradigm shift in disaster risk management offers a good opportunity to build, develop and sustain policies pertaining to social development, equity, economic growth, environmental quality and sustainable land use. Experiences have revealed that in disaster management, attention should be given to prevention and mitigation rather than to wait for the aftermath. In recognition of this, disaster managers the world over have continued to give serious attention to risk reduction strategies. NEMA, with the cooperation of the stakeholders, have embarked on a series of training, sensitisation and initiatives to promote the required consciousness on specific and general disasters. The programme emphasises on the mainstreaming of disaster management into national development policies.

viii. Awareness Generation

More than half of the disasters in the country are man-made arising from deliberate intent, error or negligence. Disasters of man-made origin can be minimised by arousing the consciousness of the people through awareness generation. The Agency utilises awareness generation against disasters as the cornerstone in building a culture of sustainable resilience to disasters.

ix. Six Zonal Offices

In its desire to decentralise disaster management and ensure active participation of the grassroots, the Agency established functional zonal offices located in each of the six geo-political zones of the country as follows: North Central (Jos), North West (Kaduna), North East (Maiduguri), South West (Lagos), South South (Port Harcourt) and South East (Enugu). The zones provide for quicker response to disaster situations before any action from the head office. The zonal offices are the extension of the Agency in their various locations having the responsibilities of implementing policies and carrying out any other assigned functions. Efforts are being made to establish additional zonal offices. Call centres are also to be established across the country for quick reaction to distress alerts. Through the zonal offices, the grassroots now have direct access to the Agency as disaster intervention is also decentralised. The intervention period has significantly improved to just about two days for assessment and delivery of relief assistance for urgent situations.

x. Disaster Response Units (DRUs)

These are units in military formations dedicated to emergency/ disaster responses. The involvement of the military is necessitated especially for situations that may require physical human strength and some military equipment. The units exist in identified Army, Navy and Air force formations across the country.

Indian

Objectives of Natural Disaster Law

Natural disaster laws and policies are aimed to achieve the following:[28]

(1) To ensure the smooth and effective implementation of natural disaster laws and management procedures for risk reduction;

(2) To see to the formation of country or national committees for the systematic and expeditious implementation of disaster management activities and programmes;

(3) To enhance sound coordination, in liaison with national and international bodies or agencies and even non-governmental organisations, government departments and organisations for natural disaster risk reduction and management;

(4) To ensure sound conservation and restoration of the environment and communities affected by disasters;

(5) To ensure and enhance the provision of sound health facilities, education programmes, social rehabilitation and other livelihood programmes to bring back better and normal living conditions for affected victims after any disaster.

Powers and Duties of National Disasters Committees

(1) To lay down rules, regulations, policies and guidelines for disaster management;

(2) To form constitute search parties and rescue work operations in event of any disaster;

(3) To draw preparedness plans to forestall potential natural disasters and facilitate reconstruction and rehabilitation programmes;

28 Natural Disaster Management Law (The Pyidaungsu Hluttaw Law No. 21,2013) See section 3

(4) To submit reports to the government with a view to declaring an area as prone or susceptible to regular or potential disaster occurrences;

(5) To assign duties to relevant agencies or ministries to ensure expeditious implementation of disaster risk reduction programmes;

(6) In liaison with relevant bodies and meteorological agencies, to study, predict and issue warning signals on potential disaster areas and impeding disasters;

(7) To coordinate relevant authorities and organisations for expeditious evacuation of the local population before or during disasters, including evacuations after issuance of early warning signals;

(8) To cooperate with all relevant national, regional or international governments or organisations for timely implementation of disaster risk reduction programmes, including their screening for assistance;

(9) To source for and provide fund for assisting affected people and relocating and even the rehabilitation of the victims;

(10) To liaise where necessary with the Police Force, Civil Defence, Fire Service, Red Cross and other non-governmental organisations to assist the victims of natural disasters;

(11) To advice, report to or assist the government in handing emergency situations and measures including issuing of press releases to local and international audience;

(12) To organise in liaison with concerned bodies, governments and agencies for providing food and other relief materials to affected people including measures to preventing the outbreak of contagious diseases in the area;

(13) To advise the government on steps to ensure expeditious resumption of normal lives and lifelines such as agriculture, commerce, electricity, communication, transportation, etc. to the affected communities;

(14) To prepare accounts and records on how relief items were utilised and issue 'thank you' messages to all donors and those who rendered any form of assistance;
(15) To establish disaster-base stations in disaster-prone areas;
(16) To assist the home government in helping and working together with other countries to stem the challenge of natural disasters in other countries including offering of international aids to affected nations;
(17) To arrange for security, law and order for internally Displaced Persons (IDP) or refugees in event of disasters;
(18) To perform any other function(s) relating to natural disaster management as may be assigned by the relevant government.[29]

Pre-Disaster Preparation

Pre-Disaster Preparation measures for risk reduction include:

(1) Identification of potential disaster-prone areas;
(2) Identification of high disaster-prone periods of the year;
(3) Identification of the likely disaster that is imminent (earthquake, flood, tsunami, landslide drought, etc.) with a view of knowing early measures to take with to stop the occurrence, where possible, or mitigate the potential harm;
(4) Issuing early warning signals and measures to take;
(5) Urgent preparation of disaster risk assessment and planning emergency management procedure;
(6) Possible evacuations where the disaster is likely unavoidable;
(7) Training of emergency and rescue personnel to handle imminent disasters, depending on the nature of the foreseeable disaster;
(8) Creation of awareness via all available media and drilling / rehearsal of community on safety strategies possible mock escapes;

29 Section 5

(9) Recruitment of volunteers and collation of interested emergency preparedness partners (both governmental and non-governmental), including international agencies; and

(10) Stockpiling of food, drugs and other relief items in readiness and to avoid being taken unawares.[30]

Hazard Reduction Measures

The ecosystem is the critical resource that provides man with all his material needs. Lowdermilk summarised this inter-relationship in his so-called "Eleventh commandment".

> Thou shalt inherit the Earth as a faithful steward, conserving its resources and productivity from generation to generation; thou halt safeguard thy fields from erosion: thy living water from drying up; thy forest from desolation; And protect the hills from overgrazing for ever; if any shall fail in his stewardship of the lands. Thy fruitful fields shall become sterile. Stony grounds for wasting gullies; and thy descendants shall decrease and live in poverty or perish from off the face of the earth.

The above commandments, thus, generally recognised that the impact of natural hazards are increasing and will continue to do so unless the world community takes concerted action. The reasons are clear: growth in population; rural-urban migration; increasing capital investment and the increasing worldwide interdependence of people in all communities. All these factors increase the world's vulnerability to disasters. Vital social, technological and scientific actions useful in mitigating the effects of disasters are as follows:

1. Construction of structures and buildings that can withstand the effects of the hazards,
2. Preventing hazards or minimising their effects,
3. Predicting hazards and issuing warning signals,
4. Identifying sites and avoiding areas where hazards are likely to occur,
5. Land use restricting in disaster volatile areas,

30 Section 15

6. Early preparation and development of emergency preparedness actions,
7. Spreading economic losses over a large population through insurance and taxation and,
8. Restructuring a community so that it is less vulnerable to hazards.

The question is how are the effects of natural hazards on community reduced? The first step is hazard and risk assessment; that is, determining the kind of disasters potential in that area, their nature, consequences and the vulnerability of the community to the hazard. The second step is disaster preparedness, showing in details of proposed response in event the disaster strikes.

The third thing is disaster mitigating. This includes such actions as building structures to withstand earthquakes, storms or tornadoes such as windbreakers, dams and other structures to reduce the effects of the disasters including drafting and enforcing grading codes to prevent possible earth tremors and landslides.

Disaster predictions and issuing early warnings are very important to hazard reduction in times of disasters.

Disaster Emergency Management Fund

Countries that have incidents of natural disasters should constitute and manage a national committee for raising Disaster Emergency Management Fund. The committee shall carry out the following duties:[31]

(1) Inclusion and defense in a national budget and management of allocation for national disasters;

(2) Advocating and receiving donations, contributions and grants from, international organisations, foreign countries and regional bodies for use in event of disasters;

(3) Advocating and receiving donations and contributions from philanthropists, civil societies and other organised private sectors;

31 Natural Disaster Management Law (The Pyidaungsu Hluttaw Law No. 21,2013)

(4) Receiving donations or contributions, issue receipts, manage and account for how the fund was used and show appreciation to contributors, nations, organisations and individuals.[32]

Offence and Penalties

It is an offence to do any act that can trigger off any natural disaster or to interfere, prevent, prohibit, assault or coerce any natural disaster officer or department from carrying out his function. It is also an offence to raise false alarm on disasters with a view to raise public fear. Punishment for natural disaster offences differ from country to country. In Myanmar for example, it can occasion as long as 3 years or both fine and imprisonment.[33]

International Legal Framework

There is a relationship between climatic science, climate change, global warming and orchestration of natural ecological problems. Evidence has shown that these phenomena triggered off and continues to enhance the frequency and intensity of ecological disasters and its incidental hazards. Undoubtedly, these disasters when not properly managed or when there is delayed or poor response has led to carnage, disruption of human activities, distortion of natural ecological balance, destruction of flora and fauna and ultimately result to forced displacement of the 'lucky' survivors. When this happens, it would lead to emergence of internally displaced persons or massive growth of refugees. That is why natural disasters have attracted international attention and that of the United Nations. That is why the United Nations High Commission for Refugees (UNHCR) was created to garner global concerted action in the control, response and aiding of countries and persons affected by disasters, inter alia. The Commission in 2009 warned that climate change will significantly contribute to

32 Section 19

33 Section 25-27 Natural Disaster Management Law (The Pyidaungsu Hluttaw Law No. 21,2013)

refugee problems and displacements if needed concerted efforts are not harnessed to check climate change which is one of the greatest contributors to ecological imbalance which ultimately lead to disasters.[34]

Similarly, the United Nations Inter-Agency Standing Committee (UN-IASC) over the years has joined the global league of synergy to check the menace of climate change triggered disasters. The Committee included disasters as a humanitarian challenge within the contemplation of universal legal human rights protection.[35] Another important legal instrument for protecting communities and peoples from the vagaries of hazards is the International Strategy for Disaster Reduction (ISDR). This instrument, guided by the internationally agreed Hyogo Framework for Action, has the objective of building the resilience of communities, peoples and nations by building their capacities to use tested tools for reducing disaster risks. In other to effectively live up to its mandate, the Hyogo Framework for Action deploys three-pronged strategic goals in disaster management namely;

1. development and strengthening of institutions, mechanisms and capacities to build resilience to hazards;
2. integration of disaster risk reduction into sustainable development policies and planning; and
3. systematic incorporation of risk reduction approaches into the implementation of emergency preparedness, response and recovery programmes.

Other legal frameworks on the right based approach to disaster management and mitigation include but not limited to International Convention on Civil and Political Rights, the International Convention on the Elimination of All Forms of Racial Discrimination and the United Nations Committee on the Elimination of Racial Discrimination. These legal instruments copiously provide for state obligations to protect rights during and after any natural disaster, disaster preparedness and timely

34 U.N. High Commissioner for Refugees (UNHCR), Climate Change, Natural Disasters and Human Displacement: A UNHCR Perspective 1 (2009)

35 2011 Tulane Environmental Law Journal, 24 Tul. Envtl. L.J. 239

response. The emphasis on global legal synergy to fight disasters as contained in these covenants and treaties was emphasised especially in the aftermath of Hurricane Katrina in the United States.[36] In the same vein, the European Court of Human Rights in landmark judgments reiterated state obligations to carry out preventive measures to protect people from risks posed by disasters as part of their fundamental right to life as contained in both national and international law.[37] Although most of these international rights treaties did not specifically provide for the protection of people displaced by disasters, nevertheless, under human rights law, countries owe a duty to all citizens and residents, the universal respect for all human rights and fundamental freedoms without distinction as to sex, race, language, creed, religion or colour.

The recent case of *Gabcikovo-Nagymaros Project Hungary v. Slovakia*[38] is very instructive and illustrative of the principles of international law on ecological problems and its impact on the environment. Here, environmental impact assessment assumed an important role in relation to ecological damage as a result of harmful and or hazardous activities resultant upon proposed projects.

Gabcikovo-Nagymaros Case

(i) The Background

This is one of the most celebrated *locus classicus* cases and international environmental law decisions to be handed down by the international court of justice. It is also the case which most directly raises issues relating to sustainable development, the equitable sharing of natural resources, and the impact of ecological damage flowing from harmful or hazardous activities.

The case arose out of a treaty signed in 1977 between Hungary and Czechoslovakia. Following the partition of Czechoslovakia in 1993, Slovakia took the place of Czechoslovakia under the treaty.

36 Advocates for Environmental Human Rights et al., The Human Rights Crisis in the Aftermath of Hurricane Katrina 1 (2010).

37 See particularly Budayeva & Others v. Russia, App. Nos. 15339/02, 21166/02, 11673/02, 15343/02, Eur. Ct. H.R. (2008).

38 ICJ (1997), 25th September 1997, General list No. 92. See compendium of judicial Decisions on Matters Related to the Environment-National Decisions volume 1.

The treaty provided for the construction and operation of a barrage system on the section of the Danube River within the two countries. This was to be a joint investment to produce hydroelectricity, improve navigation on the relevant section of the Danube and protect areas along the banks against flooding. The parties undertook to ensure that the project did not impair the quality of the water in the Danube, and that nature would be protected in the course of the construction and operation of the system.

The treaty provided for the building of two series of locks, one at Gabcikovo (in Slovak territory), and the other at Nagymaros (in Hungarian territory). The two locks were to constitute "a single and indivisible operational system of works." The cost of the joint investment was to be borne by the two parties in equal measure in the use of the system.

Work on the project started in 1978. Due to domestic criticism focusing on the economic and environmental implications of the project, Hungary suspended the works at Nagymaros in May 1989 pending the completion of the various studies. Later, in October 1989 it abandoned the works altogether. By this time, work on the Gabcikovo sector was well advanced, with the most advanced sections being 95% complete while the least advanced were up to 60% complete. On the Nagymaros sector, on the other hand, very little work had been done.

Czechoslovakia protested Hungary's suspension, and subsequent abandonment, of the works. Prior to Hungary's abandonment of the works, negotiations to find a solution commenced. These eventually proved fruitless and, in May 1992, Hungary terminated the treaty.

While negotiations were still going on, Czechoslovakia started investigating alternative solutions. One of them, "Variant C", entailed a diversion of the Danube by Czechoslovakia on its territory and the construction on its territory, of a reservoir with a storage capacity about 30% less than that of the one initially contemplated. Work on "Variant C" began in November 1991 and in October 1992; Czechoslovakia put it into operation without the involvement of Hungary.

In April 1993, the parties agreed to submit the dispute to the ICJ. They requested the court to decide;

(1) "Whether the Republic of Hungary was entitled to suspend and subsequently abandon, in 1989, the works on the Nagymaros project and on the part of the Gabcikovo project for which the treaty attributed responsibility to the Republic of Hungary."

(2) Secondly, the parties asked the court to decide whether "the Czech and Slovak Republic was entitled to proceed, in November 1989, to the provisional solution, and put it into operation in October 1992."

(3) The court was also asked to determine the legal consequences of its judgment on these questions.

(ii) Hungary's termination of the treaty

Hungary relied on a "state of ecological necessity" as justifying its termination of the treaty in 1989. It saw several ecological dangers from the works: the quality of the water would be impaired due to erosion and silting, there were risks of eutrophication and the fluvial fauna and flora would become extinct. Slovakia, on the other hand, denied the existence of a "state of ecological necessity." It argued that whatever ecological problems might have arisen, could have been remedied.

The "state of necessity" is "the situation of a state whose sole means of safeguarding an essential interest threatened by a grave and imminent peril is to adopt conduct not in conformity with what is required of it by an international obligation to another state."

The court stated that the state of necessity is a ground recognised by customary international law for precluding the wrongfulness of an act not in conformity with an international obligation. For it to be successfully invoked, the following conditions must be satisfied:

(a) it must have been occasioned by an essential interest of the state which has committed the act conflicting with one of its international obligations;

(b) that interest must have been threatened by a grave and imminent peril;

(c) that act being challenged, must have been the only means of safeguarding that interest;

(d) that act must not have seriously impaired an essential interest of the state towards which the obligation existed; and

(e) the state which has committed the act must not have contributed to the occurrence of the state of necessity.

The court accepted that Hungary's concern about the effects of the project on its natural environment related to its essential interest: safeguarding the ecological balance has come, in the last two decades, to be considered an essential interest to all states. It held, however, that Hungary's uncertainties as to the ecological impact of the project could not, alone, establish the objective existence of a peril that could justify invoking state of necessity. The existence of peril must be grave and imminent. The environmental dangers highlighted by Hungary were mostly of a long-term nature, and remained uncertain. Even if it could have been established that the project would ultimately have constituted a grave peril for the environment in the area, the peril was not imminent in 1989, the time when Hungary suspended and then abandoned the works. In any case, Hungary suspended and then abandoned the works. However, Hungary had means other than abandonment of the works, of responding to any such peril, for instance the adoption of mitigatory measures.

Therefore, on the first question, the court held that Hungary was not entitled to suspend and subsequently abandon the project, and that its notification of termination of the treaty did not have the legal effect of terminating it.

(iii) Czechoslovakia's implementation of an alternative solution

Hungary contended that "Variant C" was a contravention of the 1977 Treaty, the convention ratified in 1976 regarding the water management of boundary waters, the principles of sovereignty, territorial integrity, the inviolability of state borders, as well as the general customary norms on international rivers and the spirit of the 1948 Belgrade Danube Convention. For its part, Czechoslovakia contented that recourse to "Variant C" did not constitute an internationally wrongful act. It argued that it had a right, in the face of Hungary's abandonment of the project, to implement a solution as close as possible to the original project, i.e., the "principle of approximate application." In addition, Czechoslovakia added that it had a duty to mitigate the damage to itself resulting from Hungary's unlawful actions. Given the advanced state of the works on the Slovak side at the time of Hungary's termination of the treaty, the economic loss and environmental prejudice arising out of a failure to put the system into operation would have been immense.

The court, while acknowledging the serious problems facing Czechoslovakia because of Hungary's actions, held that "Variant C" failed to meet the cardinal condition of the 1977 Treaty that the project was to be a "joint investment constituting a single and indivisible operational system of works." This could not be carried out by unilateral action, such as Slovakia's. Moreover, the operation of "Variant C" led Slovakia to appropriate for its own use and benefit between 80 and 90% of the water of the Danube before returning them to the main bed of the river, despite the fact that the Danube is not only a shared international watercourse, but also an international boundary river. The court held that the implementation of "Variant C" by Slovakia was an internationally wrongful act.

The court considered whether the wrongfulness could be precluded on the ground that the measure was in response to Hungary's prior failure to comply with its obligations. It observed that for wrongfulness to be so precluded the countermeasure adopted must be commensurate with the injury suffered. The court held that Czechoslovakia, by unilaterally assuming control of a shared resource, thereby depriving Hungary of its right to an equitable and reasonable share of the natural resources of the

Danube failed to respect the proportionality, which is required by international law. Czechoslovakia's diversion of the river was therefore not a justified countermeasure. Therefore, the court held, on the second question that whereas Czechoslovakia was entitled to proceed with the preparatory works for "Variant C," it was not entitled to implement it.

(iv) Consequences of the court's judgment

On the issue of the consequences of the court's findings, the court held that, as the 1977 Treaty was still in force, it would continue to govern the relationship between the parties. Therefore, the parties should negotiate on how the treaty's multiple objectives could still be fulfilled. The court observed that the project's impact upon, and implications for, the environment were a key issue and that the parties should look afresh at the effects on the environment of the operation of the power plant. Whatever solution is found by the parties must take account of the objectives of the treaty, the norms of international environmental law and the principles of the law of international watercourses.

The court suggested that one way in which the parties could achieve these multiple objective would be for them to re-establish the joint regime by making "Variant C" conform to the treaty through the joint operation of the current works, and the modification to satisfy Hungary's environmental concern. The court stated further that the re-establishment of the joint regime would reflect the concept of common utilisation of shared water resources in accordance with the convention on the law of the non-navigable uses of international watercourses, and would constitute the best solution in this instance.

(v) Minority opinions

The court's decision was not unanimous: there were both separate, but concurring, and dissenting opinions.

Judge Weeranmantry gave a separate opinion focusing on the role played by the principle of sustainable development in balancing the competing demands of development and environmental protection. He also focused on the principle of continuing environmental impact assessment and the use of principles such as estoppel for the resolution of *erga omnes* problems such as environmental damage.

Judge Weeranmantry observed that had the possibility of environmental harm been the only consideration to be taken into account, Hungary's contentions would have proved conclusive. However, there were other factors to be taken into account, not least the developmental aspect. The project was important to Slovakia from the point of view of development. Therefore, the court had to balance between the environmental and developmental considerations, which it could only do through the principle of sustainable development, which Judge Weeranmantry considered to be an integral part of modern international law. This case marked the first occasion on which it had received attention in the jurisprudence of the court.

On the principle of continuing environmental impact assessment, Judge Weeramantry observed that this referred not merely to an assessment prior to the commencement of the project, but a continuing assessment and evaluation as long as the project was in operation. In this instance, the principle of EIA was incorporated into the treaty. Environmental law would read into treaties which may be considered to have a significant impact on the environment, such as this one, a duty of monitoring the environmental impacts of the project during its operation since there had been foreign international recognition of the concept of continuous monitoring as part of EIA.

On the issue of estoppel, Judge Weeramantry suggested that *inter partes* adversarial procedures might not be suitable in resolving a case involving imminent, serious or catastrophic environmental danger. He observed that international environmental law would need to proceed beyond weighing the rights and obligations of the parties within a closed compartment of individual state self-interest, unrelated to the global concerns of humanity as a whole. This case presented an opportunity for such reconsideration.

(vi) Commentary[39]

This judgment marks important milestone in the development of the ICJ's jurisprudence on international environmental law.

In the Gabcikovo-Nagymaros case, the court held that Slovakia had committed an internationally wrongful act by implementing "Variant C", though not in carrying out the preparatory works. The court equated Slovakia's act with Hungary's act in terminating the treaty. This set the stage for the ICJ's direction to the parties to negotiate a settlement on an equal footing, both having been found at fault. However, the ruling has justifiably been criticised as not supportable on the facts. Clearly, there would not have been any point in Slovakia carrying out works, which it could not put into operation. In his separate opinion, Judge Koroma expressed dissatisfaction with this aspect of the court's judgment and observed "justice would have been enhanced, had the court taken account of the special circumstances [justifying Slovakia's actions]." Other members of the court also dissented from this aspect of the judgment.

The separate opinion of Vice-President Weeranmantry is very instructive because "a rich array of environmentally related issues" including Environmental Impact Assessment arose in the case. Before his dissenting judgment, Justice Weeranmantry said:

> This case raises a rich array of environmentally related legal issues. A discussion of some of them is essential to explain my reasons for voting as I have in this very difficult decision. Three issues on which I wish to make some observations supplementary to those of the Court, are: the role played by the principle of sustainable development in balancing the competing demands of development and environmental protection; the protection given to Hungary by what I would describe as the principle of continuing environmental impact assessment; and the appropriateness of the use of inter partes legal principles, such as estoppel, for the resolution of problems with an erga omnes connotation such as environmental damage.

39 *The Law of Environmental Liability*: Theodore Okonkwo; AEDE publications 20003.p.259

RECOMMENDATIONS

For the reduction of potential damages, destructions and losses that are associated with disasters, the following are measures that could be taken in disaster-prone areas:

(1) There should be pre-disaster preparatory and preventive measures for disaster risk reduction;

(2) Measures must be put in place for emergency responses including search, evacuation, rescue and salvage in event of any during natural disaster;

(3) There should be put in place measures for post-disaster rehabilitation, rebuilding and reconstruction for improved living standards of the community as well as the conservation of the environment that are affected by natural disasters;

(4) In times of disasters, priority should be given to the rescue and protection of children, infants, the disabled, the elderly, women (pregnant and nursing mothers in particular);

(5) During evacuation, rescue, resettlement or any other disaster response activity, concerned officers and men should refrain from act that may cause injuring or demeaning of human dignity of the victims or affected persons;

(6) Other measures to prevent colossal harm by natural disasters in area it is likely to occur include but not limited to the following:

    (a) It is important to build cyclone shelters and other lifesaving sanctuaries in evacuation rough/tough terrains;

    (b) It is vital to construct embankments along waterways and coastlines in flood and erosion prone areas;

    (c) It is also imperative to plant and preserve coastal trees and mangroves along coasts and shorelines in case of water-related ecological problems such as windstorm, hurricane and blizzard.[40]

40 Section 16

Furthermore, we adopt the recommendations made by the 1994/95 Constitutional Conference as cited in the Supreme Court case of *Attorney General of Lagos State v. Attorney General of the Federation of Nigeria and 35 others.*[41]

(7) *Afforestation and Reforestation:* Concerted effort should be made in terms of afforestation and reforestation programme to help fight land degradation *and other ecological problems.*[42] Water should therefore be pumped to desert-prone areas to help combat desertification effectively, thereby intensifying efforts in afforestation and reforestation activities.

(8) *Devastation through Mining Activities:* Existing mining regulations should be enforced and adhered to strictly in order to reclaim and rehabilitate the devastated environment due to mining activities.

(i) At least 50% fund allocated to NDDC or *Niger Delta Commission.*[43] Oil Minerals Producing Areas Development Commission should be devoted to the rehabilitation of the environment of oil-producing areas.

(ii) Funds should be provided through a commission similar to NDDC or the Niger Delta Commission, preferably tagged SOMPADEC (Solid Minerals Producing Areas Development Commission), to coordinate the reclamation and the rehabilitation of the devastated environment.

(iii) Efforts should be made to complete the Liquefied Natural Gas (LNG) project to provide alternative energy source and prevent unnecessary depleting of trees for wood fuel, thereby causing severe erosion and ecological degradation.

(9) *Desertification:* Water should be conveyed to desert-prone areas to reforest and afforest and thereby control

41 Supra. P. 996-997.

42 Italics mine.

43 Italics mine.

desertification more effectively, and a Forest, Soil, and Coastal Erosion and Desert Control Commission should be established.

(10) *Disaster and Emergency Commission:* A relevant commission should be established to be responsible for the handling of National Disasters and National Emergencies.

(11) *National Environmental Standards and Regulation Enforcement Agency Act (NESREA):*[44] National Environmental Standards and Regulation Enforcement Agency Act (NESREA) should be reorganised, rescheduled and be continuously funded to perform its roles more effectively.

(12) The mitigating factors under hazard reduction above are also hereby recommended.

44 The original text had FEPA as an agency not NESREA. However, NESREA succeeded FEPA in 2007.

10

THE TORT OF CATTLE TRESPASS IN NIGERIA: AVERTING A LOOMING DISASTER

Gloria D. SHAJOBI-IBIKUNLE

INTRODUCTION

In Nigeria, the vocation of livestock grazing is largely migrant in nature and run by the fulani ethnic tribe. Previously, herdsmen lived largely in the rural parts, armed mostly with sticks, daggers and knives for protection. They rarely ventured into town and were rarely seen in the cities. For the most part, they lived peacefully with whichever host community that accomodated them. The situation has now changed for the worse. In the past ten years, clashes between herdsmen and farmers in most states of the federation have become increasingly common, vicious and have assumed dangerous levels. Farmers allege destruction of their farms and crops which they resist.The herdsmen allege cattle rustling mostly in the North and hostile communities in other parts to which they retaliate with killings, reprisal attacks, maimings, rapes and brutalisation of women and children. In severe cases, entire villages have been burnt, looted and sacked. This has led to the creation of Internally Displaced Camps (IDC) in some states. Furthermore, the already stretched Nigerian Police Force have had to be drafted in and in some cases the Nigerian Army also to maintain law and order. This crisis has assumed ethnic coloration as it is alleged livestock herdsmen are 'Fulani' by tribe and a religious connotation i.e. 'muslim' by religion. When these clashes occur in the southern part and middle belt which are largely christian, it is a combustible mix. According to the 2015 Global Terrorism Index, these Fulani militants are the fourth deadliest militant groups in the world with a record killing of 1229 people in 2014.[1]

The three possible ways to incur tortious liability through animals under ones's control are:[2]

i. Ordinary liability in tort
ii. Liability under the Scienter Rule
iii. Liability for Cattle Trespass

At common law, a possessor of cattle was strictly lalble independent of scienter for damage done when they trespassed on the land of a neighbor. Yet the tort of animal liability and more specifically tort of cattle trespass provides for liability and compensation as the case may be as civil cause of action. The descent into murder, rape etc has criminalised this tort. Yet in other jurisdictions, cattle vs farmers issues have remained largely of civil liability. It is germane to examine the full ramifications of these issues.

The Fulani people

The Fula people (Fula: Fulbe; French: Peul; Hausa: Fulani or Hilani; Portuguese; Fula; Wolof: Pel; Bambara:Fulaw); number between 20 and 25 million people and are one of the largest ethnic groups in the Sahel and West Africa, widely dispersed across the region.[3] The Fulanis generally speak the Fula language and are bound together by the Fula language and their Islamic religious affiliation, their history and their culture. A significant

1 Eyekpimi, Omawunmi, History Of Fulani Herdsmen And Farmers Clashes In Nigeria, https: //infoguidenigeria.com/fulani-herdmen-farmers-clashes, June 21st, 2016 . Last accessed 27th September 2017.

2 Winfield & Jolowicz, TORT, W.V.H. Rodger. Sweet & Maxwell, London. 17th edn, 2006, p.727.

3 Fula People, https://en.wikipedia.org/wiki/fula¬_people last accessed 02/10/2017.

number of them- a third, or an estimated 7-8 million are nomadic in nature, herding cattle, goats and sheep across the vast dry grass lands of their environment, keeping isolate from the local farming communities, making them the world's largest pastoral nomadic community in the world.[4]

They are massively spread over many countries, and are found mainly in West Africa and northern parts of Central Africa, but also in Sudan and Egypt. Generally, three different types of Fulani exist based on settlement patterns.[5]

i. The nomadic/Pastoral or Mbororo: The pastoral Fulani move around with their cattle throughout the year. They do not stay around for long stretches(not more than 2-4 months at a time).

ii. The semi-nomadic Fulani: These are either Fulbe families who happen to settle down temporarily at particular times of the year or Fulbe families who do not browse around past their immeadiate sorroundings, and even though they possess livestock, do not wander away from a fixed homestead.

iii. The Settled or 'Town' Fulani: These live in villages, towns and cities permanently; and have given up nomadic life completely.[6]

Animal Liability

An animal is any creature, fish, reptile or other creature excluding a human being.[7]Animals are companions, burden bearers, sources of food, clothing, helpers on the farm and source of alternative research material. It can be defined as any living creature that is not a plant or person.[8] The general rule of law is, that a person

4 *Ibid*

5 *Ibid*

6 *Ibid*. The process of settlement, military conquest has established them as the ruling class in parts of West Africa such as Birnin-Kebbi, Gombe, Yola, Jalingo, Mubi, Ngaoundere, Girei and Garoua in Camroun and Northern Nigeria. In Guinea, they are in Fulani towns such as Labe, Pita, Mamou and Dalaba; Kaedi, Matam and Podor in Mauritania-Bandiagara, Dori, Gorom-Gorom and Djibo in Mali and Burkina Faso on the bend of the Niger.

7 Malemi, Ese. *Law of Tort,* Princeton Publishing Co, Lagos, 1st ed, 2008, p397.

8 *LONGMAN Dictionary of Contemporary English,* Pearson Education Limited, England, 2001. 3rd ed, p 44.

keeps an animal at his own risk and is liable for any injury or damage done by it.

Animals are classfied into three:

a. Livestock also commonly referred to as cattle
b. Dangerous animals
c. Non-dangerous animals

It is said that tortious liability for damage or injury by an animal dates back to the feudal era in England. Liability for animals developed for two reasons; this was a time when livestock was the main source of wealth, they roamed freely and the fencing off of property was not common.[9] Also the situation as to cattle being a denominator of wealth may be said to be applicable to certain parts of the northern part of the country.

LIVESTOCK

The word 'livestock' is said to mean any animal reared or kept for food, wool, skin, farming or any agricultural purposes.[10] In other words, animals not in a wild state. It can be said to mean domestic animals used or raised on a farm and the term in a generic sense includes all domestic animals.[11] Cattle is said to be cows and bulls kept on a farm for their meat or milk.[12] The main focus of this article is livestock or cattle trespass. It is important to state that livestock does not include animals such as 'dogs' and 'cats'. Malemi[13] lists two reasons why this is so thus:

1. It is neither desirable nor practicable to keep them in restraint
2. The damage they cause is minimal.

Claims for damages for cats and dogs are best handled under the scienter action. This would mean a claim for harm by a wild

9 Chris Turner, Tort Law, Hodder & Stoughton, London, 2003.p.134
10 *Ibid*
11 *BLACK'S LAW DICTIONARY,* West Publishing Co, ST Paul, Minn, 1990, 6th ed, p 935
12 *LONGMAN Op cit* p200
13 Malemi p 399

or dangerous animal or under the tort of negligence for injury they cause.

Sources for Livestock Liability

There are several ways livestock liability may arise. This includes the following ways:[14]

1. By bringing in prohibited livestock into the country
2. Keeping a diseased animal likely to cause an epidemic.
3. Failure to vaccinate an animal or give any prescribed by law.
4. Failure to restrain, transport or keep an animal in a prescribed manner
5. Driving an animal intentionally on to another person's land
6. When an animal strays independently onto another person's land.
7. Carelessly leaving an animal on the highway to constitute a nuisance on another's land.
8. When an animal causes damage to crops.
9. When an animal injures another animal.
10. When an animal causes damage to chattel.

This article will mostly cover damage caused under numbers 4-10 above.

What is Cattle Trespass?

Cattle trespass was an ancient common law tort whereby the keeper of livestock was held strictly liable for any damage caused by straying livestock. Liability for the trespass was absolute. It was not necessary for the plaintiff to prove intent or negligence and it was not open to the defendant to prove its absence and escape liability. The rule on cattle trespass has had a great influence on modern law[15] and it served as the foundation of the

14 Malemi, *Op cit* p.400-401

15 Ramaswamy S. Iyer, *The Law of Torts In England & India,* Madras Law Journal Press,Madras, 1938, 2nd ed, p112.

rule in Rylands v Fletcher.[16] It comes from a very ancient stock of legal ideas as special rules for compensation for damage to crops by straying animals in ancient codes such as the 12 Tables of Rome[17] and the Code of Manu.[18] A man is answerable for not only his trespass but that of his cattle.[19]

The first recorded writ for cattle trespass was issued during the reign of King John (1199-1216). It related initially to keeping of animals but by 1353 was expanded to cover liability caused by their escape.[20] Cattle trespass includes but is not limited to straying cattle. This was the old rule before the case of Cox v Burbidge[21] was decided. The case laid the basis for liability for livestock trespass. It was enunciated by Williams J. The facts were that the horse of the defendant which was on a highway kicked the plaintiff's child who happened to be playing nearby.

> I apprehend the general rule of law to be perfectly plain. If I am the owner of an animal... I am duty bound to take care that it does not stray on to the land of my neighbor; and I am liable for any trespass it may commit, and for the ordinary consequences of that trespass...

The interest being protected is based on land. This means only a person having some acknowledged interest in land trespassed upon by cattle can bring an action for cattle trespass. A mere licensee does not have a right of action. Street submits that liability is for damage done by the livestock to the land or to any property on it and this only covers damage and not loss.[22] Susu[23] is of the view that damage to animals;[24] chattel;[25] and recovery

---

16 Rylands v Fletcher(1866) L.R 1 Exch265

17 Table V111, Clause 7: Pritchard & Nasmitn, History of Roman Law, p93

18 Chapter VIII, 240-241.

19 Blackstone, Comm, 3, 211 p384

20 Cattle Trespass, https://en. Wikipedia.org/wild/cattle-trespass. Last accessed 12th September, 2017

21 Cox v Burbridge.13 C.B.N.S. 430,32 LJCP 89.

22 Margaret Brazer, STREET ON TORTS, Butterworth, London, 1988. p361

23 Susu, Benedicta, Law of Torts, CJC Press, Lagos, 2004. p182

24 Ellis v Laftau(1874) LR 10 CP; Chattel v Cooper

25 Cooper vRailway Executive(1953)1 WLR ,223

for personal injuries[26] are all actionable under the tort. Pollock[27] is of the view that cattle liability should only be based on the natural and probable consequences of the trespass e.g. an infection of a neighbor's sheep by a trespassing diseased sheep.[28]

The owner's knowledge of their diseased states is not material, the action being a trespass. So strict is the rule that if any part of the animal to which an owner is bound to keep is over the boundary, it is a trespass. The Cattle Trespass Act of 1871 was the extant legislation applicable under English law, being a statute of general application. The common law and the 1871 Act was the operative law in most of the colonies governed by Great Britain. It is still the operative act in Nigeria. However, in the United Kingdom, Section 1(1)(c) of the Animals Act of 1971 has abolished the tort of cattle trespass, although it continues to subsist in other common law jurisdictions either in its original form as a Common Law tort or as may be modified by statute. It is noted that section 4(2) of the Animals Liability Act imposes liability on a person in possession of trespassing livestock.

Furthermore, it is the word 'stray' that is used in the act though the marginal note in the Act refers to 'trespassing.' Rodgers submitted that the word 'stray' is probably wide enough to cover 'trespass.'[29] This author respectfully disagrees and is of the view that the word 'stray' is wider than the the term 'trespass.'

Another area of change in the 1971 Act was that at common law, the remedy of self-help was available to a person harmed by straying livestock through *distress damage feasant* but the person had no power of sale. This has been abolished by Section 7. Any expenses reasonably incurred by a person in keeping livestock while it cannot be restored to the person to whom it belongs or while it is detained in pursuance of Section 7 of the Act, the owner of the livestock will be liable for damage incurred and a refund

26 Warman v Cole(1954) 1 Q.B. 614.

27 Pollock's Law of Torts: A Treatise On the Principles of Obligations Arising From Civil Wrongs In the Common Law, Stevens & Sons Ltd, London, 14th Edition, 1939. p383

28 Thoyer v Purnell(1918) 2 K.B. 333 88 LJKB 263

29 V.W.H. Rodgers footnote 80, p738

of such expenses.[30] Also any local authority operating a pound for animals found straying may properly impose a standard charge on the owner of such cattle.

EXCEPTIONS TO CATTLE TRESPASS

As stated earlier, the owner of cattle is liable for trespass committed by cattle upon trespass but exceptions exist where the owner is entitled to claim damages. These are;[31]

a. Where there is a statutory, contractual or prescriptive duty to erect a fence and trespassing animals are injured upon entering the unfenced land.

b. Where there is a defective fence which the landowner is bound to repair.

c. Where the landowner placed in his land some dangerous instruments

In *Heald v Grier*[32], a farmer brought an action against his neighbor for cost of cattle lost when the neighbor drove them off his property. The cattle were grazing on the farmer's property and strayed onto the neighbor's land. He forced them off his property onto another's property who now drove them onto the highway and some got lost.

It was held that, apart from the fact that the landownwer had no lawful fence, no law can compel animals to recognise and respect 'invisible boundary lines;' that he had the right to drive the cattle back to the farmer's property but not to drive them elsewhere to their harm. Thus, he was liable for the cost of the lost cattle.

d. Where cattle being led along the highway escape from there on to an adjacent land.[33] This is because this is regarded as a normal hazard associated with the passage

30 section 4(1)

31 Persons Liable For Trespass: https://trespass.uslegal.com/persons-liable-for-trespass. Last accessed 12th September 2017.

32 Heald v Grier (12 Mo. App 556 (Mo. Ct App 1882

33 Tillet v Ward (1882) 10 Q.B.D. 17

of traffic along the highway.[34] But if the escape is due to carelessness on the part of the person in control of the cattle on the highway, the latter will be liable in negligence.[35]

Recent Farmer v Herdsmen Clashes In Nigeria[36]

September 30th, 2012: A Fulani herdsman had been accused of murdering one Benjamin Chegue on his farm, the Director of Personal Management in the Isoko North Local Government Council In Delta State.

- April 5th, 2014: Assailants opened fire on community leaders and residents that were meeting in Galadima village. At least 200 people were killed and an unknown number were injured in the attack. Sources attributed the attack to Fulani assailants.
- February 18th, 2016: Five persons were killed by Fulani herdsmen at Okokolo village in Agatu Local Government Area of Benue State.
- March 5th 2016: About 500 persons were killed by the rampaging herdsmen following a siege on Agatu local government area of Benue state. These communities include; Aila, Akwu, Adagbo, Okokolo, Ugboju, Odugbeho, Ogbaulu, Egba and Obagaji.
- April 12th, 2016: Fulani herdsmen attacked two villages in Gashaka Local Government Area of Taraba state and killed 15 people.
- April 19th, 2016: Persons drawn from Twenty-five local government areas in Delta State grounded activities on the Benin-Asaba expressway. They reported that the herdsmen allegedly killed over 23 persons. Interestingly, the police recovered 20 AK-47 rifles, 70 dane guns, 30 double-barrel guns and over 1,000 live ammunition, mostly from Fulani herdsmen during this period.

34 Kodilinye Gilbert And Aluko, Oluwole: The Nigerian Law Of Torts, Spectrum Books Ltd, Ibadan. 1999. P.126-127.

35 Gayler & Pope v Davies(1924) 2 K.B. 212.

36 Eyekpimi, *Ibid*

- April 21st, 2016: Farmers in Lagun, Iyana Offa, Offa, Atagba, Lapata and their surrounding communities in Lagelu Local Council Area of Ibadan, Oyo State, alleged that a group of Fulani armed men attacked their communities at night, injured a guard and carted away valuables.
- April 25th, 2016: Fulani herdsmen attack seven villages in Nimbo in Uzo-Uwani Local Government Area of Enugu State. About 40 persons were reportedly killed.
- June 16th, 2016: A 45-year-old renowned farmer was shot by gunmen suspected to be Fulani herdsmen in Ossissa community in Ndokwa east local government area of Delta state.
- June 20th, 2016: At Least 59 Deaths have been recorded following recent attacks on Benue communities such as Ugondo, Turan, Gabo Nenzev – in the Logo Local Government Area by suspected herdsmen.
- 15th October, 2017. At least 29 are allegedly dead as herdsmen attack Nkiedonrwho community, Bassa Local Government Area, Plateau State.[37]

As it can be observed from the foregoing data, the clashes have spread over states like Plateau, Benue, Ekiti, Enugu, Cross River, Taraba, Edo but Plateau and Benue have been the worst hit.

CAUSES OF FARMERS VS HERDSMEN CLASHES

Several factors have been identified for the escalation of the clashes in recent times. According to DeHaan, 'destruction of crops by cattle and other property (irrigation equipment and infrastructure) by the pastoralists themselves are the main direct causes for conflicts cited by the farmers, whereas burning of rangelands, fadama and blockage of stockroutes and water points

37 Fulani Herdsmen defy Curfew, Kill 27 in Plateau Attack, *The PUNCH*, Tuesday, 17th October 2017, p 12.

by crop encroachment are important direct reasons cited by the pastoralists.'[38] Others are:

A. Land Use Conflict

Land is probably the most important resource needed by man for his day-to-day existence. All human livelihoods and activities are directly or indirectly dependent on land at varying thresholds. However, land connotes different meanings to the various user groups. Out of all user groups, agricultural production perhaps exhibits the highest form of sophistication in its use of land. Not only must agricultural land be capable of supplying crop-specific nutrients and water; soil temperature, structure, texture and pH levels are inevitable requisites in the choice of land for agricultural production activities.[39]

Yet, land is a limited, somewhat scarce, resource with both artificial and natural access and usage barriers. Non-agricultural user groups compete with agricultural user groups on one hand, while there are various levels of intra-user group competition on the other. Farmer-herdsmen conflict has remained the most preponderant resource-use conflict in Nigeria.[40] The necessity to provide food of crop and animal origin, as well as raw materials for industry and export in order to meet ever growing demands, has led to both "intensification and extensification" of land use.[41] The competition between these two agricultural land user-groups, however, has often times turned into serious overt and covert manifestation of hostilities and social friction in many parts of Nigeria. The conflicts have demonstrated high potential

38 De Haan, C. Nigeria Second Fadama Development Project (SFDP), Project, Preparation Mission Report, Livestock Component, World Bank, 2002. p1-13.

39 Adisa, Rashid Solagberu, Land Use Conflict Between Farmers and Herdsmen – Implications for Agricultural and Rural Development in Nigeria. www.intechopen.com. Last accessed 28th September 2017. P99-100

40 Fasona, M. J. and Omojola, A. S. "Climate Change, Human Security and Communal Clashes in Nigeria." Paper at International Workshop in Human Security and Climate change, Holmen Fjord Hotel, Oslo Oct. 21st-23rd, 2005.

41 Nyong, A and Fiki, C. (2005). "Droughts-Related Conflicts, Management and Resolution in the West African Sahel." Human Security and Climate change International Workshop. Oslo; GECHS, CICERO and PR20.

to exacerbate the insecurity and food crisis particularly in rural communities where most of the conflicts are localised, with reverberating consequences nationwide.[42]

B. Killer herdsmen

It is alleged that the fulani herdsmen involved in crisis rocking parts of the federation are not Nigerians but trans-border migrant fulani people.

> It is a shameful thing today that you dont even know those who are indigenous fulani or the trans-border fulani. We don't know and that is why people mix things up, you begin to suspect the fulani man that you grew up with when anything happens. He has not travelled far. Where did he get the AK 47?[43]

As stated earlier, the fulani herdsmen are categorised into two: those who are Nigerians and those foreigners from neighbouring countries like Cameroun, Chad and Niger. The Nigerian fulani insist on their constitutional right to freedom of movement within the country under section 41 of the 1999 Constitution (as amended) which states that 'Every citizen of Nigeria is entitled to move freely throughout Nigeria and to reside in any part thereof.'

However, freedom of movement is not freedom to commit actionable trespass on the land, chattel of other citizens or commit murder, rape etc which are crimes against the state.

C. Economic Community of West Africa Protocol

The Economic Community of West Africa (ECOWAS) Transhumance Protocol of 1998[44] provides for free movement of goods and West African nationals across the border of the participatory nation states. Nigeria is a signatory to the treaty. The border posts of each nation is supposed to have the appropriate regulatory agencies spanning the Customs, Immigration to regulate entry and exit of goods, persons etc. It would appear that

42 Adisa *Op cit* p101

43 Abdulgafar Alabelewe 'Southern Kaduna: Blame FG, Foreign Herdsmen-Miyetti Allah, The Nation, Sunday 16th July, 2017. p 8.

44 The Economic Community of West Africa (ECOWAS)Transhumance Protocol of 1998 A/DEC.5/10.98

the Nigeria border controls are rather lax. As stated, "we should ensure that anybody coming into the country, we know when he is coming, where he is going and control what they are coming with."[45]

D. Government Inaction

The government has been accused of inaction over the matter. The first part is its inability to put in place adequate control measures over movement of so-called non-Nigerian fulani herdsmen or fulani militia into the country.

'The porous nature of the nation's borders is responsible for free entry and exit of the killer herdsmen mostly foreigners.'[46] Related to this laxity is the undetected inflow of arms and ammunition into the country by these trans-border migrant Fulani. Speaking further, he said that for example, if you are to move from Niger Republic to Nigeria, at the border there must be what we call control posts. Benin Republic has control posts with Custom, Immigration, SSS and other related agencies so that you don't come in to that country with sophisticated weapons', which is lacking at Nigeria's border.

E. Crisis in Sub Sahara

Crisis in several African countries have also contributed to free influx of cheap weapons across the region.

F. Climate Change

The country has not been immune from effects of climate change. This can be traced to desertification, gully erosion and drying up of the Lake Chad Basin. This has reduced arable land available for grazing and pushed the herdsmen further south in search of pastures and water for their cattle.

45 Ibrahim Abdullahi, Assistant National Secretary General, Miyetti Allah Cattle Breeders Association of Nigeria (MACBAN),*The Nation,* Sunday July 16th, 2017 p 8.

46 *Ibid*

G. Fulani Culture

Fulani culture is basically migratory in nature. Nevertheless, culture is dynamic and herdsmen must fit into the dynamics of living in the 21st century.

H. Farmers

With the increasing emphasis on food production by governments and loans often obtained at high cost from banks, investment in farming activities has increased. Therefore, most farmers have a lot at stake and cannot stand idly by and see their investments lost as cattle destroy crops. They also react against invasion of their farms. Finally, the regulation that twenty percent of the fadama would need to be set aside for grazing, has not been adhered to.[47]

I. Cattle Rustling

Herdsmen allege that persons from host communities often engage in cattle rustling. This is often a trigger point for attacks by the herdsmen. It has been suggested that if the fulani cattle are branded, it would make rustling less attractive as they can easily be recovered.

J. Changing Resource Access Rights

In circumstances such as this, traditional access rights to communal grazing and water resources are being obstructed by the individual tenureship of arable farmers.[48] This is particularly severe on the traditional trek routes, which become favourite cropping sites because of their better soil fertility resulting from the concentration of animal manure from the trekking herds in these areas. Within the fadama areas, this is exacerbated by the fragmented nature of the crop plots, which makes prevention of animals straying in the crop plots difficult.

47 National Agricultural Policy 1988

48 Adisa, *Op cit* p102

K. Inadequate Grazing Resources

Increasing crop cultivation (and increasing commercialisation of the crop residues) and poor management of the existing grazing reserves have resulted in a significant reduction in available livestock feed resources, in particular in the northern states.[49]

Moreover the high value crops introduced by NFDP (tomatoes and onions) produce almost no crop-residues for livestock feeding.

L. Decline In Internal Discipline And Social Cohesion

Adherence to the traditional rules regarding grazing periods, and the authority of the traditional rulers is breaking down. This is exacerbated by increased rent seeking of the formal and traditional authorities in managing resource access.

M. Faulty Perceptions

De Haan also noted that antagonistic perceptions and beliefs among farmers and herdsmen could compound conflict situation, especially due to failing institutions and fierce competition for resources.[50] Other perceived causes of farmer-herdsmen conflicts include inequitable access to land, diminishing land resources, antagonistic values among user groups, policy contradictions, and non-recognition of rights of indigenous people.[51] Whatever the causes of farmer-herdsmen conflicts are, it is evident that the conflicts have been of great negative effects.

---

49 *Ibid*

50 DeHann, *Ibid*

51 Adisa, R. S. Management of Farmer-herdsmen Conflict in North-Central Nigeria: Implications for Collaboration between Agricultural Extension Service and Other Stakeholders. Journal of International Agricultural Education and Extension, 18 (1): 60-72. 2011

Social, Economic, Political and Human Costs of the Clashes

The incessant attacks have had a drastic effect on food security and have caused a loss of $14 billion in three years. The British Department for International Development (DFID) carried out a research between 2013 and 2016 on the causes and effects of the perennial clashes between herdsmen and farmers in Nigeria. The study also pointed out that the on-going conflict is thwarting the country's economic development to an enormous extent, and if conflicts were resolved the average household affected today could see income increase by at least 64 per cent, and potentially 210 per cent or higher.[52] Also, states affected by herdsmen-farmers conflicts lost an average of 47% of taxes (Internally Generated Revenue) during these attacks. Other economic impacts include: impeding trade practices, reduction in crop yield, displacement of farmers, loss of lives and properties, loss of products in storage and destruction of public and private buildings.

A study investigated the variables associated with farmer-herdsmen conflicts from the perspectives of farmers in Kwara State, Nigeria.[53] Most respondents suffered various effects as a result of their mutual conflict. The effects on both sides ranged from physical, economic, to socio-psychological. The study also examined gains of conflicts. Objectionable as conflicts are, the findings show that both farmers and herdsmen reported a few non-material gains. For instance, while 35% of farmers and 29% of herdsmen said they gained in knowledge, 30.0% and 17.9% respectively agreed that they gained social support as a result of mutual conflict.

The only other item that constituted a 'gain' for both parties was quality of relationship, although the figures were very low: 4.1% for farmers and 5.1% for herdsmen.These low figures indicate

52 Eiyekpeni, *Ibid*

53 Using a four-stage random sampling technique to select 300 farmers in communities contiguous with herdsmen' stock routes, data were collected with the aid of structured questionnaire and subjected to factor analysis and descriptive statistical procedures. Adisa, *Ibid*

that mutual conflict affects the quality of social relationships. Moreover, 24.6% farmers and 12.5% herdsmen indicated a loss in quality of relationship as a result of conflict.

Other non-material resources that were lost included job status (55.6% of farmers and 30.3% of herdsmen); self esteem (52.9% of farmers and 16.9% of herdsmen); and personal\family health (13.9% of farmers and 16.1% of herdsmen). The study reveals that both farmers and herdsmen suffered several negative socio-psychological consequences relating to their respective family lives.

These effects were, however, found to be more pronounced among the farmers than herdsmen, probably due to the fact that farmers suffered more losses than herdsmen.[54]

Cattle Trespass Legislationin Other Jurisdictions

South Africa

The case of *Xolisike Zondi v W. Steenburg & Ors*[55] highlighted the issue of cattle trespass. The matter hinged on the interpretation of the legislation guiding cattle trespass in the province of Kwazulu Natal.[56] Section 1 of the law provided for a landowner to seize and impound trespassing livestock. It further provided for immediate seizure of trespassing animals by a landownwer without notice to the livestock owner unless the owner is an owner of adjacent land immediate or cattle are a registered brand; assessment of damages by the cattle to be done by two disinterested persons who must be voters or landowners; and payment of impoundment fees and damages by cattle is borne by the livestock owner. Alternatively, such animals are sold if the owner is unable to pay. It should be noted that a magistrate issues instructions with regard to dispensing of vicious animals, re-offer for sale or destruction of animals that remain unsold. In Section 41, a police officer also has similar powers especially with regard to destruction of animals too vicious

54 Adisa, *Op cit* 106

55 Xolisile Zondi v W. Steenburg & Ors(2001)

56 Pound Ordinance of (Kwazulu Natal) Province of 1947, No 12 of 1947

to be driven to a pound. Section 12 provides that in the event of a death or injury to any impounded cattle, the pound keeper shall enter into his pound book, details of such injury or death. Also, Section 37 requires that where animals are unsuccessfully offered up for sale, the pound keeper is to report to the magistrate, stating the value of the animal, fees & charges incurred etc.

In all of these, the law requires for notice to be given to owners of impounded cattle which shall be by registered post. It can be seen that in South Africa system, detention places for trespassing livestock exists and the process of regulating, disposing or dealing with related matters is superintended by the judiciary and law enforcement. Thus, the matter is not left for the main parties involved to resort to self help which leads to breakdown of law and order as is the case in Nigeria.

Though the case was eventually decided on constitutional basis, that is the right to notice before an adverse decision is made is a fundamental requirement of fairness to the livestock owner; and that notice provides a person affected with opportunity to make representations. The judgment also stressed that the rights of livestock owners and landowners must be protected. This is quite instructive as it appears Government of Nigeria gives the impression the interests of the farmers and herdsmen are diametrically opposed. Any legislation dealing with same should provide for notice to affected livestock owner.

India

The statute governing cattle trespass in India is the Cattle Trespass Act of 1871 as modified by statute in some regions. Section 3, of the said Act defines cattle to include elephants, camels, buffaloes, horses, lambs and goats. Section 4 of the Act makes provision for pounds in village or groups of villages to be established by a magistrate or district officers. A magistrate has powers to fix the rates of fees, charges for feeding impounded cattle and rules for control of the pounds.[57] Pound keepers are appointed by the

57 Section 5

State and are regarded as public officers.[58] Each pound is to keep a register of: trespassing cattle showing the day and hour animals are taken into the pound; the number and description of such cattle impounded; the name and residence of the 'seizer'; and the name and residence of the livestock owner.[59] The pound keeper is by law mandated to take charge and feed the cattle until they are disposed of as provided for in Section 9. In cases of cattle damaging land, Section 10 states that the cultivator or occupier of any land or any person who has advanced cash for the cultivation of the crop or produce may seize or cause to be seized any trespassing cattle that has caused damage to any crop or produce and send them within 24 hours to the pound established by the village. The police are mandated to aid such seizures. Section 11 deals with cattle damaging public roads, canals and embankments. It provides that persons in charge of such places may seize or cause to be seized any cattle doing damage to such roads or straying livestock.

It must be mentioned that grazing land is set aside for use of such cattle by villagers. If the designated person is of the view that crops, grass or any agricultural land or grazing land is likely to be damaged, such cattle should be removed. If impounded cattle is not claimed within a week, the pound keeper reports to the officer in charge of the nearest police station; who shall issue a 'proclamation' by native means.[60] By virtue of section 19, officers and pound keepers are pre-cluded from buying such cattle offered for sale. Any person whose cattle was seized under the Act or detained under the Act, has 10 days to make complaint to the magistrate in the district.[61] Penalty for damage caused to land, crops or public roads is borne by the livestock owner under section 26.

58 Section 6

59 Section 8

60 This means a drum will be beat around the village and market place nearest to the place of seizure. A notice will also be put up in a conspicuos part of the station or with a magistrate in the office. If no one comes to claim such cattle after seven days, the cattle will be sold by public auction.-section 14

61 section 20 & section 21.

Pakistan

The applicable statute is the Cattle Trespass Act, 1871 as modified.[62] It provided for establishment of pounds in villages and its control to be under the control of a magistrate;[63] but appointment of the pound keeper to be by the provincial government;[64] and same rules as in India for register of seizures and furnishing of returns in section 8. A cultivator or occupier of land or any person who has advanced cash for cultivation of crops or produce on land, the vendee or mortgagee of such crop or produce may seize trespassing livestock and send them to a pound within 24 hours.[65] Persons in charge of public roads, canals and embankments, pleasure grounds, drainage works is authorised to seize such cattle.[66] Other provisions deal with fines for cattle impounded under section 12, sale of such cattle if not claimed within a week under section 14. An owner of cattle has power to make complaint on the basis of illegal seizure or detention of the cattle to a magistrate in the District[67] and if proven, is entitled to compensation for the illegal seizure.[68] Anyone who forcibly opposes seizure of cattle or rescuing of same after seizure either from a pound or from any person taking or about to take them to a pound is liable to pay a penalty.[69] This latter proviso is sorely required in any act to be enacted over this matter.

62 The Cattle Trespass Act, No 10 of 13th Jan 1871 of Pakistan. https://extwprlegs.fao.org/docs/pdf/pak64433.pdf.

63 section 4

64 section 6

65 section 10

66 section 11, Chapter III.

67 sections 20 & 21, Chapter V.

68 section 22

69 section 24 Chapter VI

United States of America

Nevada

Following complaints from residents on impact of cattle on city facilities in Bunkerville and Mesquite towns, the United States District Court permanently barred Cliven Bundy, a livestock owner, from grazing cattle on federal lands in the Gold Butte Area, and ordered him to so remove his cattle.[70] There were other complaints on:

a. that designated critical desert tortoise habitat have been trespassed.

b. Cattle have crushed cultural artifacts at cultural sites.

c. Damage to springs including fecal contamination, trampling soils and vegetation, and impacts on wildlife.

The court also enjoined the federal government to seize and impound any cattle still remaining on such lands by December 2013.

The enabling legislation provides that an owner of livestock is liable for all damages done by animals breaking through or over lawful fences to trespass upon the enclosed lands of another.[71] 'The animals so breaking through or over such fence may be seized as trespassing animals.'[72] The language of the statute makes a livestock owner 'strictly liable' i.e. without fault for agricultural damage' caused by their animals. The situation worsened during the drought of 2011 when livestock owners turned a blind on livestock straying into neighborhood properties and sometimes reports of gates intentionally left open.

When livestock stray into landowners's property, the principle of *'distrainment'* applies. Distraint is the seizure of someone's

70 http://www.b/m.gov.nv/st/on/fo/vfo/blm-programs/more/trespass-cattle. Last accessed 12th July 2017

71 Title 4, Section 155 of Oklahoma Statutes.

72 Jeff Todd, 'When livestock stray: Landowner Rights Under Oklahoma Fencing Laws'http;//www.mcafeeaft.com/when-livestockstray. Last accessed 12th July 2017.

property in order to obtain payment of rent or money owed. A distrainer has lien rights in trespassing livestock.[73] Where a plaintiff can recover judgment for damages caused by trespassing of animals of an owner, the judgment operates as a lien upon the stock caught trespassing so that the plainttiff may have special execution for the sale of such stock to satisfy the judgment and other costs. When livestock is distrained, the livestock owner is notified and both parties are required to try and work out amicably, payment for the damages to crops, grass, hay etc. If no agreement is reached, the Sheriff is to be notified who will then wade in as an arbiter. If the livestock owner still refuses to pay, the Sheriff is empowered by law to sell the livestock. Meanwhile, the distrainer has possession and is enjoined by law to take good care of the livestock

Where straying livestock cause damage but the animals were not distrained, an action can be filed directly against the livestock owner to recover damages and also attorney fees.[74]

Colorado

In the early 1880's, Colorado legislature passed a 'fencing' statute. It is referred to as 'open range' or 'fence out' statute.[75] Anyone maintaining in good repairs a lawful fence may recover damages for trespass from the owner of any livestock that breaks through such fence.[76] Livestock invading fenced property is not a criminal offence but civil recourse is available to the property owner. Therefore, without a lawful fence, the landowner has no civil recourse for damage done to their property by trespassing livestock. Straying livestock liability turns a great deal on fencing. A lawful fence is defined as a well constructed three barbed wire fence with substantial posts set at a distance of approximately 20 feet apart and sufficient to turn ordinary horses and cattle

73 4 Oklahoma Stat 156

74 *Ibid*

75 Open Range And Fencing.https;//www.colorado.gov/pacific/agbrands/open range-and fencing. Lst Accessed 10th August 2017.

76 CRS 35-46-102

with all gates equally as good as the fence or any other fence of like efficiency. Therein lies the challenges of its adaptability to the Nigerian environment. The peasant farmer will very likely resist the idea of fencing property which he insists is an ancestral inheritance. It is the commercial farms that may be amenable to the idea, of course the issue of financial burden this portends must be addressed. If it turns on liability and ease of claiming damages, then the idea for these class of farmers may be acceptable. When it comes to the migrant livestock, expectedly opposition would be stiff.

Class of persons must be properly enlightened that their cattle can no longer ruin crops of others under an age old belief that they have equal rights to graze cattle anywhere. Further that they should adopt modern ranching methods, which makes issue of fencing etc. then relevant. It is legal to take custody of trespassing livestock in Colorado but there is also a legal duty to ensure proper care of such stock.[77] The law requires that the Sheriff and local board Inspector be notified when livestock is held for trespass damage, but it does not mean the law does allow livestock to run around amok without penalty. The situation about livestock grazing on roads and municipalities is provided for.[78]

Duties of Livestock Owners[79]

a. Livestock should have sufficient water and pasture.
b. The owner is to quickly recover any strays or escaped stock
c. They are to inspect fences regularly
d. Maintain proper insurance
e. Work with neighbors on building and maintaining fences.

77 *Ibid*
78 CRS 35-46-105
79 *Ibid*

AVERTING FURTHER CLASHES

Mandatory Border Checks

Though a signatory to the ECOWAS protocol, Gambia requires that anyone coming into the country from any West African country to graze his animals, must have his home country write officially to the Gambian government giving a reason. The proposed route and the number of such livestock must be stated. Weapons are not allowed as the country's law enforcement are saddled with the responsibility for their security.

That is not the case with Nigeria where the migratory fulani are allowed to come in with all sorts of sophisticated weapons.[80]

> So let government do the right thing. Let us decide who comes in because it is our country, let us decide the terms for the persons coming in. Let us not because of ECOWAS protocols leave everything to fate. That is not going to help us.

Open Grazing Probition Law

Across the country, several states have promulgated laws aimed at curbing open grazing by herdsmen. These include Ekiti, Benue and Taraba States.

Establishment of Ranches

The government of Taraba State established model ranches in the three senatorial zones in the state for would-be ranchers to adopt.[81] The ranches are to be clustered for the introduction of modern technology in livestock production. This is supposed to lead to improved breeds of cattle for beef and dairy products, which in turn enhances the income of the average ranching family. Other benefits of ranches include:[82]

---

80 *Ibid.* The Assistant National Secretary General, Miyetti Allah Cattle Breeders Association of Nigeria. MACBAN)

81 h. Open Grazing Prohibition and Ranches Establishment Law of Taraba State, 2017.

82 Fanen Ihyongo, Open Grazing Probition Law will End Herdsmen- Farmers Clashes in Taraba, The NATION, Sunday 13th August, 2017. p 39.

i. Security. It will be easier for security agencies to monitor and control the influx of nomadic herdsmen into the state, thereby reducing herdsmen-farmers clashe leading to an end to the wanton destruction of lives and property.

ii. Law enforcement agencies can be deployed to critical areas that need urgent action

iii. Funds currently deployed to catering for victims of these clashes now residing in IDP's can be channelled to other developmental needs of the state.

iv. The ordinary fulani will have the opportunity to own land which they can develop for exclusive use of their cattle

v. With ranches, veterinary services will be provided to tackle animal diseases.

vi. Schools and medical facilities will be within reach of the herdsmen in ranches. Nomadic education scheme has been a failure for migrant fulani in the past.

vii Improved quality of life. When they have ownership of land, they can build houses, instead of the shacks they construct on the move.

viii. It will allow for integration into the community they live in, instead of the present situation where they dwell as strangers outside the communities.

ix. Farmers will no longer live in fear and can resume full agricultural activities hereby boosting food security for the nation.

Role of Miyetti Cattle Breeders Association of Nigeria (MACBAN)

Enlightenment and Mass Mobilisation of Herdsmen by Miyetti Cattle Breeders Association of Nigeria (MACBAN). This is vital to the success of the project. The suspicion, with which the association regards moves to curb incessant *herdsmen v farmers* clash, must end.

Role of Community

Citizens should be mobilised and enlightened for their role in ridding public space of wandering cattle. This is a nuisance and potential hazard to public safety. Payment to informers as provided for in section 5(1) of the Cattle Trespass Act of Belize will ensure strict monitoring of the law.

New Legislation

It is recommended that a new legislation be enacted to regulate herdsmen and farmers issues. It is not envisaged that this will be a matter for exclusive legislative jurisdiction of the Federal Assembly but State Assemblies' should enact their laws. Best practices from other jurisdictions laws should be deliberated on and inculcated in the state law. In particular, the following are recommended:

- It would appear that civil jurisdiction is no longer effective both in practice and theory in resolving farmer/herdsmen clashes in the nation. This is due to the violence, bestiality, lawlessness, and havoc being perpetrated by the herdsmen and its effect on security, peace and good governance in the nation. Breach of law and order under the legislation and other offences should be criminalised and state authorities should be empowered to prosecute.
- The Ecowas Transhumance Protocol of 1998 is not an excuse for lax border measures by the Nigerian Immigration & Custom officials. The Gambian model is strongly recommended for enforcement.
- Grazing Reserves is the modern way for livestock rearing. The herdsmen through the Miyetti Allah should be enlightened on this and state governments should cooperate by allocating land for this purpose.
- The proposed legislation should provide for branding of cattle being moved around the country whether by

local or migratory herdsmen. This will make notification of owner or whoever is in possession when there is an infraction easier and matters of impoundment will be easier to deal with.

- Local Governments should provide for designated places to serve as pounds where impounded cattle can be kept. As is the law in India, Pound keepers are to be appointed by the State and as public officers. Strict rules to guide operation of the pounds should be enacted as was exemplified in other jurisdictions. Also Mediation Committees made up of the traditional rulers, herdsmen, law enforcement officials should mediate.
- Any legislation dealing with same should provide for notice to affected livestock owner.
- Section 24 of the Pakistan Cattle Trespass Act is germane in that it provides that any one who forcibly opposes seizure of cattle or rescuing of same after seizure either from a pound or from any person taking or about to take them to a pound is liable to pay a penalty
- Damages for destruction of crops, property, injuries should be strictly enforced and paid for by the livestock owner or whoever is in possession of the animals. Moreover, herdsmen implicated in any wanton destruction of lives and property should henceforth be arrested, prosecuted and jailed.
- Although several jurisdictions particularly in the United States of America and the Animal Liability Act of 1971 do provide for fencing, it may not be feasible for now either by the farmers or herdsmen due to the economic burden it would impose.

CONCLUSION

Nowhere in the world has such large scale and sustained destruction of lives, property and crops, due to herdsmen and farmers clashes been witnessed, as in Nigeria today. The attendant mass exodus of survivors fleeing from their homes to become refugees is pathetic and should no longer be condoned. There has been no political will by the federal government to address this matter holistically and no legal framework developed to solve this conflict. From the incisive and deep analysis of the root causes, effects and recommendations made herein, the author hopes that government will chart a way forward towards fostering peace and security without which no progress can be made.

11

DISASTER MITIGATION IN INTERNATIONAL HUMANITARIAN LAW

Elijah O. OKEBUKOLA

ABSTRACT

The chapter examines the application of International Humanitarian Law in the mitigation of the disastrous impact of armed conflict. Disasters are either natural or man-made. Biological, geophysical and climate-related (hydro-meteorological) natural and man-made disasters have caused great damage and disruptions to human societies since recorded history. Since 1990, up to 217 million people are affected by natural disasters yearly. Despite the immensity of the population of victims of natural disasters, man-made disasters create even more victims. Armed-violence and armed conflict account for the largest percentage of victims of man-made disasters. Since 1990, about 300 million people live amidst situations of violence and armed conflict yearly. In response to both natural and man-made disasters, the international community has developed diverse mechanisms, laws, regulations and policies to protect life, limb and property. This chapter will focus on the attempts to not only reduce the number of victims of armed conflict but also to mitigate other harmful effects of armed conflict through the instrumentality of International Humanitarian Law. The chapter is divided into five sections. The first section introduces and gives an overview of the chapter. The second highlights the factual basis for recognising armed conflict as a form of man-made disaster. The third describes the role of International Humanitarian Law in mitigating the disastrous effect of armed conflict. The fourth assesses the extent to which International Humanitarian Law fails to protect certain victims from the horrors of armed violence. The fifth concludes the chapter.

Keywords: International Humanitarian Law, Armed Conflict, Man-made Disaster, Disaster Mitigation

INTRODUCTION

At the most basic level, disasters are natural or man-made. According to the Swiss Re Institute,[1] "there were 327 disaster events in 2016, of which 191 were natural catastrophes and 136 were man-made."[2] In monetary terms calculated on the basis of insured losses, the "September 11 terrorist attack in the U.S. was the costliest man-made disaster in history."[3] According to Insurance Information Institute, based on Swiss Re data going back to 1970, the September 11 terrorist attack caused $25.1 billion in insured losses (in 2015 dollars).[4] While these figures may help one to have an idea of the devastating loss that result from disasters, the emotional, social and psychological trauma caused by disasters are not quantifiable in monetary terms.

Given the devastating consequences of disasters and the enormous monetary and non-pecuniary losses that result from

1 Swiss Re is an organisation that conducts research on current and future issues affecting the re/insurance industry. See http://institute.swissre.com/about/organisation/ accessed on 16/10/2017.

2 Swiss Re, 'Natural Catastrophes and Man-made Disasters in 2016: A year of Widespread Damages' <http://media.swissre.com/documents/sigma2_2017_en.pdf> accessed on 16/10/2017.

3 *Ibid.*

4 https://www.iii.org/fact-statistic/facts-statistics-man-made-disasters accessed on 16/10/2017.

disasters, it is desirable to put an end to disasters. However, the utopian desire to end disasters has two primary obstacles. First, the chain of events that lead to most natural disasters have no connection to human activities and there is simply nothing humans can do to stop such natural disasters.

Second, to put a stop to man-made disasters will require the elimination of the incompatibilities relating to territory, government and resources that lead to armed conflict. It will also entail the elimination of limitations to human ability which allow errors or mistakes that lead to man-made disasters.

In the absence of the ability to put an end to disasters, the alternative is to develop mechanisms for disaster mitigation. In both natural and man-made disasters, law has been deployed as an important tool for attempting to reduce the devastation of disaster. The use of law as a mitigating tool is pronounced in the area of armed conflict. This is the central focus of this chapter. This first section of the chapter gives a general contextual background to the topic of discussion.

The second section of the chapter demonstrates how and why armed conflict is man-made disaster. In this regard, the section highlights the types of disasters by distinguishing man-made from natural disasters. Afterwards, the section defines armed conflict with a view to identifying the focus of International Humanitarian Law (IHL). This exercise leads to the outlining of the devastating consequences of both armed violence in general and armed conflict in particular. The discussions in this section lead to the position that armed conflict is a type of man-made disaster.

The third section starts by identifying the evolution and philosophy of International Humanitarian Law. Essentially, it highlights the approach of International Humanitarian Law in mitigating armed conflict as a man-made disaster. It ends by showing the interplay of specific principles and rules in mitigating some of the negative consequences of armed conflict which are mentioned in the second section.

The fourth section demonstrates that International Humanitarian Law suffers from the inability to mitigate the disastrous effects of violent situations that fall short of armed conflict. It also shows that it creates tensions with, and dominates other legal regimes which may afford greater protection to victims of armed conflict. The fifth section concludes and puts the chapter in perspective.

Armed Violence and Armed Conflict as Man-Made Disasters

Taxonomy of Disasters – Guns and Hurricanes

The 2015 World Disasters Report of the International Federation of Red Cross and Red Crescent Societies aptly notes the two main types of disasters by juxtaposing guns and hurricanes.[5] The symbolic elements of guns and hurricanes can serve to explain the difference between natural and man-made disasters.

Using the methodology of the Emergency Events Database (EM-DAT),[6] natural disasters can be divided into 5 sub-groups, which in turn cover 12 disaster types and more than 30 sub-types. See Figure 1.[7]

5 See http://www.emdat.be/classification accessed on 09/10/2017.

6 K. Levy Jason, 'Coping with Multiple Hazards in the Cascadia Corridor: A Transdisciplinary Disaster Management Approach' (2007) 9 Water Resources IMPACT 17 at 18.

7 See for example Rachel Schiller, 'Reconciliation in Aceh: Addressing the social effects of prolonged armed conflict' (2011) 39 *Asian Journal of Social Science* 489 where the author discusses the coexistence of natural and man-made disaster in the form of armed conflict.

Figure 11.1: Typology of Natural Disasters

Natural Disasters

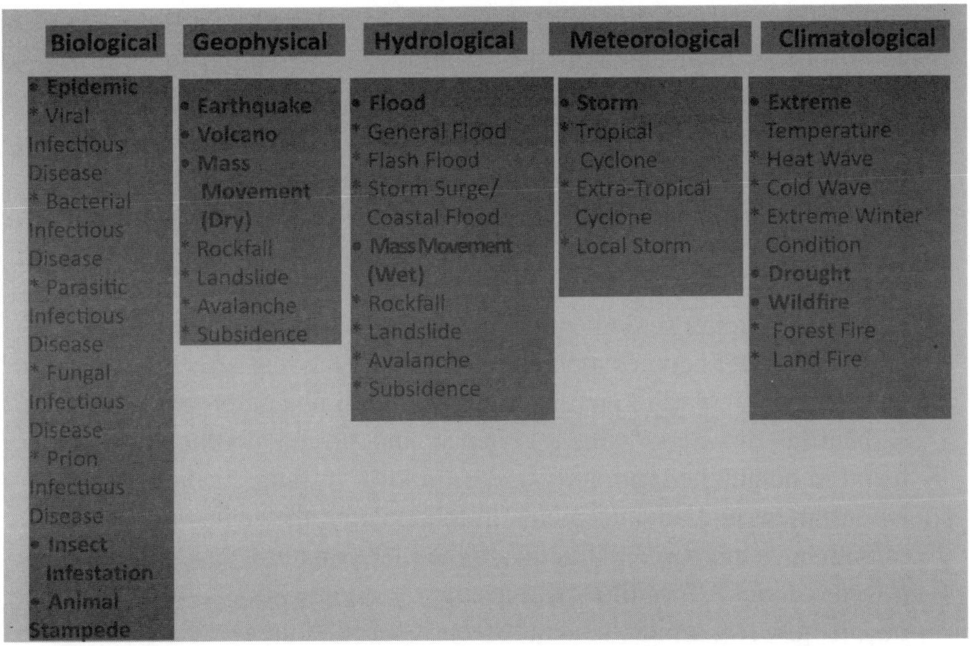

Biological disasters are caused by exposure to toxic substances from living organisms (e.g. venom) or diseases borne by living organisms. Common manifestations of biological disasters are epidemics caused by parasites, bacteria, or viruses (e.g. Ebola). Geophysical or geological disasters originate from the solid earth. They include earthquakes and dry landslides. Hydrological disasters are caused by the occurrence, movement, and distribution of surface and subsurface freshwater and saltwater. Meteorological disasters are caused by short-lived, micro- to meso-scale extreme weather and atmospheric conditions. This may be for a duration of a few minutes or several days. Climatological disasters are caused by long-lived, meso- to macro-scale atmospheric processes ranging from intra-seasonal to multi-decadal climate variability.[8]

8 Islam M Rafiqul, 'The Sudanese Darfur Crisis and Internally Displaced Persons in International Law: The Least Protection for the Most Vulnerable' (2006) 18 Int J Refugee Law 354 at 358.

Man-made disaster may be technological, terrorism-related,[9] armed violence-related or armed conflict-related.[10] The two categories of disasters are sometimes interwoven.[11] There is sometimes a correlation between climate disasters and armed conflict.[12] Both also have some similar elements. For example, vulnerability to both is heightened by population pressure in densely populated areas.[13]

The extreme natural biological, geophysical, hydrological, meteorological and climatological phenomena become disasters only when they negatively affect human societies.[14] As aptly put by James Lewis, "there are two principal constituents in a natural disaster (the extreme natural phenomenon on the one hand, and a vulnerable human settlement on the other)."[15] In the same manner, armed conflict becomes a disaster when it negatively affects human societies, particularly in terms of loss of life, limb and property.

Natural disasters, armed violence and armed conflict always result in heightened mortality and morbidity.[16] With changes in the environment and advancements in technology, the scale and scope of both have increased markedly. "Since 1990, natural disasters have affected about 217 million people every year, and about 300 million people now live amidst violent insecurity around the world."[17] Both natural and man-made disasters cause disruptions which constitute humanitarian crises.[18]

9 See generally, Carl-Friedrich Schleussner and others, 'Armed-conflict risks enhanced by climate-related disasters in ethnically fractionalized countries' (2016) 113 Proceedings of the National Academy of Sciences 9216.

10 See for example, Miao Yu, Wengong Yu and Jie Cheng, 'The Influencing Factors of Typhoon Disaster to Populated Islands and Its Defensive Countermeasures Research' (2015) Journal of Coastal Research 348 at 349 where the correlation of natural disasters and population density is highlighted.

11 James Lewis, 'Natural disaster' (1977) 3 *Built Environment Quarterly* 310.

12 *Ibid.*

13 Jennifer Leaning and Debarati Guha-Sapir 'Natural Disasters, Armed Conflict, and Public Health' (2013) 369 *New England Journal of Medicine* 1836.

14 *Ibid.*

15 *Ibid.*

16 Sylvain Vite´, 'Typology of armed conflicts in international humanitarian law: legal concepts and actual situations' (2009) 91 International Review of the Red Cross 69.

17 Mary Ellen O'Connell, 'Defining Armed Conflict' (2008) 13 J Conflict & Sec L 393 at 399.

18 ICTY, Prosecutor v. D. Tadić.

The common factor between armed violence and hurricanes, tornadoes, storms, earthquakes and other natural disasters is the damage they do to human societies. Unlike extreme natural phenomena which may occur without hurting human societies, armed violence always has a devastatingly negative impact on life, limb and property.

Definition of Armed Conflict

There is no direct definition of armed conflict in the relevant treaties on International Humanitarian Law.[19] The existence of an armed conflict is determined by the behaviour of the facts and circumstances of each situation.[20] In the Tadic case, the Appeals Chamber of the International Criminal Tribunal for the former Yugoslavia (ICTY) ruled that: "an armed conflict exists whenever there is resort to armed force between states or protracted armed violence between governmental authorities and organized armed groups or between such groups within a state."[21] In such as case, the Appeals Chamber stated that, "international humanitarian law applies from the initiation of such armed conflicts and extends beyond the cessation of hostilities until a general conclusion of peace is reached; or, in the case of internal conflicts, a peaceful settlement is achieved."[22]

There are two distinct categories of armed conflict: international and non-international.[23] In the most basic terms, the armed conflicts fought by or against Non-State Actors (NSTACs)

19 *Ibid.*

20 Natasha Balendra, 'Defining Armed Conflict' (2007) 29 Cardozo L Rev 2461 at 2469.

21 Despite this seemingly simple description of non-international armed conflict (NIAC), bona fide debates on the definition of NIACs rage on. See for example, David E. Graham, 'Defining Non-International Armed Conflict: A Historically Difficult Task Part III: Type of Non-International Armed Conflicts and the Applicabale Law: Chapter III' (2012) 88 Int'l L Stud Ser US Naval War Col 43.

22 Elijah Okebukola, 'Call for Jus Ad Bellum for Non-International Use of Force' Cambridge Law Review 2 (2017) at 109.

23 *Ibid.*

are Non-International Armed Conflicts (NIACs).[24] NIACs are distinguished from international armed conflicts (IACs) by the parties involved. Where the conflict is between a state and NSTACs or between NSTACs, it is classified as a non-international armed conflict. The geographical spread or location of the conflict does not affect the classification of the conflict. For example, the Boko Haram conflict which is taking place in the territories of Nigeria, Cameroon and Chad remains a NIAC despite involving the territories of more than one state.[25]

In addition, the participation of more than one state in a conflict does not necessarily make it an IAC unless one state is engaged in armed conflict against the other. Where the states are jointly combating a NSTAC, the conflict is a non-international armed conflict despite the involvement of several states. Using the Boko Haram example again, the involvement of the governments of Nigeria, Chad and Cameroon in the conflict does not make it an IAC since they are working together to fight the NSTACs participating in the conflict. In this regard, the multiple states engaged in a NIAC are not engaged in the use of force against themselves.[26]

IHL applies only to armed conflict and does not cover other situations of violence that fall short of armed conflict.[27] Situations of riots, criminal activities and upheavals are therefore not covered by IHL. Therefore in the proper legal context, humanitarian law or international humanitarian law applies to only situations of armed conflict. Thus, when the expression 'humanitarian' is used in relation to situations of disaster, it is used in the general sense of humanity or humaneness. If however, the expression is used alongside the word 'law', it takes on the strict legal meaning of IHL.

24 Aldo Zammit Borda, 'Introduction to International Humanitarian Law Special Issue: International Humanitarian Law and International Red Cross and Red Crescent Movement' (2008) 34 Commw L Bull 739 at 746.

25 Arthur Sementelli, 'Toward a Taxonomy of Disaster and Crisis Theories' (2007) 29 *Administrative Theory & Praxis* 497 at 508.

26 Christin Ormhaug, Patrick Meier and Helga Hernes, 'Armed Conflict Deaths Disaggregated by Gender ' <www.prio.no> accessed 09/10/2017.

27 *Ibid.*

Trends in Destruction Resulting from Armed Conflict

Disasters and crises sometimes "reveal certain longstanding, wicked problems such as poverty, crime, or other social issues."[28] This is particularly true of poverty and crime. However, it is beyond the scope of this chapter to analyse the problems revealed by disasters. It suffices, instead, to identify the loss suffered as a result of disaster.

Death is the most obvious loss that arises from armed conflict. Intuitively, it would be expected that battle-related deaths will be higher in number than non-battle related death. However, in reality, non-battle deaths tend to supersede battle deaths.[29] Battle death accounts for both civilians and soldiers or fighters killed in combat. Non-battle death relates to persons killed by other circumstances arising from the armed conflict but not direct combat. For example non-battle deaths will include one-sided violence, unorganised and criminal violence and non-violent unfavourable circumstances.[30] Table 2 and Figure 2 below clearly show the significant difference in the number of battle deaths and non-battle deaths in selected sub-Saharan African conflicts.[31]

28 *Ibid* p. 8.

29 ACAPS, *Disaster Summary Sheet Armed Conflict* (2012).

30 ICRC, 'Refugees and Displaced Persons Protected under International Humanitarian Law' 2010) <https://www.icrc.org/en/document/protected-persons/refugees-displaced-persons> accessed 10/10/17.

31 UNHCR, Refugees and Others of Concern to UNHCR, 1999 Statistical Overview. See also NHCR By Numbers, http://www.unhcr.ch/un&ref/numbers/table 1.htm.

Table 11.1: Battle-deaths versus total war deaths in selected sub-Saharan African conflicts

S/N	Country	Years	Estimates of total war deaths	Battle-deaths	Battle-deaths as a percentage of total war deaths
1.	Sudan (Anya Nya rebellion)	1963-73	250,000–750,000	20,000	3-8%
2.	Nigeria (Biafran war)	1967–70	50,000-2million	75,000	4-15%
3.	Angola	1975–2002	1.5 million	160,475	11%
4.	Ethiopia (not including Eritrean insurgency)	1976–91	1–2 million	16,000	<2%
5.	Mozambique	1976–92	500,000–1million	145,400	15–29%
6.	Somalia	1981–96	250,000–350,000 (to mid-1990s)	66,750	19–27%
7.	Sudan	1983–2002	2 million	55,000	3%
8.	Liberia	1989–96	150,000–200,000	23,500	12–16%
9.	Democratic Republic of the Congo	1998–2001	2.5 million	145,000	6%

Figure 11.2: War Deaths

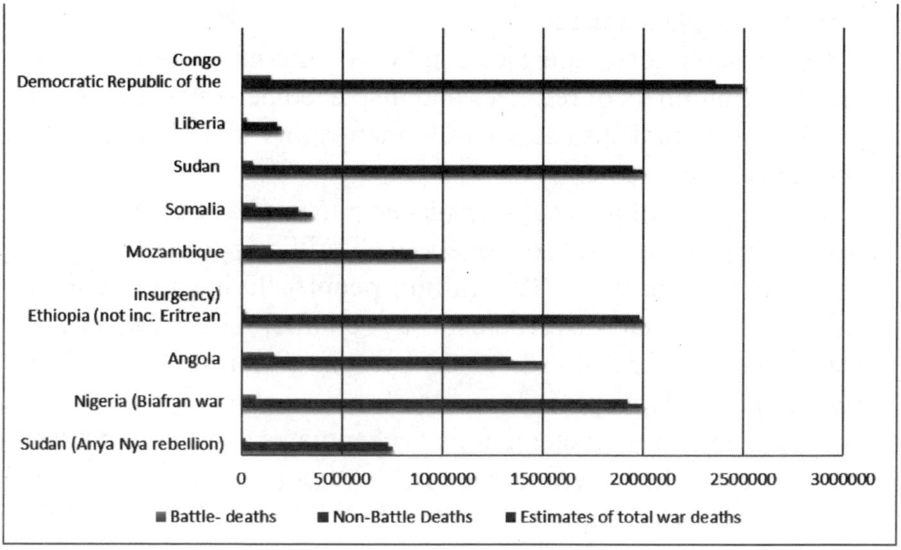

In addition to loss suffered in the form of death, families also suffer displacement during armed conflict. Armed conflicts result in large scale population movement. This movement often ends as displacement. As of December 2011, 26.4 million persons had been displaced as a result of conflict. Persons displaced by armed conflict[32] frequently end up as refugees or internally displaced persons.

Refugees and internally displaced persons are the symptoms of wars, communal violence motivated by ethnic or religious hatred, persecution and intolerance. Refugees are people who have been displaced from their homes as a result of persecution or threat of persecution in their country of origin and have crossed an international border. Internally displaced persons (IDPs), on the other hand, have not crossed an international border, but have

32 Internal Displacement Monitoring Centre, 'Displacement in the Context of Conflict and Violence' 2017) <http://www.internal-displacement.org/internal-displacement/context-of-displacement/conflict-and-violence/> accessed 22/10/17.

also fled their homes.[33] At the start of the 1990s, there were about 50 million forcefully displaced persons, of whom 30 million were internally displaced persons.[34]

Countries affected by armed conflicts or internal strife generally have large numbers of refugees and displaced persons. This is not to say that natural disasters and human rights violations do not also cause displacement.

The numbers of internally displaced persons have continued to grow. At the end of 2016, out of about 65 million people displaced by conflict, there were 40.3 million people "living in ongoing, protracted internal displacement by conflict and violence."[35] Thus, more than two-thirds of the persons displaced by conflict remain internally displaced.

Displacement is an element that accentuates conflict as a man-made disaster. In Nigeria for example, about 1.7 million people were displaced by conflict by mid-2017 while about 2,000 people were displaced by natural disasters in the same period.[36]

Armed conflict also results in food shortages and, sometimes, starvation. As aptly noted by Pejic, armed conflict disrupts all stages of human nutrition "the production, procurement, preparation, allocation, consumption and biological utilisation of food — thereby leaving malnutrition, disease and death in its wake."[37] Armed conflict undermines food security in many ways.[38] Flight and displacement will disrupt food production, trade and distribution. Implements, seeds, fertilisers, fuel and other farming

33 2016 Global Report on Internal Displacement (GRID 2016), http://www.internal-displacement.org/library/publications/2016/2016-global-report-on-internal-displacement-grid-2016/ 22/10/17.

34 Jelena Pejic, 'The Right to Food in Situations of Armed Conflict:The Legal Framework' (2001) 83 IRRC 1097.

35 Welthungerhilfe, 'More Wars, more Hunger? Poverty, Hunger, Climate Change, War' <https://www.welthungerhilfe.de/en/war-and-hunger.html> accessed 08/10/17.

36 *Ibid.*

37 BBC History, 'The Blitz September 1940-May 1941 ' <http://www.bbc.co.uk/history/events/the_blitz> accessed 10/10/17.

38 UNESCO, 'The Hidden Crisis: Armed Conflict and Education' 2011) <http://unesdoc.unesco.org/images/0019/001907/190743e.pdf> accessed 22/10/17.

inputs become scarce or fall into desuetude. "As many foods can no longer be obtained or are simply too expensive, people eat less and, mostly, a more unbalanced diet."[39]

Destruction of houses and buildings is also another constant companion of armed conflict. The Battle of Britain is illustrative of the destruction of buildings that can occur in armed conflict. "In September alone, the Luftwaffe – the German air force –dropped 5,300 tonnes of high explosives on the capital in just 24 nights."[40] The German bombing resulted not only in loss of life but also in the destruction of many industrial and non-industrial buildings.

Another consequence of armed conflict is the disruption of educational and cultural activities. A study of a group of thirty-five countries that experienced armed conflict during 1999–2008 shows that there were 28 million out-of-school children in the conflict-affected countries.[41] Armed conflict leads to the destruction of schools or the occupation of classrooms by military forces or use of schools for the accommodation of displaced people. "Teachers and students are killed, kidnapped, injured and traumatised. Even where schooling continues, conflict has a knock-on negative impact on learning and the quality of education received by children."[42]

In addition to educational disruptions, health care is also in danger in times of armed conflict.[43] In addition to direct attack on medical personnel, fighters have been known to order surgeons at gun point to give their wounded compatriots priority treatment. [44]

39 Amir Jones and Ruth Naylor, 'The Quantitative Impact of Armed Conflict on Education: Counting the Human and Financial Costs' <https://www.educationdevelopmenttrust.com/~/media/EDT/files/research/2014/r-armed-conflict-2014.pdf> accessed 20/10/17.

40 See for example, Robin Coupland and Alex Breitegger, Health Care in Danger The responsibilities of health-care personnel working in armed conflicts and other emergencies (ICRC 2012).

41 Fiona Terry, 'Violence against health care: insights from Afghanistan, Somalia, and the Democratic Republic of the Congo' (2013) 95 IRRC 23 at 26.

42 Borda, 'Introduction to International Humanitarian Law Special Issue: International Humanitarian Law and International Red Cross and Red Crescent Movement'.

43 Hilaire McCoubrey, 'The Concept and Treatment of War Crimes' (1996) 1 J *Conflict Security Law* 121-139 at 121.

44 Knut Dörmann and Robin Geiβ, 'The Implementation of Grave Breaches into Domestic Legal Orders' (2009) 7 J *Int Criminal Justice* 703-721 at 704.

The Legal Mechanism for Disaster Mitigation in Armed Conflict

Evolution Definition and Philosophy of International Humanitarian Law

Law has evolved as an instrument to mitigate the horrors of war by creating rules within which hostilities are to be conducted in times of armed conflict. International Humanitarian Law (IHL), which is also known as the 'Law of War' and the 'Law of Armed Conflict', is the body of law that determines what is acceptable or not acceptable in armed conflict. It is inspired by considerations of humanity and is centred on the protection of certain vulnerable individuals in time of armed conflict and on rendering the fighting more humane.[45]

The traditional view is that the underlying idea of IHL is to mitigate the human suffering caused by armed conflict 'once the endeavour of the *jus ad bellum,* or *jus contra bellum*, to avert such conflict has failed.'[46] As succinctly put by Knut Dörmann and Robin Geiβ, *jus in bello* seeks "to attenuate the rigours of war, to uphold humanity amidst the chaos of armed conflict and to forestall barbarianism."[47]

IHL is traditionally classified into the law of The Hague and the law of Geneva.[48] While the law of The Hague deals with the conduct of war in relation to means and methods belligerents may lawfully apply in armed conflict, the law of Geneva deals with the protection of civilians and persons ordinarily classified as combatants but who have been rendered hors de combat by reasons such as being captured by enemy forces or being sick or wounded.

IHL is perhaps the most oxymoronic aspect of international law, as it seeks to establish a legal framework within which ordinarily

45 R. Cryer, H. Friman, D. Robinson, E. Wilmshurst, p 222; A. *Cassese, International Criminal Law* (Oxford University Press 2003) p. 47.

46 Teslim Elias, *New Horizons in International Law* (Brill Publishers 1979) p. 181.

47 Nils Melzer, *International Humanitarian Law: A Comprehensive Introduction* (ICRC 2016) p. 17.

48 See Article 35.2 Additional Protocol 1 to the Geneva Conventions.

illegal activities would be conducted. This philosophy of IHL is apparently influenced by the resignation to the belief that armed conflict will always occur in and between communities.

Although the origins of contemporary IHL can be traced back to the nineteenth century, it is based on principles and practices which are much older.[49] The beginning of modern IHL is generally traced back to 1864 when the First Geneva Convention was adopted. However, the rules contained in that convention were not entirely new. Many of them were derived from existing customary international law. In addition, there have been rules protecting certain categories of victims in armed conflicts as early as 1000 BC. Although these ancient and often very rudimentary rules were not established for humanitarian reasons, but rather for purely economical purposes, their effect was humanitarian.

Modern IHL "comprises those rules of international law which establish minimum standards of humanity that must be respected in any situation of armed conflict."[50]

General Principles and Rules of International Humanitarian Law

IHL has certain general principles upon which its rules are based. These principles relate to among others, war death, destruction of buildings and disruption of education. The first of the principles is distinction. The principle of distinction demands that parties shall at all times distinguish between civilian and combatants or fighters. They shall also distinguish between civilian objects and military targets.[51] It is a cornerstone of IHL. It requires that "the only legitimate object which states should endeavour to accomplish during war is to weaken the military forces of the enemy,"[52] whereas "[t]he civilian population and individual civilians shall enjoy general protection against dangers arising

49 St. Petersburg Declaration, Preamble.
50 Art. 51(1), Additional Protocol I to the Geneva Conventions.
51 Art. 48, Additional Protocol I to the Geneva Conventions.
52 Melzer, *International Humanitarian Law: A Comprehensive Introduction* p. 19.

from military operations."[53] Therefore, the parties to an armed conflict must "at all times distinguish between the civilian population and combatants and between civilian objects and military objectives and accordingly shall direct their operations only against military objectives."[54]

The parties to an armed conflict must take precautions in attack. This entails a duty to avoid or, at least, minimise the infliction of incidental death, injury and destruction on persons and objects protected against direct attack.[55] Accordingly, IHL requires that, "[i]n the conduct of military operations, constant care shall be taken to spare the civilian population, civilians and civilian objects."[56] This applies both to the attacking party, which must do everything feasible to avoid inflicting incidental harm as a result of its operations *(precautions in attack),*[57] and to the party being attacked, which, to the maximum extent feasible, must take all necessary measures to protect the civilian population under its control from the effects of attacks carried out by the enemy *(precautions against the effects of attack).*[58]

Where the parties to an armed conflict cannot avoid incidental harm on civilians or civilian objects, they are mandated to apply the principle of proportionality. This entails that, those who plan or decide on an attack must refrain from launching, or must suspend, "any attack which may be expected to cause incidental loss of civilian life, injury to civilians, damage to civilian objects, or a combination thereof, which would be excessive in relation to the concrete and direct military advantage anticipated."[59]

Moreover, IHL demands that all persons who have fallen into the power of the opposing forces must be treated humanely regardless of their status and previous function or activities.

53 Art. 57, Additional Protocol I to the Geneva Conventions.

54 *Ibid.*

55 Art. 58, Additional Protocol I to the Geneva Conventions.

56 *Ibid.*

57 Art. 3(1) Geneva Convention I-IV, 1949.

58 Melzer, *International Humanitarian Law: A Comprehensive Introduction* p. 20.

59 Rule 129 Customary IHL Study, ICRC.

Common Article 3, to the Geneva Conventions of 1949 provides that:

> Persons taking no active part in the hostilities, including members of armed forces who have laid down their arms and those placed hors de combat by sickness, wounds, detention, or any other cause, shall in all circumstances be treated humanely, without any adverse distinction founded on race, colour, religion or faith, sex, birth or wealth, or any other similar criteria.[60]

The entitlement to humane treatment applies to both persons deprived of their liberty and to all the inhabitants of territories under enemy control.[61]

In addition to minimising the horrors of armed conflict in the area of war deaths and destruction, IHL also seeks to prevent needless displacement. The ICRC found in its study of customary international humanitarian law that parties to an armed conflict may not deport or forcibly transfer the civilian population of an occupied territory, in whole or in part, unless the security of the civilians involved or imperative military reasons so demand.[62]

Under the Charter of the International Military Tribunal (Nuremberg), "deportation to slave labour or for any other purpose of civilian population of or in occupied territory" constitutes a war crime.[63] In addition, according to the Fourth Geneva Convention and Additional Protocol I, it is a grave breach of IHL to deport or transfer the civilian population of an occupied territory, unless the security of the civilians involved or imperative military reasons so demand.[64] Under the Statute of the International Criminal Court, "the deportation or transfer [by the Occupying Power]

---

60 Article 6(b), IMT Charter.

61 Article 147, Fourth Geneva Convention; Article 85(4)(a) , Additional Protocol I.

62 Article 8(2)(b)(viii), ICC Statute.

63 UN General Assembly, Res. 2675 (XXV) (adopted by 109 votes in favour, none against and 8 abstentions).

64 UN General Assembly, Res. 3318 (XXIX) (adopted by 110 votes in favour, none against and 14 abstentions).

of all or parts of the population of the occupied territory within or outside this territory" constitutes a war crime in international armed conflicts.[65]

In 1970, the UN General Assembly affirmed that "civilian populations, or individual members thereof, should not be the object of ... forcible transfers."[66] In a resolution on the protection of women and children in emergency and armed conflict, adopted in 1974, the UN General Assembly declared that "forcible eviction, committed by belligerents in the course of military operations or in occupied territories, shall be considered criminal."[67] Instances of forced displacement in international armed conflicts and non-international armed conflicts, have been condemned by the United Nations in several conflicts including in Bosnia and Herzegovina, Burundi and Sudan.[68]

There are however exceptions to the rule against displacement. In cases where the security of the civilians involved or imperative military reasons (such as clearing a combat zone) require the evacuation of civilians, they may be made to leave that place for as long as the conditions warranting the movement exist. This exception is allowed by the Fourth Geneva Convention and Additional Protocol II.[69]

The Guiding Principles on Internal Displacement prohibit the "arbitrary" displacement of persons, which is defined as including displacement in situations of armed conflict, "unless the security of civilians involved or imperative military reasons so demand."[70] The exception of "imperative military reasons" does not justify the removal of the civilian population for the purpose of persecution.[71]

---

65 UN Security Council, Res. 752 and Res. 819; UN Security Council, Statement by the President; UN General Assembly, Res. 55/116; UN Commission on Human Rights, Res. 1995/77 and Res. 1996/73.

66 Article 49, Fourth Geneva Convention; Article 17(1), Additional Protocol II.

67 Principle 6(2), Guiding Principles on Internal Displacement.

68 Article 45, Fourth Geneva Convention.

69 Article 49, Fourth Geneva Convention.

70 Article 17(2), Additional Protocol II.

71 Principle 5, Guiding Principles on Internal Displacement.

The Fourth Geneva Convention further specifies that evacuations may not involve displacement outside the bounds of the occupied territory "except where for material reasons it is impossible to avoid such displacement."[72] In non-international armed conflicts, evacuations must never involve displacement outside the national territory.[73]

There is also a duty on parties to a conflict to prevent displacement caused by their own prohibited acts. The Guiding Principles on Internal Displacement requires that "all authorities and international actors shall respect ... prevent and avoid conditions that might lead to displacement of persons."[74]

Connected to the prohibition of displacement is the prohibition of starvation of the civilian population as a method of warfare. The Commission on Responsibility set up after the first world war identified deliberate starvation of civilians as a violation of IHL subject to criminal prosecution. The prohibition of starvation as a method of warfare is codified in Article 54(1) of Additional Protocol I. Similarly, the Statute of the International Criminal Court provides that "intentionally using starvation of civilians as a method of warfare" is a war crime.[75]

The prohibition of starvation as a method of warfare does not mean that parties to an armed conflict cannot engage in siege warfare. Rather, siege warfare is not prohibited as long as the purpose is to achieve a military objective and not to starve a civilian population. Similarly, the prohibition of starvation as a method of warfare does not prohibit the imposition of a naval blockade as long as the purpose is to achieve a military objective and not to starve a civilian population. If the civilian population is inadequately provided for, the blockading party must provide for free passage of humanitarian relief supplies.[76]

72 Article 8(2)(b)(xxv), ICC Statute.

73 George Alfred Mudge, 'Starvation as a Means of Warfare' (1969) 4 Int'l L 228.

74 Articles 24–26 , First Geneva Convention; Article 36, Second Geneva Convention; Article 20, Fourth Geneva Convention.

75 Article 15, Additional Protocol I.

76 Article 8(2)(b)(xxiv), ICC Statute.

Certain principles of IHL specifically protect medical personnel. Medical personnel exclusively assigned to medical duties must be respected and protected in all circumstances. They lose their protection if they commit, outside their humanitarian function, acts harmful to the enemy. This rule is established in the First, Second and Fourth Geneva Conventions of 1949.[77] It applies to both civilian and military medical personnel in all circumstances. Under the Statute of the International Criminal Court, intentionally directing attacks against medical personnel using the distinctive emblems of the Geneva Conventions is a war crime.[78]

The term "medical personnel" refers to personnel assigned, by a party to the conflict, exclusively to the search for, collection, transportation, diagnosis or treatment, including first-aid treatment, of the wounded, sick and shipwrecked, and the prevention of disease, to the administration of medical units or to the operation or administration of medical transports. Such assignments may be either permanent or temporary.[79]

Other persons performing medical duties are not medical personnel but enjoy protection against attack as civilians, as long as they do not directly participate in hostilities. Since they are not medical personnel, they have no right to display the distinctive emblems.

There are several other principles and rules of IHL that help to mitigate the disaster of armed conflict. These include the prohibition of use of: weapons which are by nature indiscriminate; hostage taking; human shields; children as soldiers; [80] and pillage.

77 Article 8(c), of Additional Protocol I.

78 Elijah Oluwatoyin Okebukola, 'Training Children for Armed Conflict - Where does the Law Stand?' (2014) 14 *International Criminal Law Review* 588.

79 Jacques Meurant, 'Inter Arma Caritas: Evolution and Nature of International Humanitarian Law' (1987) 24 *Journal of Peace Research* 237.

80 Medard R. Rwelamira, 'The Role of International Humanitarian Law in Internal Disturbances and Tension Situations: Some Reflections' (1987) 20 Comp & Int'l LJ S Afr 175.

Limits of International Humanitarian Law in Disaster Mitigation

The objective of IHL to mitigate disaster caused by man's violence is watered down by the non-application of IHL to extremely violent situations that fall short of armed conflict. The challenge is exacerbated by the legal difficulty which arises in determining if a situation of armed conflict exists.[81] For example, violence in the form of terrorism is one of the most glaring manifestations of man-made disaster,[82] yet many situations of terrorisms fall short of the strict legal definition of armed conflict.

It is true that human rights law operates in situations short of armed conflict. However, human rights law "does not address itself directly to the limitations and restrictions on confrontations between warring groups within states or between tile dissidents and central authority."[83] This shortcoming is particularly glaring in situations that involve armed violence inflicted by NSTACs against the civilian population without any effective intervention or prevention by the governmental authorities or agents.[84] This challenge is well illustrated by violence emanating from drug cartels which may control vast territories and kill hundreds of people.[85]

As aptly noted by Moreillon:

> Humanitarian law has followed the evolution of violence in order to try and temper its effects. This law only struggles against the violence of war. There are many other types of violence to which we have to give our attention. This is the case with regard to internal disturbances and tensions which escape the

81 Jason, 'Coping with Multiple Hazards in the Cascadia Corridor: A Transdisciplinary Disaster Management Approach' at 18.

82 Rwelamira, 'The Role of International Humanitarian Law in Internal Disturbances and Tension Situations: Some Reflections' at 183.

83 Claude Bruderlein and Jennifer Leaning, 'New Challenges for Humanitarian Protection' (1999) 319 BMJ: British Medical Journal 430.

84 See for example, Michael T. Wotherspoon, 'Mexico's Drug War, International Jurisprudence, and the Role of Non-International Armed Conflict Status' (2012) 3 J Int'l Human Legal Stud 291.

85 Jacques Moreillon, 'Contemporary Challenges to Humanitarian Law' (1987) 24 Journal of Peace Research 215 at 217.

> formal fields of application of humanitarian law. The protection of human rights law is insufficient and thus further normative regulation of these situations needs to be undertaken.[86]

Although humanitarian law and human rights law overlap, they are not coterminous.[87] [88] This therefore sometimes leads to conflict. The conflict is premised on the contemporaneous application of both legal regimes. As emphasised by the International Court of Justice (ICJ), in the cases of the *Construction of the Wall in the Occupied Palestinian Territory, and Armed activities case,*[89] human rights treaties continue to apply in wartime. They apply together with humanitarian law.[90]

Thus, what happens in the event of conflict between human rights law and international humanitarian law? The general position formulated by the International Court of Justice is that the international humanitarian law provides for the *lex specialis* "that derogates more stringent standards under any applicable international human rights law."[91] The fallout from this is that a party to an armed conflict may get away from conduct which may be atrocious and unacceptable in human rights, but is consistent with IHL.[92] In effect, "some peacetime human rights may disappear altogether in wartime."[93]

86 Legal Consequences of the Construction of a Wall in the Occupied Palestinian Territory , Advisory Opinion of 9 July 2004, General List No. 131 [2004] ICJ Rep 136 at 178, para. 106.

87 Robert Cryer, 'The Interplay of Human Rights and Humanitarian Law: The Approach of the ICTY' (2009) 14 J *Conflict Security Law* 511 at 513.

88 Case Concerning the Armed Activities on the Territory of the Congo (Democratic Republic of the Congo v. Uganda) , Judgment of 19 Dec. 2005, General List No. 116, at para. 216.

89 Alexander Orakhelashvili, 'The Interaction between Human Rights and Humanitarian Law: Fragmentation, Conflict, Parallelism, or Convergence?' (2008) 19 *Eur J Int Law* 161.

90 Claus Kreß, 'Some Reflections on the International Legal Framework Governing Transnational Armed Conflicts' (2010) J *Conflict Security Law* krq009 at 17.

91 William Abresch, 'A Human Rights Law of Internal Armed Conflict: The European Court of Human Rights in Chechnya' (2005) 16 *Eur J Int Law* 741.

92 Allan Rosas and others, 'The Frontiers of International Humanitarian Law' (1987) 24 *Journal of Peace Research* 219 at 226.

93 Allan Rosas and others, 'The Frontiers of International Humanitarian Law' (1987) 24 *Journal of Peace Research* 219 at 226.

Conclusion

Some occurrences in nature have such a catastrophic impact on human life, society and the environment that they are regarded as disasters. Therefore, in ascribing the label of disaster on an occurrence, the primary consideration is the depth of its negative impact on man. It follows from this proposition that a natural phenomenon such as a thunderstorm or even an earthquake may not be a natural disaster if it does not negatively impact on human life, society and the environment.

It is because of its negative consequences on humanity that armed conflict is viewed as a disaster. Just as certain provisions of environmental law are made to mitigate the negative impact of natural disasters, international humanitarian law is applied in mitigating the disaster of armed conflict. IHL is however not a perfect regime for disaster mitigation because of the inherent limitation in the very essence of IHL. The most prominent of these is the non-applicability of IHL to situations of extreme violence which are as disastrous as armed conflict but fall short of the legal definition of armed conflict. This is an area that calls for more scholarly work aimed at filling the gaps and mitigating the disastrous impact of violence short of armed conflict.

In all, despite the areas of violence not covered by IHL and other gaps in the application of IHL, it is advantageous to humanity to have IHL rules that aim to mitigate the barbarity of armed conflict. Certainly, armed conflict would be more disastrous if parties can carry on without any restraints.

12

CONNECTING THE DOTS: CLIMATE CHANGE, DISASTER AND SOCIAL SECURITY UNDER THE AFRICAN REGIONAL HUMAN RIGHTS SYSTEM

Ademola O. Jegede & Untalimile C. Mokoena

Abstract

Global climate change will significantly affect the frequency and intensity of hazard occurrence, in that the number of people facing climate change related disasters will rise as Greenhouse Gases (GHGs) continue to accumulate in the atmosphere. However, the development will occasion more negative consequences of human rights implications (including social security) in Africa than in other regions, despite the fact that the continent contributes little to the causation of climate change. Scholars of regionalism have shown that some potential exists at the regional level to solve cross border challenges associated with climate change. The role of regional institutions in shaping positive national level response on environmental challenges has also been discussed. However, while the problem of climate change raises multi-dimensional complex issues for multidisciplinary attention, its link to disaster management as well as social security and the potential at the African regional level to shape domestic approaches have received scanty attention. Yet, in terms of its applicable instruments and interpretation of treaty monitoring mechanisms, the African regional human rights system can significantly contribute to climate change and disaster law at the national level.

This contribution seeks initially to demonstrate the nexus of climate change, disaster management and social security. Thereafter, it argues the normative as well as the jurisprudential trend under the African regional human rights system as potentials that can shape climate change and disaster management law at the domestic level in Africa.

Keywords: Climate change, disaster management, social security, human rights

INTRODUCTION

According to the United Nations Framework Convention on Climate Change, climate change may be referred to as a change of climate which is attributed directly or indirectly to human activity that alters the composition of the global atmosphere and which is in addition to natural climate variability observed over comparable time periods.[1] This is reiterated in the scientific Report of the Intergovernmental Panel on Climate Change (IPCC) which defines climate change as any change in climate over time, whether due to natural variability or as a result of human activity. Authors such as Banholzer *et al* argue that global climate change is one of the key causes of numerous hazard occurrences.[2] Examples of such occurrences associated with climate change include hurricanes, floods, droughts, forest fires among others. The implication of this, according to Banholzer *et al.* has also been the increase of natural disasters globally. A disaster is defined by the United Nations as a serious disruption of the functioning of a community or a society involving wide spread human, material,

1 United Nations Framework Convention on Climate Change (UNFCCC) (1992) ILM 851, art 1(2)

2 S Banholzer *et al* "The Impact of Climate Change on Natural Disasters' in A Singh and Z Zommers (ed) *Reducing Disaster: Early Warning Systems For Climate Change* (2014, Springer) 21-49.

economic or environmental losses and impacts, which exceeds the ability of the affected community or society to cope using its own resources.[3]

Literature has shown that Africa, as a continent, has not been immune from climate change or its associated disasters. Toulmin is of the view that in as much as climate change poses a major challenge for everyone, it represents a particular threat for African countries.[4] The IPCC Report affirms that Africa is likely to be hit the most by climate change despite its relatively small contribution to the global greenhouse gas production.[5] According to Stock,[6] fragile conditions of most of Africa's states facing numerous development challenges like high rates of population growth, ecosystem degradation, regional poverty, and so on, aggravate the continent's vulnerabilities. Literature also indicates that the African continent is exposed to disaster risk from diverse natural causes, especially those arising from hydro meteorological hazards.[7] Authors such as Lukamba,[8] and Schlein[9] analysed statistical surveys for natural disasters in different regions and found that disaster frequency in Africa is increasing and such growth is a consequence of climate change. Although, the International Federation of Red Cross and Red Crescent Societies,

3 United Nations International Strategy for Disaster Reduction (UNISDR), Geneva, Switzerland, May 2009.

4 C Toulmin *Climate Change in Africa* (2009, Zed Books) 1.

5 IPCC *Climate Change 2014: Synthesis Report. Contribution of Working Groups I, II and III to the Fifth Assessment Report of the Intergovernmental Panel on Climate Change* (2014, Cambridge University Press) 151

6 R Stock *Africa South of the Sahara: A Geographical Interpretation* (2012, The Guilford Press) 145.

7 International Strategy for Disaster Reduction (ISDR) (2004, ISDR) 149; G Malugeta et al 'Natural and Human-induced Hazards and Disasters in sub-Saharan Africa'International Council for Science (ICSU) Report (2006, ICSU); MT 'Lukamba Natural disasters in African countries: what can we learn about them?' (2010) 6/1*The Journal for Transdisciplinary Research in Southern Africa* at 478 – 495.

8 Lukamba *Natural disasters in African countries,* above at note 7

9 L Schlein 'UN: Most Deaths From Natural Disasters Occur in Poor Countries' *VOA News* 13th October 2016, available at https://www.voanews.com/a/un-says-most-deaths-from-natural-disasters-occur-in-poor-countries/3548871.html. (last accessed 16th October 2017).

mainly advocates for a well-prepared regulatory framework to assist in times of disasters, it also affirms the presence of a close link between climatic factors and the natural disasters outbreak.[10] Despite the lack of establishing the link between natural disasters and climate change, Spooner indicates that disasters need not be gigantic to cause serious harm but small, recurring natural disasters that constantly destroy critical public infrastructure, housing and the means of production are just as detrimental.[11]

Scholarship on regionalism has shown that focusing on legal development at the regional level is helpful in addressing national challenges. In particular, in terms of global environmental issues, Alagappa explains that regional institutions are a significant 'component of the global architecture for environmental governance.'[12] Explaining the different waves of regionalism in the context of environmental protection also, Hettne notes that regionalism is useful in solving trans-boundary environmental problems, particularly those that 'were not effectively tackled at the national level.'[13] Regional institutions are helpful, according to Birdsall and Lawrence, in addressing challenges to environmental protection.[14] Considering that it operates between the national and global level, Katzenstein notes that regional developments 'as in the story of Goldilocks, are neither too hot, nor too cold, but just right.'[15]

---

10 International Federation of Red Cross and Red Crescent Societies, Geneva, 2011.

11 S Spooner 'Africa's most costly disaster zones' *Mail & Guardian* 10th July 2015.

12 M Alagappa 'Energy and the environment in Asia-Pacific: Regional co-operation and market governance' in PS Chasek (ed) *The global environment in the twenty-first century: Prospects for international co-operation* (2000, United Nations University Press) 255-270

13 B Hettne 'Beyond the 'new' regionalism' (2005) 10 *New Political Economy* at 543, 549

14 N Birdsall & RZ Lawrence 'Deep integration and trade agreements: Good for developing countries?' in I Kaul, I Grunberg & MA Stern (eds) *Global public goods: International co-operation in the 21st Century* (1999, Oxford University Press) 128-51

15 PJ Katzenstein 'After the global crises: What next for regionalism?'(September 1999) available at http://www2.warwick.ac.uk/fac/soc/csgr/events/conferences/1999_conferences/3rdannualconference/papers/bowles.pdf(last accessed 19th April 2014)

While emphasising the need for the connection of dots of climate change and disaster, political leaders particularly in Africa have urged the continent to take a significant action on the issue.[16] Hence, it is not surprising that there have been disparate attempts at the regional level in Africa to address climate change and disaster occurrences. For instance, there is significant normative development. In 1985, the African Ministerial Conference on the Environment (AMCEN) was established with an intention of advocating for environmental protection in Africa. AMCEN has greatly influenced the manner in which environmental issues are tackled in the region. In 2014 the African Ministers partnered with European Union in an endeavour to address climate change related issues.[17]

The 12th AMCEN advocated for an integration of Africa's existing and new climate change programmes under a consolidated framework.[18] Africa has also established the Conference of African Heads of State and Government on Climate Change (CAHOSCC) whose objective is to give the highest political attention to climate-change-related issues in Africa.[19] The New Partnership for Africa's Development (NEPAD) addresses several issues amongst which Africa's significant role with regard to the critical issue of environmental protection is highlighted. One of its key thematic areas target agriculture and food security, as well as climate change and natural resource management. The continent prides itself in its extensive resources which include among others, rain forests, carbon dioxide-free atmosphere above the continent and the nominal existence of toxic effluents in the rivers and soils that interact with the Atlantic and Indian Oceans, and the Mediterranean and Red Seas.[20] NEPAD devices strategies of nurturing the said resources for the advantage and development

16 UN Climate Summit (2014).

17 EU-Africa Ministerial Statement on Climate Change (2014).

18 Southern Africa Sub- Regional Framework of Climate Change Programmes Adaptation (2010).

19 Draft African Union Strategy on Climate Change (2014)

20 African Union NEPAD (2001)

of the continent and the world at large.[21] Furthermore, NEPAD acknowledges that the agrarian systems of Africa face a lot of challenges including climate change, which then lead to poverty in the continent. As such, a healthy and productive environment is seen as one of the key elements for NEPAD. Investing in a healthy environment results in a reduction of poverty and greatly contributes to employment as well as social and economic empowerment.[22]

In terms of normative development, article 13(1) (e) of the Constitutive Act[23] provides that the Executive Council shall coordinate and take decisions on policies in areas of common interest to the Member States, including environmental protection, humanitarian action and disaster response and relief. The environmental protection envisaged in the Act may be taken to include even issues of climate change. Notable normative development at the regional level also include the African Charter on Human and Peoples' Rights (African Charter),[24] which in its article 24 guarantees the right to a healthy environment. Also, the African Union Convention for the Protection and Assistance of Internally Displaced Persons in Africa (the Kampala Convention) has as its main intention the protection and assistance of internally displaced people as a result of climate related causes. State actors are, in particular, mandated to assist and protect people internally displaced by natural or human made disasters, including those stimulated by climate change.[25]

Despite the foregoing, neither the relationship between climate change, disaster management and social security nor the regional legal response that can shape domestic disaster management practices has been adequately clarified.Yet, social security, defined by International Labour Organisation as the protection that a

21 As above

22 As above

23 2000 Constitutive Act of the African Union

24 Organisation of African Unity (OAU), *African Charter on Human and Peoples' Rights ("Banjul Charter")* 1981.

25 Article 5(4) of the Kampala Convention.

society provides to individuals and households to ensure access to health care and to guarantee income security, particularly in cases of old age, unemployment, sickness, invalidity, work injury, maternity or loss of a breadwinner,[26] is important in constructing an appropriate legal response to disaster associated with climate change. This chapter demonstrates the link of climate change, and disaster to social security as well as the implication of this link for regional human rights. It then explores how the legal response at that level can shape domestic disaster management law particularly in Nigeria. Following this introduction, Section II demonstrates the link amongst climate change, disaster and social security. Section III deals with implications of the link for human rights while Section IV focuses on how development at the regional level can shape legal response in Nigeria.

The link of climate change and disaster to social security

Regarded as the single biggest challenge for humanity over the coming centuries,[27] climate change is among other influences, the lead driving force of different natural disasters globally. It influences for instance, hurricanes, droughts, fires and so on in a number of ways. Though scientists generally have not shown that each and every particular disaster is a result of climate change, they attest to the likelihood of most of catastrophes being influenced by climate change.[28] Climate change is perceived to have various dire repercussions, some having effect over a longer period, such as accelerated sea level rise, drought and spread of

26 ILO 'Facts about social security' available at http://www.ilo.org/wcmsp5/groups/public/---dgreports/---dcomm/documents/publication/wcms_067588.pdf(last accessed 12th October 2017)

27 P Huntjens and K Nachbar 'Climate Change as a Threat Multiplier for Human Disaster and Conflict' (2015, Working paper, The Hague Institute for Global Justice)

28 A Khan 'Fires, droughts and hurricanes: What's the link between climate change and natural disasters?' Los Angeles Times, 10th September 2017 available athttp://www.latimes.com/science/sciencenow/la-sci-sn-climate-change-natural-disasters-20170907-htmlstory.html (last accessed 10th October 2017)

diseases, while some impact like flooding are immediate. With Africa and the world at large currently making headlines and characterised by extreme weathers, an outcry for establishing the connection of these disasters to climate change has become necessary.[29] Different predicaments are faced by the region, and these include but are not limited to prolonged and intensified droughts in eastern Africa; unprecedented floods in western Africa; depletion of rain forests in equatorial Africa; and an increase in ocean acidity around Africa's southern coast. Africa being a continent that is mainly dependent on agriculture,[30] may be said to be greatly impacted by climate change as immensely changed weather patterns and climate extremes negatively impact agricultural production and food security, health, water and energy security, which in turn undermine Africa's ability to grow and develop.[31]

The Sendai Framework for Disaster Risk Reduction 2015-2030 (Sendai Framework) is the first major agreement of the post-2015 development agenda, with seven targets and four priorities for action as endorsed by UNGA.[32] It links with climate change is that it provides that disasters, many of which are exacerbated by climate change and which are increasing in frequency and

29 J Edelson 'Thousands evacuate as California wildfires kill 10' Daily Maverick, 10 October 2017 available at https://www.dailymaverick.co.za/article/2017-10-10-thousands-evacuate-as-california-wildfires-kill-10?utm. (last accessed 10th October 2017); J Ruttle 'Bankrupt, hungry and homeless: Life after Madagascar's cyclone' available at https://www.theguardian.com/global-development/2017/may/16/after-cyclone-enawo-madacasgar-village-forging-new-life-from-ruins (last accessed 10 October 2017; J Cronje ' Western Cape natural disaster death toll rises to 11' News24 11 June 2017 http://www.news24.com/SouthAfrica/News/western-cape-natural-disaster-death-toll-rises-to-11-20170611 (last accessed 10th October 2017); J Anderson 'Climate Change and Natural Disasters: Scientific evidence of a possible relation between recent natural disasters and climate change' available athttps://www.ecologic.eu/sites/files/project/2013/Brief_CC_and_natural_disasters_scientific_evidence_of_relation_Jan_2006_EP_version.pdf. (last accessed 10th October 2017)

30 Agriculture in Africa: : Transformation and outlook, NEPAD (2013)

31 As above

32 'United Nations Office for Disaster Risk Reduction Sendai Framework for Disaster Risk Reduction' available at http://www.unisdr.org/we/coordinate/sendai-framework (Sendai Framework)(last accessed 10 October 2017)

intensity, significantly impede progress towards sustainable development.[33] Hence, it is urgent and critical to anticipate, plan for and reduce disaster risk in order to more effectively protect persons, communities and countries, their livelihoods, health, cultural heritage, socioeconomic assets and ecosystems, and thus strengthen their resilience.[34] Equally, it is important to strengthen the design and implementation of inclusive policies and social safety-net mechanisms, including through community involvement, integrated with livelihood enhancement programmes, and access to basic health-care services, including maternal, newborn and child health, sexual and reproductive health, food security and nutrition, housing and education, towards the eradication of poverty, to find durable solutions in the post-disaster phase and to empower and assist people disproportionately affected by disasters.[35] Phillip *et al*[36] compiled evidence among major stakeholders including lawmakers, researchers and members of communities on the relationship between climate change and natural disasters. The survey required respondents to identify whether or not they believe climate change has an effect on the frequency, intensity, and/or duration of various natural disasters such as tornadoes, hurricanes, and earthquakes. It was found that there is a strong relationship between natural disasters and climate change. Researches also found that 'a changing climate leads to changes in frequency, intensity, spatial extent, duration and timing of extreme weather and climate events, and can result in unprecedented extreme weather and climate events.'[37]

Climate variations that spearhead disasters can cause enormous negative effects on societies, the environment, and economic

33 Sendai Framework, para 4

34 Sendai Framework, para 5

35 Sendai Framework, para 30 j

36 MCK Phillips et al 'The Effect of Climate Change on Natural Disasters: A College Student Perspective' (2015) 7 *Weather Climate Change and Society* at 60

37 SI Seneviratne and N Nicholls 'Changes in Climate Extremes and their Impacts on the Natural Physical Environment' in V Barros et al *Managing the Risks of Extreme Events and Disasters to Advance Climate Change Adaptation* (2012, Cambridge University Press) 111

wealth of the affected countries. In such events, agriculture, tourism, and water, often face a great burden by extreme events as they are closely related to climate.[38] The impact of disaster resulting from climate change is not just environmental in nature, but also economic and social in that it alters people's lives in a great way. It can cause poverty and food shortages and occasion displacement with varying consequences for the old, women and children.

The examination of the link of climate change and disaster to social security with clarity is scanty in literature. Yet, the General Comment No 19 of 2008 on the right to social security,[39] made in pursuant to article 9 of the International Covenant on Economic, Social and Cultural Rights (ICESCR) is relevant for the articulation of such linkage.[40] This is because the components of the general comment offer a comprehensive framework to examine the linkage in a functional manner. The General Comment No 19 makes provision for nine main branches of social security, and these are health system, cash benefits for the sick, old age family and child support, maternity, unemployment, employment injury, disability, and lastly, survivors and orphans.[41] These components provide a flexible framework that link disasters resulting from climate changes to the universal standards of the provision of social security measures. In other words, in terms of its components, the general comment is useful in amplifying the linkage of climate change, disaster and social security as a contingency deserving a proper address by social security measures.

Functional health system as a component of social security is expected to guarantee adequate access to health services for

38 n 2 above

39 UN Committee on Economic, Social and Cultural Rights (CESCR), *General Comment No. 19: The right to social security (Art. 9 of the Covenant),* 4th February 2008, E/C.12/GC/19 (General Comment No.19)

40 International Covenant on Economic, Social and Cultural Rights Adopted and opened for signature, ratification and accession by General Assembly resolution 2200A (XXI) of 16 December 1966 entry into force 3rd January 1976, in accordance with article 27

41 General Comment No.19, section 2

all.[42] This is connected to climate change and associated disaster in that climate change can lead to health disasters. For instance, increasing concentrations of greenhouse gases can lead to an increase of both average and extreme temperatures which may escalate death and illnesses from heat.[43] Hotter days than the average seasonal temperature in the summer for instance, may result in intensified levels of illness and death by discrediting the body's ability to regulate its temperature or by inducing direct or indirect health complications (extended exposure to high temperatures is associated with increased hospital admissions for cardiovascular, kidney, and respiratory disorders).[44] According to the general comment, states are to give cash benefits to those who are unable to work by the reason of ill health to cover their loss of earnings.[45] This suggests that those who are sick due to disaster associated with climate change are entitled to cash benefits to cover their loss of earnings. This may be made feasible by implementing a system related to the South African social security system which has amongst its different social security strands; social relief of distress. Social relief of distress entails short term measures undertaken by the state to help people during community or individual crises that have made the affected people to be unable to meet their most basic needs. Governments should be prepared to provide relief for different crises arising from natural disasters.[46]

General Comment 19 also emphasises that special attention should be given to populations in vulnerable circumstance, in particular, in the context of old age, family and child support, and maternity.[47] The instrument calls upon the state to establish social security schemes that provide benefits to older persons

42 General Comment No.19, para 13

43 GP Kenny et al'Heat stress in older individuals and patients with common chronic diseases' (2010) 182/10CMAJ : *Canadian Medical Association Journal* at 1053.

44 A Crimmins 'The Impacts of Climate Change on Human Health in the United States: A Scientific Assessment' (2016, US Global Change Research Program)

45 General Comment No.19, para 14

46 EML Strydom *Essential Social Security Law* (2015, Juta)

47 General Comment No.19, paras 15,18 and 19

and indicates that benefits for families are crucial for achieving the rights of children and other dependents. Also, assistance is required for pregnant women in that disaster associated with climate change will affect more disproportionately vulnerable groups such as older persons, children and pregnant women. Older people are more vulnerable to the effects of temperature extremes and may have a significantly higher mortality risk in extreme weather events.[48]

Owing to biological and psychological factors, research also shows that the foetus and children are vulnerable to the direct and indirect effects of climate change and related disaster. These effects include malnutrition and infectious disease, physical and psychological trauma from extreme weather-related disasters, heat stress, and respiratory disease.[49] According to the World Health Organisation, more than 88 per cent of the existing global burden of disease due to climate change occurs in children less than five years of age.[50] Climate change and its resulting disaster will worsen environmental hazards that threaten the health of pregnant women.[51] Consequently, in terms of the general comment, social security will be essential to older persons, children and pregnant women due to the disproportionate burden of climate change and disaster which they continue to bear. It is apparent that disasters emanating from climate change may affect peoples' employment in that land degradation, rise in sea level and damage to infrastructures may result to loss of jobs. Hence, State Parties are expected to make provision of benefits to cover inability to obtain or maintain suitable employment.[52]

48 Climate change in an ageing world https://reliefweb.int/sites/reliefweb.int/files/resources/COP21_HelpAge_PositionPaper_Final_0.pdf

49 F Perera 'How Climate Change Impacts Children' available at https://www.ecowatch.com/how-climate-change-impacts-children-1881782625.html (last accessed 12th November 2017)

50 As above

51 'Climate Change and the Health of Pregnant Women' available at https://www.cmu.edu/steinbrenner/EPA%20Factsheets/pregnant-health-climate-change.pdf(last accessed 12th November 2017)

52 General Comment No.19, para 17

Referring to General Comment No. 5 of 1994, on persons with disabilities,[53] the General Comment No 19 endorses the importance of the provision of adequate income support to persons with disabilities, in particular, to those who have temporarily lost, or received a reduction in their income, have been denied employment opportunities or have a permanent disability.[54] This is of particular relevance in climate-related disaster considering the vulnerability of people with disabilities in disaster and emergency situations associated with climate–change. People with disabilities are more likely to be forgotten or abandoned during evacuation.[55] Hence, as part of social security measures required of states, the concerns and peculiarities of people with disabilities should be integrated in disaster risk reduction.[56]

According to General Comment 19, survivors or orphans must not be excluded from social security schemes but should be given assistance in accessing the schemes.[57] This envisages that for these categories of people, non-contributory measures of social security are necessary. Owing to its consequences, disaster associated with climate change may leave communities devastated and as result confronted with different needs of survivors and orphans. Hence, this connotes that states must put in place a non-contributory social assistance to the survivors and dependants of climate change and its associated disaster. In all, climate change is connected to disaster and raises concerns of relevance to social security. Arguably, the connection has implications on regional human rights in Africa.The vulnerability of the continent of Africa suggests, according to Masipa, that

53 UN Committee on Economic, Social and Cultural Rights (CESCR), *General Comment No. 5: Persons with Disabilities*,1994

54 General Comment No.19, para 20

55 E Lockwood 'Climate Change and Persons with Disabilities' available at http://blog.cbm.org/climate-change-and-persons-with-disabilities/. (last accessed 23 October 2017)

56 The Impact of Climate Change on People with Disabilities Report of the e-discussion hosted by The Global Partnership for Disability & Development (GPDD) and The World Bank (Human Development Network - Social Protection/ Disability & Development Team) July 8, 2009, 2

57 General Comment No.19, Para 21

climate change will have dire consequences on social security.[58] This was reinforced by the Intergovernmental Panel on Climate Change (IPCC) which further reported that in sub-Saharan Africa, agricultural productivity will decline from 21% to 9% by 2080,[59] a development confirmed by different authors who contend that climate change has dire consequences on agricultural land, which ultimately affect food security.[60] In particular, Liliana states that about two-thirds of Africa's arable land is expected to be lost by 2025 because of the lack of rainfall and drought.[61] The foregoing confirms that climate change is affecting Africa's social security.

Implications for regional human rights

Arguably, a gap or deficit of social security focus on disaster occurrences associated with climate change has implications for human rights instruments guaranteed at the African regional level to which State Parties are signatories. Core human rights instruments at the regional level such as the African Charter to which 53 members of the African Union are State Parties,[62] and the Kampala Convention to which 45 members are currently State Parties,[63] guarantee socio-economic rights of relevance to

58 T Masipa 'The impact of climate change on food security in South Africa: Current realities and challenges ahead'(2017) 9/1Jàmbá: *Journal of Disaster Risk Studies* at 1-7.

59 Intergovernmental Panel on Climate Change (IPCC), 2007, *Intergovernmental panel on climate change fourth assessment report* (2007, Cambridge University Press)

60 H Liliana The food gaps: *The impacts of climate change on food production: A 2020 perspective* (2005, Universal Ecological Fund); C Vogel 'Usable Science: An assessment of long-term seasonal forecasts among farmers in rural areas of South Africa'(2005) 82/2 *South African Geographical Journal* at 107–116.

61 Liliana *The food gaps*, note 60 above.

62 'Ratification Table: African Charter on Human and Peoples' Rights' available at http://www.achpr.org/instruments/achpr/ratification/(last accessed 12th October 2017)

63 African Union Convention for the Protection and Assistance of Internally Displaced Persons in Africa (Kampala Convention), adopted by the Special Summit of the Union held in Kampala, Uganda, 23 October 2009; on the status of ratification available at https://au.int/sites/default/files/treaties/7796-sl-african_union_convention_for_the_protection_and_assistance_of_internally.pdf(last accessed 12th October 2017)

the link of climate change to disaster and social security. There are soft instruments at the regional level such as Resolutions 153 of 2009,[64] Resolution 271 of 2014,[65] and Resolution 342 of 2016, 'the latter urging member states to improve regional cooperation for climate action that respects human rights and tasks, the African Commission on Human and Peoples' Rights (the Commission), the treaty monitoring body of African Charter to investigate the impact of climate change on human rights in Africa.[66] Although an elaborate connection has not been officially articulated at the African Union level, it is not impossible to link a deficit of social security focus in disaster occurrences associated with climate change to human rights, in particular, the rights to health care, environment, water, and food, housing and the right to work.

The fact that health systems can be impacted adversely by climate change and associated link to social security is of implications to the right to health under the African Charter. The right to health is guaranteed in article 16(1) of the African Charter. Applying article 16(1) of the African Charter in *Social and Economic Rights Action Centre (SERAC) and Another v Nigeria,*[67] the Commission ruled that environmental depreciation of the region had a negative impact on the right to health.[68] This connotes that since disaster emanating from climate change is connected with environmental depreciation such as rise in sea level, drought and desertification, except there is an adequate social security attention, such occurrences will negatively impact

64 African Commission of Human and Peoples' Rights, ACHPR/Res153(XLVI)09: Resolution on Climate Change and Human Rights and the Need to Study its Impact in Africa, November 25th, 2009 (Resolution 153),preamble.

65 African Commission on Human and Peoples' Rights '271: Resolution on climate change in Africa', adopted at the 55th ordinary session of the African Commission on Human and Peoples' Rights held in Luanda, Angola, from 28th April-12th May 2014

66 '342: Resolution on Climate Change and Human Rights in Africa' adopted at its 58th Ordinary Session, held in Banjul, The Gambia, from 6th to 20th April 2016-ACHPR/Res. 342(LVIII) (ACHRP Resolution 342).

67 *Social and Economic Rights Action Centre (SERAC) and Another v Nigeria* (2001) AHRLR 60 (ACHPR 2001).

68 *SERAC case* para 52.

on human right to health guaranteed in the African Charter. A similar argument can be made for the right to healthy environment guaranteed by article 24 of the African Charter. In the SERAC case, the Commission reasoned that article 24 imposes clear obligations upon a government to take reasonable and other measures to prevent pollution and ecological degradation, to promote conservation, and to secure an ecologically sustainable development and use of natural resources.

Hence, it recommends that government should ensure adequate compensation to victims of the human rights violations, including relief and resettlement assistance to victims of government-sponsored raids, and undertaking a comprehensive clean-up of lands and rivers damaged by oil operations.[69] Although, climate change and disaster as well as relationship with social security was not up for consideration in the decision, it is useful in terms of its provision dealing with compensation and the provision of relief and resettlement assistance which resonates well with the position of the General Comment No 19 that special attention should be given to populations in vulnerable circumstance, in particular, in the context of old age, family and child support, and maternity in assessing the need for social security. Failure of government to make such provisions in that context will have grave implication on the right of populations to a healthy environment.

In Free Legal Assistance Group case, the Commission affirmed that the failure by governments to provide basic services, including safe drinking water, could constitute a violation of article 16 of the African Charter.[70] Disaster associated with climate change conditions such as drought, flooding and disrupted rainfall will constrain or even deprive populations of their water sources, thereby constituting a threat to the right to water under the African Charter. Water is a basic need for all living creatures, as such a substance that no human being should be denied of.

69 General Comment No.19, paras 52 and 71

70 *Free Legal Assistance Group and Others v Zaire* (2000) AHRLR 74 (ACHPR 1995) para 47.

In an endeavour to adequately manage and conserve water in Africa, the African Water Vision for 2025 was born.[71] One of the intentions of the vision is to move Africa from where it is presently to where it intends to be in future in order to ensure that accessible water is adequate and sustainable. Water and socio-economic development are closely related and dependent to each other. Scarcity of water in Africa would highly affect the continent's social security as the continent mostly depend on agriculture.[72] Climate change often affects sustainability of water resources in Africa thereby infringing upon Africans right water. Africa's main environmental component remains the variation of the climate that result in droughts, flooding, desertification and other natural disasters. Therefore, as the rise in sea level, drought and desertification adversely impact water, it is required of states to earmark concrete social security measures to increase the access of populations to water in the event of climate change. The failure of the state to take such social security measures will undermine the realisation of the right to water.

Failure to respond to climate change and associated disaster with an appropriate social security measure may also have implication for the realisation of the regional right to food. Although the African Charter does not expressly refer to the right to food, the Commission acknowledges that the Charter recognises it through the right to life as well the right to health.[73] States should be seen to refrain as well as protect against contamination of food sources. It is apparent that climate change may threaten the right to food. Unpredictable rain patterns, extending droughts and floods are some of the environmental factors affecting food security.[74] This

71 United Nations, UN-Water/Africa (Agency), African Union, and African Development Bank *The Africa Water Vision for 2025: equitable and sustainable use of water for socioeconomic development* (2009, Economic Commission for Africa)

72 L Bounoua 'Climate change is hitting African farmers the hardest of all' *The Conversation* 12th May 2015, available at http://theconversation.com/climate-change-is-hitting-african-farmers-the-hardest-of-all-40845 (last accessed 23rd October 2017)

73 African Charter, articles 4 and 16.

74 O De Schutter 'What is the impact of climate change on the right to food?' available at http://www.srfood.org/en/climate-change-2 (last accessed 21st October 2017)

implication was confirmed in the SERAC case [75] where it was said that the right to food is inherent in the Charter's protection of the rights to life, health and the right to economic, social and cultural development. The court held that Nigeria violated the right to food of the Ogoni people by failing to safeguard their land from environmental damage in the Niger Delta. The right to food is inseparable from the fulfilment of other rights, thus, one of the significant human rights. It is equally guaranteed to women by the Protocol to the African Charter on Human and Peoples' Rights on the Rights of Women in Africa (Maputo Protocol) which calls upon State Parties to ensure that women have adequate food.[76] In the Endorois' case, the Commission has noted that the eviction of indigenous peoples from their land can hinder their access to food.[77] In a similar manner, in affecting peoples' employment and occasioning land degradation, climate change may lead to mass loss of jobs, displacement and thereby undermine the right to food of populations. The possibility of this reality requires an appropriate response by government in the form of social security assistance to ensure that the right to food of populations experiencing disaster associated with climate change is not endangered.

Failure to respond to climate change and associated disaster with an appropriate social security measure may also have an effect on the right to adequate housing. Climatic incidents such as sea level rise and flooding can bring about destruction of infrastructure including houses. This development, no doubt, poses a challenge to the right to adequate housing which principally aims at providing and setting standards for adequate shelter. Although not directly guaranteed by the African Charter, it was highlighted in the SERAC case that rights to housing are protected through the integration of provisions protecting the

75 *Social and Economic Rights Action Centre (SERAC) and Another v Nigeria* (2001) AHRLR 60 (ACHPR 2001).

76 African Charter, article 15.

77 *Centre for Minority Rights Development & Others v Kenya* (2009) AHRLR 75 (ACHPR 2009) (Endorois case);

right to property, the right to enjoy the best attainable standard of mental and physical health, and the protection accorded to the family.[78] General Comment 7 on the right to housing, deals with the issue of forced evictions[79] defined as:

> Permanent or temporary removal against their will of individuals, families and/or communities from the homes and/or land which they occupy, without the provision of, and access to, appropriate forms of legal or other protection.[80]

Disaster associated with climate change can result in forced evictions which, does not only interfere with the right to housing, but with several other rights of the ICESCR, and with rights enshrined in the International Covenant on Civil and Political Rights (ICCPR).[81] These rights include the right to life, the right to security of person, the right to non-interference with privacy, family and home and the right to the peaceful enjoyment of possessions.[82] All the elements in the right to adequate housing, namely, legal security of tenure, availability of services, materials, facilities and infrastructure, accessibility, location, affordability, habitability, and cultural adequacy are disturbed when populations are displaced or forcefully evicted from their shelter in the event of disaster associated with climate change. Hence, where the state fails to put in place adequate security to assist affected populations, this can constitute a breach of the right to housing as grounded in articles 14 and 16 of the African Charter.

---

78 African Charter, articles 14, 16 and 18(1).

79 'United Nations General Comment No. 7: The right to adequate housing, art 11 (1) of the Covenant: Forced evictions' (1997) 16thsession (United Nations General Comment No. 7)

80 United Nations General Comment No. 7 para 3

81 International Covenant on Civil and Political Rights (ICCPR), Dec. 16th, 1966, 9 U.N.T.S. 171

82 United Nations General Comment No. 7 para 4

According to the International Labour Organisation, there are risks and negative impacts of climate change on employment.[83] Disaster resulting from change in climate in the form of droughts, heat waves, heavy precipitation, and increased occurrence and intensity of cyclones, or from more gradual climatic processes such as sea level rise, rising temperatures and changing rainfall patterns can result in job losses as they can displace workers or result in damage to business assets, transport and industrial infrastructure.[84] Those employed in the rural markets may also lose their jobs due to a greater incidence of heavy precipitation and/or extreme heat and resulting damage to agricultural crops.[85] The foregoing scenarios will offend article 15 of the African Charter which provides that every individual shall have the right to work under equitable and satisfactory conditions, and shall receive equal pay for equal work. In *the Nubian Community in Kenya vs The Republic of Kenya,*[86] the complainants argued that the discrimination to which Kenya Nubians are subjected result to the violation of their rights to equal access to work guaranteed under article 15 of the African Charter.[87] The Commission upheld the argument of the complainants that the significant burdens are placed by the state on the process of acquiring ID documents frustrate the ability of the Nubian community to enjoy the right to work under equitable and satisfactory conditions.[88] In finding the violation of article 15, the Commission urged the state to establish objective, transparent and non-discriminatory criteria and procedures for determining citizenship. This decision shows that the failure of the state to adopt appropriate measures can result in an infringement of the right to work. Similarly, the absence of

83 'International Labour Organisation 'Frequently Asked Questions on Climate Change and Jobs' available at http://www.ilo.org/global/topics/green-jobs/WCMS_371589/lang--en/index.htm(last accessed 12th October 2017)

84 As Above

85 As Above

86 Communication 317/06 The Nubian Community in Kenya vs The Republic of Kenya February 28th, 2015 (Nubia case)

87 Nubia case, para 93

88 Nubia case, para 132

specific measures to avert the loss of jobs in the event of disaster associated with climate change can undermine the right to work guaranteed to the African populations under the African Charter. As has been shown, the lack of an appropriate measure in terms of social security for disaster occurrences associated with climate change can have implications for human rights, in particular, the rights to health care, environment, water, and food, housing and the right to employment. It remains to be seen how that can shape climate change and disaster management law in Nigeria.

WAY FORWARD FOR NIGERIA

Nigeria should take into consideration its obligations, in particular, under the regional framework to shape appropriate legal response to disaster associated with climate change and formulate an appropriate social security measures in response. Nigeria ratified the UNFCCC in 1994,[89] it is represented at COP, which as earlier observed, is the highest decision making body of the UNFCCC that takes decisions necessary to promote the effective implementation of climate mitigation actions.[90] It acceded in 2004 to the Kyoto Protocol,[91] and ratified the Paris Agreement on 6th May 2017.[92] In response to the obligation under the UNFCCC, Nigeria has submitted her First National Communication in 2003[93] and a Second National Communication in February 2014 in which she lists climate change mitigation as a priority.[94] She deposited the instrument of ratification of Kampala Convention

89 It ratified the Convention on 29th August 1994, see 'Status of Ratification of the Convention' available at <http://unfccc.int/essential_background/convention/status_of_ratification/items/2631.php>(last accessed 19th October 2017).

90 UNFCCC Conference of parties (n 47).

91 n 81.

92 Paris Agreement 'Status of Ratification' available at http://unfccc.int/paris_agreement/items/9444.php (last accessed 19th October 2017)

93 Federal Republic of Nigeria 'Nigeria's First National Communication under the United Nations Framework Convention on Climate Change' (2003, Federal Ministry of Environment).

94 Federal Republic of Nigeria 'Nigeria's Second National Communication under the United Nations Framework Convention on Climate Change'(2014, Federal Ministry of Environment).

on 22nd May 2012,[95] and for the Maputo Protocol on 18 February 2005. Like the foregoing instruments, the Kampala Convention and Maputo Protocol are not yet domesticated in line with Article 12 (1) of the Constitution which provides that 'no treaty between the Federation and any other country shall have the force of law to the extent to which any such treaty has been enacted into law by the National Assembly.' Also, soft instruments which are of significance to disaster management and social security such as the Sendai Framework is not binding.

However, it is trite principle of international law that a state should not act in a manner inconsistent with its obligations, and even if not yet domesticated, states should still ensure that treaties influence the application of domestic law, policy, and judicial decisions.[96] This is not surprising as a contrary reasoning will undermine article 27 of the Vienna Law of Treaty that recognises that 'a party may not invoke the provisions of its internal law as justification for its failure to perform a treaty.'[97] This is more so as, although the Constitution has no provision on socio-economic rights but only retains socio-economic objectives as directive principles of state policy (DPSP),[98] they are safeguarded under the African Charter which is a part of the law in Nigeria.[99]

Accordingly, in so far as the provisions of the above instruments reflect how social security measures can be employed to legally response to disaster associated with climate change, they can influence domestic law, policy, and judicial decisions. They can be useful in applying domestic law and policy dealing with climate change and disaster management. In particular,

95 'Status of Ratification' available at https://au.int/sites/default/files/treaties/7796-sl-african_union_convention_for_the_protection_and_assistance_of_internally.pdf/ (last accessed 19th October 2017)

96 F Viljoen *International human rights law in Africa* (2012, Oxford University Press) 25

97 Vienna Convention on the law of treaties (with annex). Concluded at Vienna on 23rd May 1969 art 27.

98 Nigeria 1999 Constitution chapter II.

99 African Charter (Ratification and Enforcement) Act (Cap 10, Laws of Federation of Nigeria 2004).

considering that foreign decisions are not binding on Nigerian courts,[100] decisions of the African Commission in so far as they relate to climate change and disaster management can be useful to courts in Nigeria in their analysis of issues around disaster associated with climate change.

100 Elioclin Nig. Ltd v. Mbadiwe (1986) 1 NWLR (Pt. 14) 47

13

PROTECTING REPRODUCTIVE AND SEXUAL HEALTH RIGHTS THROUGH LAW DURING DISASTER

Folake M. Tafita

Abstract

Reproductive health is an important component of health that must be protected, even in disaster. Globally and in recent decades, natural disasters such as hurricanes Catherina, Harvey and the 2004 Tsunamis and other man-made disasters such as war and other conflict situations have seriously impacted the world, increasing the problem of disaster risks and impact on victims.

The impact of these disasters on sexual and reproductive health of disaster victims raises a cause for concern. While Disaster Risk Reduction (DRR) has received increased global attention and support, Sexual and Reproductive Health (SRH) is yet to be systematically included in global and national efforts to reducing disaster risks, and remains a critical gap.

The reproductive and sexual health aspects and effects of the disaster, particularly on women and girls, are little acknowledged or addressed. During disasters, women and girls are often mostly affected and their reproductive and sexual health needs neglected and marginalised.

The chapter discusses disaster risk management and violations including reproductive and sexual ill health of disaster victims from a global perspective but with a major focus on Nigeria.

From a legal and human rights approach, it elucidates the obligation of society, government and health/humanitarian workers in the protection and promotion of reproductive and sexual health rights in disaster situations, calling also for the integration sexual and reproductive health rights (SRH) into national disaster risk reduction and management.

It is imperative that governments enact national laws and policies that will better protect disaster victims' reproductive and sexual health rights.

Keywords: Disaster risks, reproductive and sexual health, disaster victims, disaster risks protection in Nigeria

INTRODUCTION

In the last decade, reproductive health has become a topical area of interest. It is not just a major health issue but also a development issue, and a human rights issue.[1] The International Conference on Population Development in Cairo, defined women's reproductive health as 'a state of complete physical, mental, and social well-being, and not merely the absence of diseases or infirmity, in all matters relating to the reproductive system and to its functions and processes.'[2] Reproductive health is an important component of health that must be protected. Globally, there has been a dramatic

1 Rebecca Cook., Bernard Dickens, and Mahmoud Fathalla, *Reproductive Health and Human Rights: Integrating Medicine, Ethics and Law* (O U P, New York 2003) p.8

2 UN, Population and Development, Programme of Action Adopted at the International Conference on Population and Development, Cairo, 5-13 September 1994 (New York: United Nations, Department for Economics and Social Information and policy Analysis, ST/ESA/SER.A/149,1994), para.7.2.

increase in the number of people affected by natural disasters [3] and man-made disasters.[4] In recent decades, natural disasters such as hurricanes Catherina, Harvey and the 2004 Tsunamis and other man-made disasters such as war and other conflict situations have affected disaster victims' men, women, boys' and girls' reproductive health.[5] Disasters are often unanticipated, unpredictable external events that influence human survival including control over one's reproductive health.[6] However, disasters affect women, girls, boys and men differently. Women and girls victims of conflicts situations or other natural disasters are more susceptible during natural and man-made disasters due to their elaborate reproductive system that is vulnerable to dysfunction, disease and other social dysfunctions such as sexual violence and abuse that impact their physical, mental and social health.[7] Studies have also shown that both domestic violence and sexual violence (assault, sexual abuse, and exploitation) child/early marriages and trafficking increase following disasters.[8] Disasters cause impoverishment, which can induce some people to adopt negative coping strategies, including

3 The Women's Refugee Commission & DRR: Promoting Sexual and Reproductive Health in Emergencies; 'Towards a post-2015 framework for disaster risk reduction. Geneva, United Nations International Strategy for Disaster Reduction 2012' http://www.unisdr.org/files/25129_towards a post¬2015framework for disaster. pdf accessed 17th September 2017

4 A disaster can be defined as an unexpected event that results to human casualty. Generally there are two types of disaster-Natural and Man made. Major natural disaster include Flood, Cyclones, Cold waves, Drought, Thunderstorm, Earthquake, Heat waves, Mud slides and Storms among others. www.gamji.com/article8000/news 8296.htm accessed 12th September 2017

5 Integrating sexual and reproductive health into health emergency and disaster risk management, WHO/RHR/12.32 World Health Organization 2012 www.who.int/disasters accessed 25th October2017

6 R Swatzyna and V Pillai, 'The Effects of Disaster on Women's Reproductive Health in Developing Countries' [2013] 5 Glob J Health Sci (4) 106–113 https://www.ncbi.nlm.nih.gov/pmc/articles/PMC4776806/ accessed 26th October 2017

7 Rebecca Cook., Bernard Dickens, and Mahmoud Fathalla, *Reproductive Health and Human Rights: Integrating Medicine, Ethics and Law* (n1)

8 International Federation of Red Cross and Red Crescent Societies, 'Unseen, unheard: Gender-based violence in disasters' (Global study) (Geneva 2015) http://www.ifrc.org/Global/Documents/Secretariat/201511/1297700_GBV_in_Disasters_EN_LR2.pdf accessed 23rd November 2017

transactional sex.[9] The impact of disasters as well as the sexual and reproductive health impact on disaster victims therefore raises a cause for concern.[10] Protecting reproductive health and human rights must be central to disaster risk reduction and management in any national legislation and policy on disaster.

Disaster risk reduction (DRR) has received increased international, regional and national attention and support. National governments are working to develop plans to address priority areas for DRR as established in the Hyogo Framework for Action.[11] However, sexual and reproductive health (SRH) is yet to be systematically included as a priority area in global and national efforts to reducing disaster risks, and remains a critical gap.[12]

Unlike other countries, not until recently, Nigeria was not at high risk of natural disasters such as flood, spontaneous earthquakes and other natural disasters.[13] The frequency of man-made conflicts in Nigeria far outweighs those from natural disasters. Due to the constantly heated polity, Nigeria has experienced a major civil war and continues to experience other sporadic skirmishes of internal armed conflicts. Historically, Nigeria has had to contend with a series of violence, beginning from the period she attained her political independence in 1960. Notable among such violent conflicts were the Aba women riot of 1929 in the East, and the Iseyin-Okeho uprising in the West.[14] The first major violence occurring at the inception of a civilian regime occurred in May 1999 in the Niger Delta region between local militias, who emerged to put an end to the long years of deprivation and neglect, led to the displacement of more than 60,000 people and particularly

9 *Ibid.*

10 WHO, 'Reproductive health during conflict and displacement; a guide for program managers' (Report)

11 The Women's Refugee Commission & DRR Promoting Sexual and Reproductive Health in Emergencies (n3)

12 *Ibid.*

13 Unlike Asia and the Americas where natural disasters are a common occurrence

14 A Adesola and A Peters, 'A Historical Analysis of Violence and Internal Population Displacement in Nigeria's Fourth Republic 1999-2011'(2015) 2 IJPCS p.15

affected the people of the Odi community.[15] Also significant is the fact that between 2003 and 2007, many communities in Nigeria experienced a number of religious, ethnic and political violence and by the time post-election violence abated in 2009, a new and more dreadful conflict arose between Boko Haram Islamic sect and government security agencies.[16]

> ...the advent of Boko Haram terrorism, herdsmen attacks, flooding (which now occurs almost every year due, mainly, to climate change and poor drainage, especially in major urban areas), collapsed buildings, oil spills, extreme coastal and gully erosion, fire disasters, communal violence, among others, have become major sources of emergency displacement of hapless citizens. Nigerians faced with this gamut of disasters suddenly find themselves homeless, hapless and destitute.[17]

Nigeria continues to experience conflicts; a chronic threat of political/religious instability which therefore creates the imperative for preparedness and protection of would be victims of both natural and man-made disasters.

Disaster contexts often generates an environment in which reproductive and sexual health needs are compromised. According to varied reports, in most refugee and displacement camps established following disaster situations, more attention is given and priority placed on providing food and other relief materials, while reproductive and sexual health concerns are often ignored.[18] Women and girls often displaced as refugees due to conflicts or other natural disasters are more vulnerable to sexual abuse by refugee camp guards, security personnel and

15 *Ibid.* This conflict is popularly referred to as the Odi Crisis in Bayelsa State.

16 Internal Displacement Monitoring Committee (IDMC), 2009 (Report) www.internal-displacement.org accessed 10th October 2017.

17 *Vangaurd* Newspaper, 'Need for Emergency Shelters' https://www.vanguardngr.com/2017/09/need-emergency-shelters/ accessed 10th October 2017

18 Rebecca Cook., Bernard Dickens, and Mahmoud Fathalla, *Reproductive Health and Human Rights: Integrating Medicine, Ethics and Law* (n1) p. 228; Human Rights Watch, 'Seeking Protection: Addressing Sexual and Domestic Violence in Tanzania's Refugee Camp' (New York Report 2000).

humanitarian aid workers.[19] The reproductive and sexual health aspects and effects of the disaster particularly on women and girls are little acknowledged or addressed.[20] During disasters, women and girls are often mostly affected and their reproductive and sexual health needs neglected and marginalised.[21] Similarly, post disaster and in the management of disaster victims, there is an increase in reproductive and sexual health rights violations, these includes incidences of rape, sexual assault, sexual harassment, unwanted pregnancy and abortion, forced prostitution, sex trade and sexual exploitation, sexual slavery, sexually transmitted diseases, forced marriages and maternal deaths.

Sexual and reproductive health services available to victims and survivors of disasters are grossly inadequate. There is lack of access to contraception or emergency reproductive health services including maternity and child delivery, and when services are available, there is lack of privacy, hindering access to reproductive and sexual health services. These attendant reproductive and sexual health risks associated with disasters, thus informed the imperative for legal and human rights approach to disaster risks management.[22] The imperative to integrate SRH into national laws and policies on disaster risk management has received global support through the UN International Strategy on Disaster Risk Reduction.[23] As stated in the outcome document of the Rio+20

19 Arrows, 'Sexual and Reproductive Health And Rights: Key To Building Disaster Resilience' The Asian-Pacific Resource & Research Centre For Women (Report 2017) www.arrow.org.my accessed 10th October 2017.

20 International Federation of Red Cross and Red Crescent Societies, Geneva, 2015, Unseen, unheard: Gender-based violence in disasters Global study (n8).

21 Arrows, 'Sexual and Reproductive Health And Rights: Key To Building Disaster Resilience'(n19)

22 WHO, 'Disaster Risk Management for Health; Sexual and Reproductive Health' Disaster Risk Management for Health Fact Sheets Global Platform, May 2011 https://www.who.int/hac/events/drm_fact_sheet_sexual_and_reproductive_health.pdf accessed 23 November 2017 ; A Hutchinson and others 'Understanding processes of risk and protection that shape the sexual and reproductive health of young women affected by conflict: the price of protection' [2017] 11Confli Health https://www.ncbi.nlm.nih.gov/pmc/articles/PMC5559816/ accessed 23rd November 2017

23 UN International Strategy on Disaster Risk Reduction

United Nations Conference on Sustainable Development, sexual and reproductive health is a significant public health need in all communities, including those facing emergencies. Universal access to reproductive health, including family planning and sexual health, is needed and should be integrated into national strategies and programmes.[24]

This chapter therefore discusses ways of integrating sexual and reproductive health rights (SRH) into national disaster risk reduction and management, calling also for national laws and policies that will better protect disaster victims' reproductive and sexual health rights.

The chapter begins by explaining the concept of reproductive and sexual health, why reproductive and sexual health is legitimate, compelling the attention of society, government, institutions, professionals and other allies in other fields. The aim is to inform the reading audience generally about ethical, legal and human rights dimensions of reproductive and sexual health. Furthermore, the chapter explains the effect of disaster on reproductive and sexual health, the relationship between disaster risks and reproductive and sexual ill health. The various forms of violations of reproductive and sexual health rights in disaster situations are also elucidated.

It further discusses disaster risk management and violations including reproductive and sexual ill health of disaster victims from a global perspective but with a major focus on Nigeria. Furthermore, the chapter elucidates the obligation of society, government and health/humanitarian workers in the protection and promotion of reproductive and sexual health rights in disaster situations.

Finally, a number of case studies provide a practical application of principles of ethics, law and human rights to a series of hypothetical and also fact situations in disaster management and reproductive and sexual health. The purpose is to showcase the

24 Integrating sexual and reproductive health into health emergency and disaster risk management, WHO/RHR/12.32 World Health Organisation, 2012(n5).

protection of reproductive and sexual health of disaster victims through the approach of human rights, ethics and law in responding to similar scenarios in disaster risk management.

The overall aim is to inform students, government, non-governmental bodies, humanitarian aid workers and other stakeholders involved in disaster management on the need to protect and promote the reproductive and sexual health rights of disaster victims.

Defining Reproductive and Sexual Health

Professor B. Dickens, defined reproductive health in relation to the positive definition of health in the constitution of the World Health Organization (WHO) stating that reproductive health is 'a state of complete physical, mental and social well-being and not merely the absence of disease or disorder of the reproductive processes.'[25] The 1994 ICPD Programme of Action defines reproductive health as complete physical, mental and social well-being in all matters related to the reproductive system. This discussion extends the above definition to include the right to reproductive health services like access to treatment for reproductive health complications due to disaster and conflicts, provision of adequate and emergency reproductive health facilities for the treatment of disaster victims. According to ICPD Programme of Action, reproductive rights are the rights of 'couples and individuals to decide freely and responsibly the number, spacing and timing of their children, have the information and means to do so,' attain the highest standards of health, and make decisions on their reproductive health 'free of discrimination, coercion and violence.'[26]

25 Rebecca Cook., Bernard Dickens, and Mahmoud Fathalla, *Reproductive Health and Human Rights: Integrating Medicine, Ethics and Law* (n1); F Tafita and O Bamgbose 'Access to Justice for Reproductive and Sexual Health Rights of Women through Law Faculty Clinics' (2015) 5 JPIL 82-109.

26 UN, Population and Development, Programme of Action Adopted at the International Conference on Population and Development, Cairo, 5-13 September 1994 (n2).

Sexual health is relatively new and distinct from reproductive health, though there is a strong link between the two concepts.[27] One definition of sexual health which has been widely accepted different from the 'terse' inclusion of a definition under reproductive health is the elaborate and more inclusive WHO definition of sexual health and sexual rights which states:

> Sexual rights embrace human rights that are already recognised in national laws, international human rights documents and other consensus statements. They include the rights of all persons, free of coercion, discrimination and violence, to: (1) the highest attainable standard of sexual health, including access to sexual and reproductive health care services; (2) seek, receive and impart information related to sexuality; (3) sexuality education; (4) respect for bodily integrity; (5) choose their partner; (6) decide to be sexually active or not; (7) consensual sexual relations; (8) consensual marriage; (9) decide whether or not, and when, to have children; and (10) pursue a satisfying, safe and pleasurable sexual life. The responsible exercise of human rights requires that all persons respect the rights of others.[28]

The above definition buttresses the opinion of writers and experts on reproductive and sexual health rights that

> ... definition of sexual health should include the ability to enjoy mutually fulfilling sexual relationships, freedom from sexual abuse, coercion, or harassment, safety from sexually transmitted diseases, and success in achieving or in preventing pregnancy.[29]

The 1994 ICPD Programme of Action defines sexual health as healthy sexual development, equitable and responsible relations, sexual fulfilment and freedom from illness, disease, disability, violence and other harmful practices related to sexuality. This definition has in recent times been extended to cover contemporary reproductive health issues.

---

27 E Durojaiye, 'Realizing Access to Sexual Health Information and Services for Adolescents through the Protocol to the African Charter on the Rights of Women' (2009) 16 WASH & LEE J.C.R & SOC. JUST 135.

28 WHO, www.who.int/reproductive-health/gender/sexualhealth.html#4 accessed 10 October 2017

29 Rebecca Cook., Bernard Dickens, and Mahmoud Fathalla, *Reproductive Health and Human Rights: Integrating Medicine, Ethics and Law* (n1) p. 8

Reproductive and sexual health rights are among the most important but controversial human rights in international human rights law. These rights are intrinsically based on other well recognised and established human rights. It includes the right to autonomy and self-determination without any form of discrimination, stigma, coercion, or violence. This includes the right of individuals to enjoy and express their sexuality, be free from interference in making personal decisions about sexuality and reproductive matters, and to access sexual and reproductive health information, education, services, and support. These rights have been identified to be based on the right to equality and equal protection of the law, the right to life and liberty and to be free from torture and cruel and inhuman treatment, and the right to privacy.[30] Generally, reproductive and sexual health rights rest on the recognition of the right to attain the highest standard of reproductive and sexual health.

Too often reproductive health has been considered as relevant only to childbearing women of reproductive age (15-44 years). It is true that women bear by far the greatest burden of reproductive health problems[31] and that biological, social, cultural and economic factors increase a woman's vulnerability to reproductive ill-health. However, reproductive and sexual health is about more than just the reproductive organs, and more than just reproduction.[32] It is about how social and sexual behaviours and relationships affect health and create ill-health. It is relevant to both men and women, and to persons of all ages. Reproductive health has to be understood within the context of relationships between men and women, communities and society, since sexual and reproductive behaviours are governed by complex biological, cultural and psychosocial factors.[33]

30 ICPD –"United Nations Report of the International Conference on Population and Development" Cairo, Egypt, 5-13 September 1994. United Nations Doc. N.Y.A/ Conf. 171/13 Rev. 1 U. N Sales No.9, XIII. 18.

31 Rebecca Cook., Bernard Dickens, and Mahmoud Fathalla, *Reproductive Health and Human Rights: Integrating Medicine, Ethics and Law* (n1) p. 8

32 WHO, Reproductive health during conflict and displacement; a guide for program managers (n5)

33 *Ibid.*

Reproductive health covers a wide range of issues: pregnancy and childbirth; the protection of women, children, adolescents and men from emotional, physical and sexual abuse; family planning counselling and services to prevent unwanted pregnancies, unsafe abortion, the treatment and prevention of sexually-transmitted diseases including HIV/AIDS; and the discouragement of harmful traditional practices. Recognising the importance of sexual and reproductive health, the provision of reproductive health services should be based on the needs of the population, with particular attention being paid to vulnerable groups, such as women and adolescents.[34] Reproductive health is not just a major health issue; it is a development issue, and a human rights issue,[35] particularly within a complicated environment such as disaster and conflict situations.

Disaster risks impacting Reproductive and Sexual Health

The first decade of the millennium has begun with a global upsurge of disasters including armed conflict. There has also been a dramatic increase in frequency and severity of natural disasters over the last two decades. However, in spite of growing instability resulting from disasters, studies on reproductive health ignore their impact.[36]

During disaster and the aftermath, reproductive and sexual health is at risk. Vulnerable populations such as women and adolescents are exposed to more disaster risks compromising and affecting their reproductive health. Women are affected adversely by natural disasters and war. They suffer poor health, and economic and social consequences of war.[37] Disasters both natural and socially generated disasters such as war and conflict are often unanticipated, unpredictable external events that influence human

34 *Ibid.*

35 Rebecca Cook., Bernard Dickens, and Mahmoud Fathalla, *Reproductive Health and Human Rights: Integrating Medicine, Ethics and Law* (n1)

36 R Swatzyna and V Pillai , 'The Effects of Disaster on Women's Reproductive Health in Developing Countries'(n6)

37 *Ibid.*

survival including women's control over her reproductive health.[38] Reproductive health disaster risks include rape, sexual assaults, and risk of unwanted pregnancy, sexually transmitted diseases (STDs), including HIV infection, child prostitution, sexual abuse, enforced prostitution, forced termination of pregnancies and forced sterilisation. Four immediate effects on women are poor birth outcomes, sexualised violence, displacement and trafficking. Other disaster risks peculiar to conflict situations include all foregoing and others, for example forced marriage and sexual slavery. Six of the ten countries with the highest under five mortality rates in the world are war torn.[39] Populations affected by disaster, particularly internally displaced populations lack access to reproductive and sexual health service and intervention. Kissinger *et al,* in their research reported the effect of Hurricane Katrina disaster on displaced persons in metropolitan New Orleans. The study observed and described changes in sexual behaviour and access to reproductive care pre- and post-rapid displacement among a cohort of young women receiving family planning services before displacement. In the wake of Hurricane Katrina, there was limited access to contraception and provision of reproductive health care, and this led to many thousands of unintended pregnancies among young women.[40] In another study on the effects of disaster on reproductive health, Zotti *et al* also found that there is an association between disasters and reproductive health outcomes highlighting the need for further studies to clarify associations.[41] Claudia Felten-Biermann in her

38 *Ibid.*

39 *Ibid.*

40 Kissinger and others, 'The Effect of the Hurricane Katrina Disaster on Sexual Behavior and Access to Reproductive Care for Young Women in New Orleans' [2007] 34 Sexually Transmitted Diseases 883-886 http://journals.lww.com/stdjournal/Abstract/2007/11000 accessed 26th October 2017.

41 M Zotti and others, 'Post-Disaster Reproductive Health Outcomes' [2013] 17 Maternal & Child Health Jour .https://link.springer.com/article/10.1007/s10995-012-1068 accessed 26th October 2017.

article states sexualised violence as a consequence of natural disasters for women in the aftermath of the 2004 Asian Tsunami.[42]

Similarly, the effects of war and conflicts on reproductive health of women are well documented in the 2009 conflict of the Democratic Republic of Congo (DRC) and the 1991 armed conflicts in former Yugoslavia. These wars were characterised by rape and other forms of sexual violence. According to reports,[43] in the conflict in Kivus (DRC), cases of sexual violence registered at health facilities near the areas of the conflict in North and South Kivu doubled or tripled. Human Rights Watch reported that combatants of the Democratic Forces of the Liberation of Rwanda (FDLR) were responsible for nearly half of all the rapes. Human Rights Watch also documented 42 cases of rape by coalition soldiers who were deployed against the FDLR in January and February 2009.[44]

During disasters and the aftermath, women and girls in Nigeria as well as other parts of Africa are prone to sexual abuse and other forms of sexual violence. In Rwanda, Uganda and Nigeria, cases of abuse have been well documented.[45] There is a growing literature and attention to sexual violence and rape as a strategy of warfare. After the incidences of sexual violence, women and girls are often rejected by family or community.[46] The raping of women in armed conflicts is a means for the aggressor to symbolically and physically humiliate the defeated men. In some contexts, the

42 C Felten-Biermann, 'Gender and Natural Disaster: Sexualized violence and the tsunami' [2006] 49 Development. 82–86 https://link.springer.com/article/10.1057/palgrave.development accessed 26th October 2017.

43 Human Rights Watch, 'Stopping Rape as a Weapon of War in Congo' (Report) https://www.hrw.org/news/2009/09/18/stopping-rape-weapon-war-congo accessed 24 November 2017 ; 'Democratic Republic of Congo: Ending Impunity for Sexual Violence New Judicial Mechanism Needed to Bring Perpetrators to Justice' (Report) https://www.hrw.org/news/2014/06/10/democratic-republic-congo-ending-impunity-sexual-violence accessed 24th November 2017.

44 *Ibid.*

45 UNICEF, 'Sexual violence as a weapon of war' http://www.unicef.org/sowe96pk/sexviol.htm accessed 23 November 2017; WHO, 'Sexual violence' World Report on Violence and Health http://www.who.int/violence_injury_prevention/violence/global_campaign/en/chap6.pdf accessed 23rd November 2017.

46 WHO, 'Sexual violence'

actors of the conflict use sexual violence as a punishment and a general warning to the female population within the community under control. Rape or threat of rape is also used to drive communities off lands or to heighten the terror during attacks.[47]

Women and girls have been held against their will and repeatedly raped until they conceive, as in the Bosnia and Herzegovina[48] and Rwanda.[49] During the East Timorese conflict, women and adolescent girls were abducted and forced into sexual slavery, slave labour and other forms of abuse,[50] and the Chibok girls of Nigeria who were abducted were forced into different forms of slavery including sexual slavery, forced marriages, and various forms of abuses.[51] Amnesty International reported widely alleged cases of rape by security forces engaged in peacekeeping operations in Ogoni land.[52]

The wars that raged Bosnia, Rwanda, Sierra Leone, Liberia, DRC and Darfur have made the logic of mass rape undeniable.[53] An average of 40 women are raped daily in South Kivu, DRC; between 20,000 and 50,000 were raped during the war in Bosnia in the early 1990s; in Sierra Leone, between 50,000 and 64,000

---

47 M Kudakwashe, 'Causes of armed conflicts and their Effects on Women' [2015] 2 *International Journal of Research in Humanities and Social Studies* 80

48 United Nations, ' Women, Peace and Security' Study submitted by the Secretary General pursuant to Security Council resolution 1325 (2000) https://www.un.org/ruleoflaw/files/womenpeaceandsecurity.pdf accessed 24 November 2017.

49 Human Rights Watch, 'Shattered lives: Sexual Violence during the Rwandan Genocide and its Aftermath' (Human Rights Watch (Women's Rights Project 1996) https://www.hrw.org/reports/1996/Rwanda.htm accessed 24 November 2017.

50 United Nations, 'Women, Peace and Security'(n48)

51 Human Rights Watch, 'Those Terrible Weeks in Their Camp' Boko Haram Violence against Women and Girls in Northeast Nigeria (Report 2014) https://www.hrw.org/report/2014/10/27/those-terrible-weeks-their-camp/boko-haram-violence-against-women-and-girls accessed 24 November 2017 ; M Kudakwashe, 'Causes of armed conflicts and their Effects on Women' (n47)

52 L Okereafor, 'Nigeria Government Must Demonstrate a Commitment to Stop Violence against Women' http://www.accesstojustice.ng.org accessed 13 October 2017.

53 Sexual Violence Against Women and Children in Armed Conflict, UN Parliamentary Hearing Session II

internally displaced women suffered sexual assault at the hands of combatants; the Rwandan genocide memorial notes that 500,000 women were raped during the 100 days conflict.[54] Apart from the rape of women, one area that has received little attention is sexual violence against men in armed conflicts. Reports of sexual violence by men against men emerge from numerous conflicts, from ancient conflicts to the most recent conflicts in Iraq and DRC. Male rape and male sexual violence unlike the rape of women is grossly under reported due to the shame, confusion, guilt, fear and stigma.[55] The most thorough investigation of male sexual violence in armed conflict is that of the atrocities committed in the conflict of the former Yugoslavia. During and after that conflict, cases of male sexual violence were found at all stages of the investigative process.[56] These reports undeniably establish the fact that men also experience associated disasters risks affecting men's reproductive health. Sexual violence during armed conflicts affects men as well as women. However, the associated risks for women are more for women than men. For a woman, there are the added risks of pregnancy, rejection by husband and families and stigmatisation of children born of rape.[57] Rape also affects women's eligibility to marry.[58] In February 2006, Amnesty International interviewed some of the women who had been raped, abducted and forced into sexual slavery during the Odi crisis; many continue to suffer with physical and psychological consequences.[59]

54 *Ibid.*

55 Shanks and others, 'Responding to Rape' (2001) 357 The Lancet 304; Coxcell and others, 'Lifetime Prevalence, Characteristics and Associated Problems of Non-consensual Sex in men: Cross Sectional Survey' (1999) 318 Brit Medical J 864.

56 Amnesty International, 'Bosnia-Herzegovina: Rape and Sexual Assault by Armed Forces'(Report 1993) cited in Jones, 'Gender and Ethnic Conflict in ex-Yugoslavia'(1994) 17 Ethnic and Racial Studies 115 at 132.

57 S Rouhani and others, 'Stigma and Parenting Children Conceived from Sexual Violence' (2015) 2 Pediatrics e1195–e1203.

58 BBC Reports, 'Indian woman marries her alleged rapist' http://www.bbc.com/news/world-asia-india-31483956 accessed 14th November 2017.

59 Amnesty International, 2000, 'Nigeria Rape: The Silent Weapons' accessed 15 October 2017 http://www.amnesty/library/afr44/020/2006.

DISASTER RISKS, VULNERABILITY AND VIOLATION OF REPRODUCTIVE AND SEXUAL HEALTH

Natural disasters and socially made disasters like armed conflict have marked human existence throughout history and have always caused peaks in mortality and morbidity.[60] Natural disasters do not affect people equally. Women and girls' social vulnerability has a greater bearing on the impact of disasters on their reproductive and sexual health than female biology. However, in a few instances, female biology has a role to play on exposure to reproductive health risks during disasters.[61] The two pertinent questions are, first, what determines vulnerability to disaster risks? Second, how do these affect reproductive and sexual health?

According to Neumayer and Plumper,[62] women are more vulnerable in natural disasters than men. In their research, they tried to establish the link between disaster and the factors that determine vulnerability. They identified three factors that make women more vulnerable in disasters than men as follows: biological and physiological differences, social norms and role behaviours, and discrimination in access to resources and breakdown of social order. The authors states that "... it is the socially constructed gender-specific vulnerability of females built into everyday socioeconomic patterns that lead to the relatively higher female disaster mortality rates compared to men."[63] A vulnerability approach to disasters would suggest that inequalities in exposure and sensitivity to risk as well as inequalities in access to resources, capabilities, and opportunities systematically disadvantage certain groups of people, rendering them more vulnerable to the impact of disasters.[64]

60 J Leaning and D Guha-Sapir, 'Natural Disasters, Armed Conflict, and Public Health' (2013) 369 N Engl J Med 1836-1842.

61 R Murthy 'Review of Sexual and Reproductive Health Rights in the Context of Disasters in Asia, Kuala Lumpur, Malaysia' (2009) The Asian Pacific Resource and Research Centre for Women https://www.researchgate.net.publication/ accessed October 27 2017.

62 E Neumayer and T Plumper, 'The Gendered Nature of Natural Disasters: Impact of Catastrophic Events on the Gender Gap in Life Expectancy, 1982-2002'. (2007) 97 *Annals of the Association of American Geographers* 551-556

63 *Ibid.*

64 *Ibid.*

In disasters, the vulnerable population most affected are women, adolescents and the aged. In the study of specific disasters, researchers have found out that consistently a wide range of factors related to gender difference and socially constructed gender power increases women's vulnerability.[65] Gender power imbalance reduces the ability of women to meet the challenges of disaster and recover from its effects, for example the right to make and freely take a decision. A typical case of gender constructed power and women's vulnerability in disaster is the case of Monia, a Bangladesh woman who died in a cyclone because she refused to leave her home to take shelter in a relatively safe shelter which is more socially accessible to men of her strictly segregated male community.[66] It was reported that her husband said "she was a good woman; she did not venture out of the home alone"[67] Women's reproductive right is violated in disaster when gender subordination affect women's fundamental human rights directly linked to and interrelated with reproductive health.

Specific human rights of women were recognised during the 1993 World Conference on Human Rights as "inalienable, integral and indivisible" women's fundamental human rights in disaster which are an integral part of reproductive and sexual health include: right to life and survival; right to liberty and security of the person; right to be free from inhuman and degrading treatment; right to decide the number and spacing of one's children and the right to private family life; right to marry and found a family; right to health (availability, accessibility, acceptability and quality) the right to education, right to family planning, provision of obstetrics and child care... All of these rights must be interpreted broadly in the context of disasters.[68]

---

65 *Ibid.*

66 R Dhungel and R Ojha, 'Women's Empowerment for Disaster Risk Reduction and Emergency in Nepal' (2012) 20 Gender & Development, Post-Disaster Humanitarian Work.

67 *Ibid.*

68 E Enarson and M Fordham, 'From Women's Needs to Women's Rights in Disasters' [2001] 3 Environmental Hazards Human and Policy Dimensions 133-136 www.earthscan.co.uk/journal/chaz accessed 27 October 2017.

Disaster has a disproportionate effect on the poorest and most vulnerable, particularly women and children. The right to life is the most obvious right affected in disaster. Pregnant women are at the risks of dying in childbirth where basic health care infrastructures have been severely damaged by disaster, due to lack of essential or emergency obstetrics care. Where obstetrical care is reduced the number of miscarriages increases, as does maternal and infant mortality.[69] Eight of the ten countries with the highest maternal mortality ratios in the world are in fragile circumstances and are affected by current or recent conflicts. Maternal mortality due to neglect violates the right to life and health.[70] Also, disaster victims particularly women that have been raped or sexually assaulted and who are in need of emergency care must be given priority attention. Life-saving treatments for severe assault and women that have been gang-raped are essential obstetrics care.

Right to liberty and the security of the person is severely compromised and curtailed in disaster and conflicts situations. In the case of natural disasters in communities where there are social norms and taboos restricting women's freedom to express their needs and seek help in disaster situations, women die because they are inhibited from seeking help. Many women risk and lose their lives while trying on their own to protect themselves or escape natural disasters.

In situation of conflict, right to liberty and security of the civilian population living in the affected communities are usually affected. In a situation where a community is occupied by government security forces, militants or other armed group, there may be curtailment or restriction of movement. Pregnant and expectant mothers who are in need of emergency delivery may be delayed at road blocks mounted by security forces requesting identification. In similar situations of rape and sexual assault perpetrated by security forces in an occupied community, victims may be afraid to report or seek help from appropriate authority.

---

69 E Neumayer and T Plumper, 'The Gendered Nature of Natural Disasters: Impact of Catastrophic Events on the Gender Gap in Life Expectancy, 1982-2002'(n62).

70 Rebecca Cook., Bernard Dickens, and Mahmoud Fathalla, *Reproductive Health and Human Rights: Integrating Medicine, Ethics and Law* (n1) p.162

In disaster and conflict situations, humanitarian workers in displacement camps, security forces and other militant groups sometimes violate the right to be free from inhuman or degrading treatments. For example, the Inter-American Commission found the government of Mexico responsible for the illegal detention and interrogation by Mexican military officers of three sisters and their mother. They also found Mexico responsible for the acts of rape committed by the military officers against the girl and their mother. The court also held that the acts of rape violated their rights to respect of their privacy and bodily integrity. The rape amounts to torture, inhuman and degrading treatment.[71]

In disaster risks reduction, the right to education and information is an imperative for the protection of lives. Vulnerable persons such as women, adolescents and the aged must be informed on how to identify early warning signs. Education on how to heed the local warning system measures can greatly reduce the vulnerabilities of women.[72] In areas where natural disasters such as flood are a common occurrence, women and girls must be informed on the dangers of using contaminated water for bathing and washing purposes. A study on the health impact of the 1998 floods in Dhaka slums notes that flood waters mixed with waste and faecal matter entered the vagina and urethra of women and adolescent girls and increased the risk of urinary tract infections.[73] For effective and meaningful protection pre- and post-disasters, women and girls must be educated on the likely reproductive health effects following natural disasters. In the aftermath of 2004 Indian Tsunami, there were reports that some Thailand adolescent girls and women in the reproductive age group

71 Ana, Beatriz and Cecelia Gonzalez Peres v Mexico, Report No. 53/01, Case 11, 565, 4 April 2001, para.27 Rebecca Cook., Bernard Dickens, and Mahmoud Fathalla, *Reproductive Health and Human Rights: Integrating Medicine, Ethics and Law* (n1) p.170.

72 Project On Internal Displacement Protecting And Promoting Rights In Natural Disasters In South Asia: Prevention And Response, 2009. Summary Report : All India Disaster Mitigation Institute The Brookings Institution, University Of Bern www.brookings.edu/idp accessed 27 October 2017.

73 R Murthy, 'Review of Sexual and Reproductive Health Rights in the Context of Disasters in Asia, Kuala Lumpur, Malaysia' (n51) p. 23.

reported that their menstruation stopped suddenly due to sudden shock and only resumed after a few months. There was no pre- or post-enlightenment by the government or health personnel.[74] The foregoing are a few of instances of disaster risks, vulnerability and reproductive health.

Human Rights Approach, Core obligations and Reproductive and Sexual Health in Disaster

Traditionally, disaster risk management had focused on the emergency response and provision of relief materials to populations affected by disaster, the current challenge is to take a more proactive approach that protect and prevent reproductive and sexual health violations and abuses during and at the aftermath of disasters as well as prepare for reproductive and sexual health emergencies in advance with timely and effective response and recovery services.[75] A multi-sectorial and multidisciplinary approach to disaster risk management protects better, public health, particularly reproductive and sexual health, reducing morbidity and mortality.

Governments, individuals, health care professionals and society in general are obliged to respect, protect and promote reproductive health through human rights, law and policies that integrate reproductive and sexual health rights into the provision of reproductive health services.

Law is a powerful instrument because of its influence on how people think and act. Policy is a plan or purpose that a government, an institution, or individual may adopt in order to achieve a particular goal.[76] Reproductive health may therefore be advanced by law, human rights and policy, where legislations made are based on a clearly articulated reproductive health policy.[77]

74 *Ibid.*

75 "Integrating sexual and reproductive health into health emergency and disaster risk management" (n3).

76 Rebecca Cook., Bernard Dickens, and Mahmoud Fathalla, *Reproductive Health and Human Rights: Integrating Medicine, Ethics and Law* (n1) p.95.

77 *Ibid.*

There are a number of human rights instruments, setting standards and obligations that government, government agencies and other stakeholders must observe for effective protection of reproductive and sexual health in disaster and conflicts. Although some authors[78] have argued that some of these human rights instruments fall short of provisions for the protection of reproductive and sexual health rights in disaster, it is our argument in this chapter that even where these human rights instruments do not expressly contain provisions on protecting reproductive and sexual health rights in disaster, there is an implied ethical and legal obligation created by law which can be inferred in these provisions to make them applicable. For example the right to life as contained in the UDHR[79] and the right of children to special protection under the Convention of the Rights of the Child (CRC).[80]

The following are some of human rights instruments containing specific provisions on the protection of reproductive and sexual health rights in disaster.

i) The Fourth World Conference on Women, 1994 ICPD Programme of Action

The ICPD specifically mentioned the provision of reproductive health services in the context of conflict and disasters. It states for "measures to ensure that internally displaced persons receive basic education, employment opportunities, vocational training and basic health care services, including reproductive health

78 R Murthy, 'Review of Sexual and Reproductive Health Rights in the Context of Disasters in Asia, Kuala Lumpur, Malaysia' (n51) The author's position in this article is that there is no separate UN Convention on the protection of the rights of people affected by disasters or rights of people living in disaster-prone areas or affected by disasters. The author went further to examine some specific human rights instrument for provisions protecting these rights and concluded that these instruments fall short of the provisions for protection.

79 Universal Declaration of Human Rights (adopted 10th December 1948 UNGA Res 217 A(III) (UDHR).

80 Convention on the Rights of the Child (adopted 20th November 1989).

services and family planning." According to Murthy,[81] it does not spell out what reproductive health and family planning services are to be provided in disaster situations. Neither did it state the persons to whom the services should be rendered. The terms 'sexual health' or 'reproductive and sexual' was not referred to or defined. In our opinion, where an instrument falls short of a relevant provision of human rights, provisions of other instruments enunciating similar rights can be read in conjunction with that instrument to compel compliance with the obligations. In this particular instance, the provisions of the ICPD programme of Action can be read in conjunction with the core obligation of states; 'to ensure the right of access to health facilities, goods and services on a non-discriminatory basis, especially for vulnerable and marginalised groups', 'to ensure reproductive, maternal (pre-natal as well as post-natal and child health care.'[82] States have a compelling obligation to meet with these core obligations.

ii) The Beijing Platform for Action

The Beijing Platform for Action reached at the end of the Fourth World Conference on Women, 1995, recognises that women suffer considerably due to disasters and often lose their livelihoods. It calls for establishing gender-disaggregated data based on the impact of disasters on women, expanding the role of women in disaster risk reduction and response, and addressing the specific needs of women displaced by disasters. However, there is no mention of reproductive and sexual health issues in disaster contexts. Unlike the ICPD Programme of Action, the Beijing uses the terms 'reproductive and sexual health' and 'reproductive rights.'

Health is a fundamental human right which is indispensable for the exercise of the other human rights. Every human being is entitled to the enjoyment of the highest attainable standard of

81 R Murthy, 'Review of Sexual and Reproductive Health Rights in the Context of Disasters in Asia, Kuala Lumpur, Malaysia' (n51).

82 Core obligations of states as contained in the Alma-Ata Declaration.

health conducive to living a life of dignity. The realisation of the right to health which embodies reproductive and sexual health, and reproductive rights can be interpreted through numerous human rights instruments, health policies and programmes. In this context, Economic, Social and Cultural Rights provision on the right to the highest attainable standard of health can be interpreted to include reproductive and sexual health and reproductive rights.

iii) The Minimum Initial Services Package (MISP) for Reproductive Health in Crisis Situations

Recognising the impact of disaster on vulnerable members of populations, the need to protect women and girls, and the gap for reproductive health services in disaster risk reduction and emergency preparedness, a summit[83] was held on the need for more proactive response to reproductive health in crisis situation. This summit led to the initiative on "The Minimum Initial Services Package (MISP) for Reproductive Health in Crisis Situations."

The Women's Refugee Commission (WRC) has been at the forefront of advocating for the sexual and reproductive health needs of women and girls in crises, and is widely recognised as the "go-to" expert on reproductive health care for displaced populations. The Women's Refugee Commission has been advocating for the inclusion of the Minimum Initial Service Package (MISP) for reproductive health in related Disaster Risk Reduction (DRR) strategies at the global, national and community level.[84] The Minimum Initial Services Package (MISP) describes the key sexual and reproductive health priorities that are expected in emergencies to include:

a) Identify an organisation(s) and individual(s) to facilitate the coordination and implementation of the MISP.
b) Prevent and manage the consequences of sexual violence.

83 Reproductive Health Response in Crisis Consortium, 2012

84 The Women's Refugee Commission & DRR: Promoting Sexual and Reproductive Health in Emergencies (n3)

c) Reduce HIV transmission.

d) Prevent excess maternal and neonatal morbidity and mortality.

e) Plan for the provision of comprehensive reproductive health services, integrated into primary health care, as soon as possible.

Similarly, the Hyogo Framework for Action[85] and the Granada Consensus[86] recognises priority actions that must be taken by government and other stakeholders before an emergency, and these must be across all sectors; these initiatives recommends priority reproductive and sexual health activities during an emergency; which are; incorporating sexual and reproductive health into multi-sectorial and health emergency risk management policies and plans at national and local levels. Second priority is to integrate sexual and reproductive health into disaster risk assessment and provide early warning for communities and vulnerable groups. Thirdly, create an environment of learning and awareness. Fourth, identify and reduce risks for vulnerable communities and sexual and reproductive health services by reducing underlying risk factors, and fifth, prepare existing sexual and reproductive health services to absorb impact, adapt, respond to and recover from emergencies.[87]

85 Integrating sexual and reproductive health into health emergency and disaster risk management (n3).

86 Granada Consensus on Sexual and Reproductive Health in Protracted Crises and Recovery, statement developed at the Consultation on Sexual and Reproductive Health in Protracted Crises and Recovery in Granada, Spain, 28-30 September 2009http://www.who.int/hac/techguidance/pht/reproductive_health_protracted_crises_and_recovery.pdf

87 *Ibid.*

APPLICATION OF ETHICAL, LEGAL AND HUMAN RIGHTS PRINCIPLES TO PRACTICAL CASES

CASE STUDY 1[88]

Sexual Violence in Conflict

Security forces involved in counter-insurgency operations during a conflict have committed rape with impunity, according to reports, 15 cases of rape by the security forces were documented. The victims were women, girls and boys. They were all examined by gynaecologists with medical evidence obtained. The matter was reported to the government but the authorities fail to prosecute the case, dismissing the testimony and medical evidence and accusing the women as sympathizers of the terrorist forces.

Suggested Approach and Measures

- There must be immediate response by appropriate authorities to re-open investigation into the report and allegations of sexual abuse against the security forces.
- Sexual and reproductive health care must be given to the victims. The victims must be given preventive treatment against risk of pregnancy, sexually transmitted diseases (STD's) and HIV.
- Government should set up a panel to look into the medical evidence that was obtained by the health care team.
- Women and girls in the area of conflict and other members of the community must be encouraged to give testimony before a panel, truth committee or fact-finding body.
- Society or members of the affected community should be educated on the need to protect and provide support for victims. Victims should not be ostracised or discriminated against.

88 This hypothetical Case Study was adapted from Marco Sassoli, Antoine Bouvier and Anne Quintin, How does Law Protect in War? https://casebook.icrc.org/case-study/india-press-release-violence-kashmir accessed 13 October 2017.

- There is need for training of police, military personnel, security and peacekeeping forces on the rights of people and the need to ensure protection of members of the community to which they are assigned during conflicts.
- They must refrain from rape and other forms of sexual violence. They must be sensitized on the damage that sexual violence causes to both victims and perpetrators.
- Where the investigation reveals that acts of sexual violence has been committed by security forces, the perpetrators must be prosecuted and punished.
- Non-governmental organisations advocating reproductive health must engage in the education of militias and other armed groups on the need to respect and protect women and girls against sexual violence.
- Government must address sexual violence by taking all legislative and judicial measures. There must be domestication of international human rights treaties and enactment of national laws prohibiting sexual violence in disaster and conflict situations.
- Government must provide rehabilitation and reparation to victims and communities for sexual violence by its security forces.
- Government must ensure justice for the victims of sexual violence.
- The government must enlighten people on disaster risks and the necessary measures to take for protection. Emphasis must be laid on the risks relating to reproductive and sexual health.
- National laws, legislations and policies must be put in place for the protection of persons particularly women and girls against abuse or any form of violence, particularly sexual violence. There must be sanctions against perpetrators.

CASE STUDY 2[89]

Disaster Victim and Sexual Abuse

A community was hit by an earthquake in which many houses were destroyed and several people injured. An unmarried girl who was rescued from the rubbles of her home suffered severe injuries and was evacuated to a hospital in the district, without accompaniment. She was raped by a doctor, but the family was not keen on pursuing the case for fear of being ostracised by their community.

Suggested Approach and Measures

- Appropriate authority must investigate the allegation of sexual abuse perpetrated by the health personnel against the girl.
- The body of health professionals should ensure that there are ethical codes of guidance for health practitioners. There must be practice direction in specific cases and treatment procedures.
- Health care professionals found guilty or liable for such conduct must face disciplinary sanctions
- The community must be encouraged to report cases of sexual abuse by health care practitioners and support victims, particularly the family affected. Community should refrain from shielding or protecting perpetrators of sexual abuse.
- Human rights advocate groups should launch campaigns against sexual abuse of adolescent girls and women. They must not hesitate to press charges where the family or a victim is intimidated by community pressure.
- Government must observe the core obligation to protect and ensure the rights of people living in disaster prone zones. Evacuation programmes should be embarked on as soon as there are early warning signs.

89 Hypothetical case adapted from R Murthy, 'Review of Sexual and Reproductive Health Rights in the Context of Disasters in Asia, Kuala Lumpur, Malaysia' (n51).

- To prevent future occurrence, adolescent girls must be informed and encouraged to report cases of rape or sexual abuse by health care personnel.
- Hospital procedures should include measures to protect adolescent girls seeking health services from further abuse. For example, examination and treatment of female patients by male health personnel should be witnessed by female health attendants. Adolescent girls must not be attended to alone.

Conclusion

Populations affected by disaster, particularly vulnerable populations lack access to reproductive and sexual health service. This chapter has examined the impact of disasters and conflicts on the sexual and reproductive health and rights of women, adolescents and other vulnerable persons, the need for a multidisciplinary approach to disaster risks reduction and protection of reproductive and sexual health. The discussion exposes violation of reproductive health during disasters, human rights and legal approaches to addressing the various forms of violations. The obligation of government, health care professionals, society and other stakeholders; and good practices on addressing reproductive and sexual health issues in disaster contexts were elucidated. Based on this, the case studies provide suggestions on integrating reproductive and sexual health rights in disaster risk reduction and management.

This chapter discussed how disasters impact women, adolescents and other vulnerable person's reproductive and sexual health. It argued for a multi-sectorial approach to disaster risks reduction, particularly disaster risks impacting reproductive and sexual health. The multi-sectorial approach to ensuring reproductive and sexual health rights is feasible.

Government, society and other stakeholders are under obligation to ensure, protect and promote respect for reproductive and sexual health rights in disaster situations and disaster risks management. Disaster risks associating reproductive sexual health can be addressed through the effective application and interpretation of human rights core obligations and a combination of other existing international standards, such as the Minimum Initial Service Package (MISP) for reproductive health disaster.

National laws and policies on disaster risk management should include reproductive and sexual health as an effective and holistic approach to preventing and combating disaster risks in any form whether natural or man-made.

14

DEALING WITH CORPORATE FAILURE IN TIMES OF ECONOMIC CRISIS: A RESCUE-ORIENTED APPROACH

Damilola O. Odetola

Abstract

Nigeria, the most populous nation in Africa, has been recognised as one of the fastest growing economies on the continent. Unfortunately, the country struggles to sustain this growth because of economic volatility and instability. Like physical disasters, economic crises can have far-reaching effects on businesses and leave a lot of destruction in their wake. During an economic crisis, there is a need for responsive, multi-pronged mechanisms that can step up to the demands of urgent rescue. And when the storm passes, there remains a need for frameworks and mechanisms to help rehabilitate businesses to recovery. This chapter critically discusses corporate rescue frameworks and coping strategies, emphasising the need for a corporate rescue culture generally, but more specifically during and after an economic crisis. Drawing from the experience internationally, it considers how Nigerian insolvency and restructuring frameworks can become better equipped to handle economic volatility.

Keywords: economic crisis; financial distress; corporate insolvency; corporate rescue.

INTRODUCTION

The financial distress of a company in itself can be rightly described as a form of disaster. When this distress is caused or accompanied by widespread economic crisis, the disaster is two-fold. The effect of economic crisis on businesses can manifest in different forms. Companies may become unable to carry on operations, lose customers and employees, and/or become unable to meet their debt obligations.[1]

Each financial crisis teaches the world new things about insolvency law and policy. One vital lesson that the world has learnt from financial crises is the importance of rehabilitative frameworks and corporate rescue.[2] The management of financial distress becomes more challenging during a financial crisis and its aftermath, because the formal insolvency regime and institutions easily become too burdened to effectively deal with the volume of distress in the economy.[3] As a result of changing times and markets, there is a need to routinely revisit policy and adapt the law to changing realities.[4] Many jurisdictions have sought to develop appropriate responses to financial distress and have had considerable success in adopting viable insolvency alternatives.[5] For regions that are highly unstable and for which financial crises

1 It is noted that companies and businesses in this chapter refer to companies other than financial institutions. Because of their systemic significance, distress, failure and financial institution regulation generally is closely monitored and implemented by the state through bodies like the Central Bank of Nigerian and the Nigerian Deposit Insurance Corporation.

2 Thomas Laryea, Approaches to Corporate Debt Restructuring in the Wake of Financial Crises (IMF Staff Position Note, 26th Jan 2010).

3 *Ibid*

4 Vanessa Finch, *Corporate Insolvency Law: Perspectives and Principles* (2nd edn CUP 2009) 28; The American Bankruptcy Institute, Commission to Study the Reform of Chapter 11 (2012-2014) Final Report, 18

5 Finch, 'Corporate Rescue in a World of Debt' (2008) JBL 756

are constantly on the horizon, corporate rescue is a valuable means of avoiding the collapse of potentially viable businesses.[6]

While the phenomenon of corporate rescue has become recognised transnationally and has inspired debates among scholars and practitioners, it has remained largely untested in African economies.[7] This chapter makes two vital contributions to the literature on corporate rescue:[8] it expands the discussion of corporate rescue to specifically deal with the issue of economic crisis and applies this analysis to the widespread problem of economic instability in Africa. The chapter examines formal and informal corporate rescue mechanisms across jurisdictions. It makes a case for the adoption of corporate rescue mechanisms as an effective insolvency alternative that can cushion the adverse impact of economic crisis. The discussion is particularly relevant at a time when Nigeria is recovering from an economic recession. The discussion on corporate rescue also contributes to ongoing developments in Africa as many countries in Africa (including Nigeria) are in the process of reforming their insolvency laws to reflect the goal of corporate rescue.

The chapter is in four parts. Part 1 considers the concept of economic crisis, its meaning, causes and symptoms. It also examines how economic crisis affects businesses and companies. Using selected corporate rescue frameworks and mechanisms, the second part discusses the concept of corporate rescue, highlighting

6 *Ibid*

7 See Finch, 'Corporate Rescue in a World of Debt' (2008) JBL 756; Gerard McCormack, *Corporate Rescue Law, an Anglo-American Perspective* (Edward Elgar Publishing, 2008). For a commentary on current insolvency law reforms in Africa, Odetola 'Corporate Insolvency Reform in Emerging Africa: The Need, Challenges and Prospects' (2017) ICCLR.

8 For some discussion on the relevance of corporate rescue in economic and financial crises, see Thomas Laryea, *Approaches to corporate debt restructuring in the wake of financial crises* (International Monetary Fund 2010); Stephan Haggard, Wonhyuk Lim, and Euysung Kim (eds), *Economic crisis and corporate restructuring in Korea: Reforming the Chaebol* (Cambridge University Press, 2003); See Bolanle Adebola, *Corporate rescue and the Nigerian Insolvency System* (Diss. University College London 2013) for a general discussion of corporate rescue in Nigeria.

its vital components and importance in an economic crisis. Part 3 focuses on Nigeria, describing the effect of economic crisis on Nigerian businesses and outlining the existing frameworks for resolving corporate financial distress. It also considers the coping strategies that stakeholders adopt and question whether they make up for the lapses in the insolvency regime. The chapter concludes with some thoughts for reform as the nation moves towards adopting a rescue-oriented culture in distress resolution.

Economic crisis: meaning and causes

Every country has its challenges, at every stage of its development. Economic crises are therefore not a strange occurrence to any nation. Like physical disasters, economic crises occur for different reasons and have varying degrees of impact. For this reason, the term 'economic crisis' does not lend itself to easy definition. Scholars tend to describe an economic crisis in relation to its causes and indicators, which also presents some difficulty as there is no consensus as to the causes and manifestations of an economic crisis. Economic crisis has been described as a sudden downturn in the economy which can be characterised by a fall in GDP, rising or falling prices, reduction in purchasing power, little financial movement in the system, absence of lending, collapsing sales, unemployment, declining profit rates amongst others.[9] An economic crisis has also been characterised with the use of terms

9 Adam Blanden, 'What is an Economic Crisis' Critical Theory (June 18 2014) <http://www.critical-theory.com/what-is-an-economic-crisis/>;Business Dictionary<http://www.businessdictionary.com/definition/economic-crisis.html>accessed 10 October 2017

like recession,[10] depression,[11] collapse[12] or slump.[13] Without getting bogged down by the debates around how an economic crisis should be defined, what is most valuable for the purpose of this chapter, is an understanding of some of the causes and symptoms of an economic crisis and their effects on corporate entities in the economy.

One common cause of economic crisis is financial crisis. A financial crisis is localised to the financial sector and characterised by a reduction in the value of financial assets such as stocks and securities. If left unchecked, a financial crisis would eventually culminate in an economic crisis, which is more pervasive and affects every aspect of the economy.[14] Because of the close causal relationship between a financial crisis and an economic crisis, it is useful to analyse some of the causes of a financial crisis in some detail. Based on historical experience, financial crises can be caused by monetary excesses, toxic financial innovations, and loose monetary policies that 'first yield a deceptive boom and then result in a dramatic bust.'[15] The global financial crisis of 2008 which originated in the US has been attributed to the boost

10 Described as a period of general economic decline, defined usually as a contraction in the GDP for six months (two consecutive quarters) or longer. Marked by high unemployment, stagnant wages, and fall in retail sales. A recession generally does not last longer than one year and is much milder than a depression Business Dictionary <http://www.businessdictionary.com/definition/recession.html>accessed 10 October 2017

11 'Lowest point in an economic cycle characterized by reduced purchasing power, mass unemployment, excess of supply over demand, falling prices, or prices rising slower than usual, falling wages, or wages rising slower than usual, and general lack of confidence in the future. *Ibid* <http://www.businessdictionary.com/definition/depression.html>accessed 10 October 2017

12 A sudden and dramatic slowdown in economic activity, resulting in a steep drop in prices with a consequent fall in level of employment. *Ibid*

13 A period of economic collapse with high unemployment and loss of trade. *Ibid*

14 G. Perera, 'Difference Between Financial Crisis and Economic Crisis' PEDIAA (16 September 2016)<http://pediaa.com/difference-between-financial-crisis-and-economic-crisis/> accessed 10 October 2017

15 John B Taylor, *The financial crisis and the policy responses: An empirical analysis of what went wrong* (No. w14631. National Bureau of Economic Research, 2009)

in the housing sector caused by subprime mortgages[16] which were funded by complex securitisation,[17] credit default swaps and re-securitisation.[18]

Almost a decade after, analysts continue to seek to understand the 2008 crisis as financial systems still reel from its effects and deliberate on how to prevent a recurrence. Some lessons have emerged; financial crisis can build up gradually until they explode into a disaster of epic proportions. They can leap across country borders like wildfire and the likelihood for this increases as countries and markets become more interconnected.[19]

In Europe, the Eurozone debt crisis which was at its worst between 2009 and 2012 has been described as the world's greatest threat in 2011.[20] The greatest sufferers from that crisis were Greece, Portugal, Italy and Ireland. Overspending, mismanagement, over indebtedness and low production have been identified as some of the culprits in this crisis.[21]

In emerging jurisdictions, the East Asian crisis is one of the significant economic catastrophes occurring in arguably the fastest growing region of the world. Before the crisis, the growth

16 They were financial instruments packaged to enable a lender and borrower take advantage of the housing sector boom in the US to make quick returns.

17 Securitisation is the process in which certain types of assets are pooled so that they can be repackaged into interest-bearing securities. The interest and principal payments from the assets are passed through to the purchasers of the securities. Jobst Andreas, 'What is Securitization' (2008) 45(3) Finance & Development.

18 Financial institutions made quick profits and took advantage of the short-term boom coming from untested financial products with investors willing to pay for securities backed by sup-prime mortgages while ignoring the risk factors.

19 The global financial crisis spread to other parts of the world as a result of the globalisation of credit markets, and the original loss of several hundred billion dollars in the mortgage market quickly amplified into a full blown economic crisis. Pol Eduardo, Understanding the global financial crisis (9th Global Conference on Business & Economics, Cambridge University 2009) 1.

20 Kimberly Amadeo, 'Eurozone Debt Crisis: Causes, Cures and Consequences' *The Balance* (23 July 2017)https://www.thebalance.com/eurozone-debt-crisis-causes-cures-and-consequences-3305524

21 Simone Foxman, 'The Eurozone Crisis for Dummies' *Business Insider* (30 December 2011) <http://www.businessinsider.com/updated-the-eurozone-crisis-for-dummies-2011-12?IR=T>

of East Asia was described by some as miraculous.[22] In retrospect, analysts have blamed the crisis on 'mismanaged exchange rate policies, badly regulated financial markets, lack of transparency, wasteful investment, current account deficits, and inadequate corporate governance.'[23] East Asia is particularly relevant to Africa because of its transition from a developing region into a fast growing one, with some of the countries in the region now leading in many sectors globally. It is remarkable that the same issues that were lauded as significant to the growth of East Asia soon became the alleged causes of its economic downturn.[24] A fast growing country like Nigeria could learn useful lessons on managing its growth and rapid changes in the economy from the experience in the East Asian region.

A drop in the price of valuable commodities like oil can also lead to economic crisis. For oil-dependent economies, a lack of diversification of the economy can quickly cause the financial constraints caused by the drop in oil price to turn into a full blown economic crisis. Finally, natural disasters which are acts of God that are clearly outside the contemplation or control of people, can lead to economic crises. While having strong institutions in place can help to contain the devastation caused by natural disasters, their effects can still be far-reaching. Hurricanes, floods, fires, volcanoes, drought and earthquakes can cause such physical destruction and disruption that have the effect of bringing economic activities to a complete halt, resulting in famine and extreme poverty.

---

22 Furman, J., Stiglitz, J. E., Bosworth, B. P., & Radelet, S, 'Economic crises: evidence and insights from East Asia' (1998) 2Brookings papers on economic activity 1-135

23 *Ibid.* These are common causative factors in emerging African economies; Central Bank of Nigeria, Financial Stability Report, January 2009 – June 2010

24 *Ibid* 11

The Effect of Economic Crises on Businesses

Economic crisis, as stated earlier, has a pervasive effect on every aspect of the economy. For companies specifically, it often means less access to funds, inability to pay back debts, reduced production and ultimately, death. The revenue of companies also shrinks as a result of reduced public spending. Many companies are highly dependent on financial institutions to fund their operations, funds which become unavailable or unaffordable during a crisis. For companies that already owe money to banks and other creditors, they become unable to meet their debt obligations because of shrinking profits and operations. Companies struggle under the weight of obligations to banks, employees, suppliers, regulators, the government, among others. This financial distress can eventually lead to their death, if no intervention is made. This has a ripple effect because businesses (including financial institutions) are interdependent in the services they provide and as a group, they form important pillars in the architecture of an economy. A situation where all these entities simultaneously face the threat of insolvency[25] could spell disaster for the economy.

From the discussion above, the causes of economic crisis are widely varied in nature, sometimes uncontemplated and unavoidable. While it may be impossible to completely prevent economic crises from occurring, what is within the power of countries to do is to build strong institutions that are able to weather the storm and cushion the effects for corporate bodies.

25 Insolvency is generally understood as the inability of a company to pay its debt as they fall due (cash flow insolvency) or where the liabilities of a company exceeds its assets (balance sheet insolvency) CAMA, IA 1986

CORPORATE RESCUE AND WHY IT IS IMPORTANT IN ECONOMIC CRISIS

The global financial sector has evolved such that modern insolvency regulations increasingly adopt the constructive approach of rehabilitation of businesses where possible.[26] During an economic crisis, corporate rescue mechanisms can avert a mass death of companies that are crucial to the survival of the economy. Also, corporate rescue frameworks help to protect companies from the effect of creditor enforcement actions that can potentially push a struggling company into insolvency.[27] These enforcement actions can become more destructive in a crisis as the dire economic situation triggers more desperate and aggressive creditor actions against a distressed company. Corporate rescue not only offers the company some protection from creditor actions, it also serves to protect the interests of creditors. Creditors, particularly financial institutions, stand to gain more from keeping companies alive for future business than from killing them for immediate recovery. In addition, ensuring that viable companies are kept alive benefits the society by protecting jobs and improving 'confidence in enterprise and healthy entrepreneurial risk taking.'[28] Corporate rescue mechanisms provide for early intervention and allow distressed companies to be rescued in a timely manner.[29]

The presence of corporate rescue procedures also provides certainty regarding the steps that can be taken when a company is in trouble. Without frameworks that provide clear, definitive guidelines for rescue and restructuring, stakeholders are thrown into confusion when a company begins to sink. Even where the creditors of a distressed company have the best intentions to help

26 Goode, Roy M, 'Principles of Corporate Insolvency Law', 4th edition (Sweet & Maxwell, 2011) 382: Cork Report of the Insolvency Law Review Committee,Insolvency Law and Practice Chapter 9

27 *Ibid* 862-864.

28 Vanessa Finch, 'Re-invigorating corporate rescue' (2003) Journal of Business Law 527-557.

29 Jennifer Payne, 'Debt Restructuring in English Law: Lessons from the US and Need for Reform, Legal Research Paper Series, February 2014. Paper no 89/2013

rehabilitate the company, the absence of legal frameworks and mechanisms to provide clear directions as to what actions can or cannot be taken, may frustrate their efforts. Thus, corporate rescue frameworks can provide a degree of sanity in addressing crisis-induced corporate distress.

Furthermore, introducing rescue-oriented laws helps to promote a culture of rescue. While a change in formal rules on its own cannot guarantee an immediate change in the actions of stakeholders, context-specific formal rules can help to gradually shape the culture of actors in the corporate world.[30] Finally, corporate rescue mechanisms are vital to the recovery process after a crisis has passed. The adverse effect of an economic crisis can linger for years after the crisis has passed. Corporate rescue may be even more vital to nurse companies to recovery at this stage, where other emergency responses like government bailouts may have been withdrawn.[31]

The Components of Corporate Rescue

Deciding to rescue a distressed company

The starting point of any corporate rescue exercise is making the decision to rescue. Whether during an economic crisis or outside it, not every company deserves to be rescued. It must be determined whether the company is economically or financially distressed. Corporate rescue prescribes that businesses be offered a second chance when it is clear that they are still economically

30 Terence Halliday and Bruce Carruthers, Bankrupt: *Global Lawmaking and Systemic Financial Crisis* (Stanford University Press, 2009) 3-20. This is especially true when the actors themselves are undergoing a re-orientation that allows them to acknowledge the value in adopting a second-chance approach to the resolution of financial distress.

31 Bailout funds are often provided as an emergency, one-off response to distress. They would be more effective if paired with corporate rescue mechanisms such as debt restructuring and/or corporate reorganisation to provide long term rehabilitative support until the company's financial position improves.

viable.[32] The distinction between the two is important because attempting to rescue an economically distressed company under any circumstances is a wasteful effort. This distinction is even more important in an economic crisis because resources are strained and must be judiciously deployed.[33]

A financially distressed business is one that is suffering from liquidity issues but remains viable, i.e. has prospects of making profits. On the other hand, an economically distressed business lacks profit making capacity and deserves to be liquidated immediately before its market value deteriorates any further.[34] For such a company, its assets are worth more if broken down than if the business were to be kept together as a going concern.[35] Corporate rescue must be strategically directed to companies whose underlying businesses remain viable and capable to contributing to the economy as a going concern.[36]

Proposing a Rescue Plan

After the assessment has been made and a company is adjudged as deserving of rescue, other elements of the rescue process include the rescue plan and rescue finance. Before embarking on the rescue of a distressed company, a plan that highlights the strategy to be adopted considering the overall circumstance of the company, must be made. The plan helps to assess important matters like cost and legal structure. Because of the sense of emergency that is present during a crisis, it may be tempting for actors to dig in and act immediately, without first making a well thought out plan. However, the sense of direction and clarity that

32 See Baird, D. G. (1998). 'Bankruptcy's uncontested axioms.' *The Yale Law Journal,* 108(3), 573-599.;Goldschmid, P. M. (2005). 'More phoenix than Vulture: The Case for Distressed Investor Presence in the Bankrupcty Reorganization Process.' Colum. *Bus. L. Rev.,* 191, 198.

33 The decision makers are typically the entities that have a major stake in the company, the majority creditors and/or controlling shareholders.

34 Baird, 'Bankruptcy's Uncontested Axioms (n 32)

35 Payne (n 29)

36 From this point, reference to distressed companies in this chapter should be understood to mean financially distressed companies only.

a rescue plan brings amid a crisis can be vital to the success of the corporate rescue. Debt restructuring often forms an important part of a corporate rescue plan. It is a process by which the liabilities of a company in financial difficulties are rearranged and/or modified to enable the company pay off its obligations while preserving its value as a going concern.[37] The financially distressed company has a better chance of survival if the creditors jointly agree not to pursue enforcement actions against the company. In addition, they can agree to reduce the interest rate on the debt, increase the tenure of the loan, write off some of the principal and pump in more money into the company to fund the operations of the company.

Apart from restructuring the debts of a distressed company, it may also become necessary to restructure the company itself. A corporate restructuring takes different forms which may result in the business of the company, rather than the company itself being saved. The rescue of a business as opposed to the company in its current form is still a desirable result where it has become impracticable to keep the company as is. This could mean disposing some parts of the business that place unnecessary strain on the company's efficiency, a reorganisation of the company's operations, change in management, staff layoff, a merger with a stronger company or a sale of the business.

Rescue Finance

A rescue plan must also be accompanied by rescue finance, which is a vital element of corporate rescue. The most significant feature of financial distress, especially during an economic crisis, is a dire shortage of funds. Thus, the rescue of a distressed company necessarily demands that funds are made available to finance the rescue process and the continued operation of the company. Rescue finance may come in the form of additional funds from lenders, equity contributions from existing and new shareholders,

37 Chris Howard and Bod Hedger, *Restructuring Law and Practice* (Butterworths-LexisNexis, 2008) 1.

third party financiers like angel funds, private equity funds and distressed debt investors, the sale of some of the company's assets and concessions or credit from the company's major suppliers.

These components are important considerations in any rescue exercise, regardless of the framework or mechanism being used. We now turn to examine corporate rescue mechanisms and legal frameworks in selected jurisdictions and how these mechanisms can play a crucial role in times of economic crisis.

CORPORATE RESCUE: FRAMEWORKS AND MECHANISMS

The United Kingdom

In England, following the observation that creditors were often too eager to pursue aggressive enforcement actions and put viable companies into liquidation at the earliest sign of trouble, the English Insolvency Act and the Enterprise Act of 2002 aimed to introduce rescue-oriented reforms through the administration procedures and the Companies Voluntary Arrangement (CVA) mechanism. The CVA allows an ailing company to make an arrangement with its creditors which may include a compromise by the creditors to give the company some space to trade out of its difficulties.[38]

The administration procedure provides a mechanism for the rescue of distressed companies. A company or its creditors can appoint an administrator when the company runs into financial difficulty and there is a risk of insolvency. In carrying out his duties, the first concern of the administrator is to aim to rescue the company as a going concern. If this is impracticable, he aims to achieve a better result for the company's creditors as a whole than would be likely if the company were wound up. It is only if these two outcomes fail, that the administrator goes on to realise the property in order to make a distribution to one or more secured or preferential creditors.[39] The administration procedure

38 In the CVA, the manager stays in place while an administration displaces existing management.

39 Schedule B1 Para 3, UK Insolvency Act 1986.

is particularly useful for protecting a distressed company from aggressive creditors by providing for a stay. When a company goes into administration, all enforcement actions against the company by both secured and unsecured creditors are stayed.[40]

Apart from these mechanisms, existing devices in company law have also been used to achieve the goals of corporate rescue. One of such is the English scheme of arrangement. Schemes of arrangement are very important legal tools that can be used at any time in the company's life to achieve the company's aims, but most importantly in times of distress to bring the company out of the woods. It provides a mechanism for the company and its creditors to meet in classes and vote on proposals that affect the rights of the voting members.[41]

Other Jurisdictions

In the US, the Chapter 11 reorganisation procedure provides a means of restructuring for companies in distress. The Chapter 11 reorganisation procedure is often regarded as a 'debtor-oriented' mechanism because it allows the management of a distressed company to stay in place (subject to supervision) during the reorganisation process.[42] This contrasts with the UK administration procedure that displaces the management once the company enters the process. Chapter 11 has served as inspiration for many jurisdictions in the adoption and or design of rescue and rehabilitative procedures. Most developed jurisdictions have their own version of corporate rescue procedures which are variants of

40 See <https://www.gov.uk/government/collections/insolvency-service-official-statistics #history> for statistics of the use of administration and other insolvency procedures in the UK

41 In the UK, the court has actively contributed to the widening of the ambit of schemes, especially with the increased inclination towards corporate rescue. Jennifer Payne, *Schemes of Arrangement: Theory, Structure and Operation* (Cambridge University Press, 2014) Ch2

42 Chapter 11 of the US Bankruptcy Code U.S.C

the US Chapter 11 procedure.[43] Many emerging jurisdictions have recently established corporate rescue provisions patterned after US chapter 11 reorganisation, the UK administration procedure or a cross between both approaches.[44] Other developing jurisdictions are in the process of overhauling their insolvency regimes to accommodate corporate rescue.[45]

Corporate Rescue Frameworks in Times of Crisis

As stated earlier, in times of crisis, the frameworks and institutions that would ordinarily be able to handle failing companies become so overwhelmed by the sheer volume of demand on them. In developing countries, the situation is even worse. This is because even at the best of times, institutions like the courts operate sub-optimally.[46]

The frameworks considered above are to a large degree, dependent on working courts and efficient institutions, something that may be difficult to find in a crisis. For instance, the

43 Honohan, Patrick, and Luc Laeven, eds, *Systemic Financial Crises: Containment and Resolution* (Cambridge University Press, 2005) 233-234; other notable rescue procedures include the French conciliatory procedure and its use of special agents; the German self-administration procedure and prepackaged plan. Takagi, S, 'Recent Developments of Restructuring Schemes and Practices in the US, the UK, Germany and France: Comparative Study' (2011) iiiglobal

44 The system adopted in South Africa is an example. The business rescue law which came into effect in May 2011 contains elements of both the US Chapter 11 and UK administration procedure. ENS Africa <https://www.ensafrica.com/news/SouthAfricasnewbusinessrescuelawthecourtsview?Id=1412&STitle=insolvency%20ENSight>

45 As discussed later in the chapter, African countries like Ethiopia, Ghana, Mozambique, Kenya, Malawi, Cape Verde, Tanzania, Nigeria among others are at various stages of reform.

46 The creation of special courts for commercial matters has been offered as a potential solution to this. However, it is noted that this requires significant resources and specialised training of judges to effectively handle complex matters like those that arise in insolvency and restructuring. See M.B Idris, 'Insolvency and Judicial Capacity: Challenges of he African Courts' (Paper Delivered by Honorable Justice M. B. IDRIS at ART INSOL conference 2010, Nigeria) <https://www.insol.org/_files/Africa%20Round%20Table/2015/Papers/INSOLVENCY%20AND%20JUDICIAL%20CAPACITY%20PAPER%20BY%20JUSTICE%20M%20B%20%20IDRIS%20NIGERIA.pdf>

administration procedure, like the Chapter 11 and schemes of arrangement, requires court sanction and a degree of supervision. These procedures can also be very costly and complex; 'luxuries' that are in short supply during an economic crisis.[47]

Thus, while enacting corporate rescue frameworks like the ones mentioned above is a crucial starting point in proliferating a sustainable rescue culture, it may become important to explore additional options to provide extra support to corporate rescue frameworks. One device that is present in almost every jurisdiction are informal (private work out) and semi-formal mechanisms. A work out is a contractual agreement between entities that have a stake in the interests of a company and the power to determine the fate of the company to work out the difficulties of the company. The aim of the agreement is likely to achieve a restructuring of the company's debts and stakeholders' interests.

A private work out does not typically involve courts and regulatory authorities. It can be done within as little time and as privately as possible, thereby preserving the goodwill of the company while seeking efficient ways to keep it alive. These qualities make private work out a pragmatic choice and appropriate emergency response. The London Approach is a widely renowned example of the use of private work out and has influenced the informal debt restructuring strategy of many countries.[48] INSOL International[49] has developed the London

47 The American Bankruptcy Institute, Commission to Study the Reform of Chapter 11 (2012-2014) Final Report, 18

48 The London Approach is a set of principles developed in England in the 1970s by the Bank of England and London Banks on how to informally work out corporate financial distress. Armour, J., & Deakin, S. (2001). Norms in private insolvency: the "London approach" to the resolution of financial distress. *Journal of Corporate Law Studies,* 1(1), 21-51; Thomas Laryea, 'Approaches to Corporate Debt Restructuring in the Wake of Financial Crises,' IMF Staff Position Note, Jan 26,2010, p. 17.

49 INSOL International is a global association of practitioners (lawyers and accountants) who specialise in insolvency and turnaround.

Approach further through the INSOL Global Principles for Multi-Creditor Workouts.[50] In addition to their formal insolvency frameworks, East Asian countries like Thailand, Malaysia, Indonesia, Hong Kong, Korea and Japan have been inspired by the London Approach and the INSOL Guidelines to create rules for out-of-court workout restructuring (OOCR).[51]

The major challenge with work outs is that there is no binding mechanism or stay on the action of creditors. This means that the efforts of all the parties may be jeopardised by just one dissenting party, at any stage of the process. To get around these constraints, a hybrid between the court supervised restructuring and informal work outs can been used. This comes in the form of pre-packaged and pre-negotiated deals that seek to combine the advantages of the court supervised and informal work outs.[52] Typically, the company negotiates with its creditors and reaches an agreement, the court is then consulted to give effect to the pre-agreed or pre-negotiated plan.[53] This allows the parties to informally work out an agreement at their pace and on their terms and for such deals to be subsequently backed by the power of the courts. Thus, the restructuring process is over almost as soon as it begins. An instance is the use of pre-packaged administration in the UK, informed by the US pre-packaged reorganisation.[54] The varied use of expedited corporate debt restructuring in Latin American jurisdictions like Brazil also provides useful illustration of how these mechanisms have evolved over time.[55]

---

50 In 2000, INSOL further built on the London Approach and developed eight principles as statements of best practice for all multi-creditor work outs globally.

51 The Guideline for Out-of-Court Multi- Financial Creditors Workout ('Guideline') was created in Japan in 2001 by the Bankers' Association and other relevant associations to clean up NPLs that ballooned after the burst of the bubble in the early 1990s.Dr Shinjiro Takagi, Debt Restructuring by Out-of-Court Workouts in East Asia' (2012)3 9(3) *International Corporate Rescue* <http://www.chasecambria.com/site/journal/article.php?id=634>

52 Olivares-Caminal and Frigerio, Expedited Debt Restructuring in Latin America; a Regional Overview, IDB Working Paper Series, No 113.

53 *Ibid*

54 This involves the drafting and agreement of the reorganisation plan before the company goes into reorganisation formally. Section 1126(b) of the US Bankruptcy Code

55 Olivares Caminal (n 52)

Distressed debt investment is also increasingly playing a role in the resolution of financial distress and providing an alternative to traditional insolvency mechanisms. Distressed debt investors (DDIs) flood the market and provide liquidity for everything ranging from 'publicly traded debt, bank loans, trade claims, tort claims to rejected executory contract claims.'[56] DDIs usually seek to exploit the restructuring surplus in a troubled company.[57] They typically buy such debt at discounted rates, restructure the company and sell at a profit after the restructuring. Others may bank on the difference between the value of the company if sold immediately and the value after time has passed. They may in the meantime deleverage the company, increase its operational capacity or simply wait till the market has recovered before selling or floating the company.[58] While DDI is neither an insolvency nor corporate rescue mechanism by itself, it can play a role in determining the fate of a distressed company by providing much needed finance. It is noteworthy that there are divergent views about the value of distressed debt investment for corporate rescue.

On one side are those who argue that these investors are vultures who hover around distressed companies, ready to pick the company apart and make a quick profit rather than making any real efforts to save the company. On the other hand, it has been argued that it is often the case that DDIs are turnaround experts who possess specialist knowledge about valuation and restructuring. They are therefore in a good position to make the important decision of whether a company should be sold in

---

56 Paul M. Goldschmid, 'More Phoenix Than Vulture: The Case for Distressed Investor Presence in The Bankruptcy Reorganization Process' (2005) Colum. *Bus. L. Rev.* 191,297

57 Sarah Paterson, 'Rethinking the Role of the Law of Corporate Distress in the Twenty-First Century' (2014) LSE *Law, Society and Economy Working Papers* 27/2014 11 (adopting explanation given by LM Lopucki and WC Whitford, in 'Corporate Governance in the Bankruptcy Reorganization of Large, Publicly Held Companies' (1993) 14 *University of Pennsylvania Law Review* 669, 752, of restructuring surplus as the difference between the value of the business if it were to be sold at the time of the restructuring and a sale after the restructuring of the existing capital structure)

58 *Ibid* 11

piecemeal, as a going concern or restructured.[59] Whatever side of the arguments is true, the fact remains that distressed debt investment is on the rise and becoming increasingly prominent on the insolvency scene, even in emerging economies. In an economic crisis, foreign distressed debt investors can provide some succour by providing rescue finance that may be otherwise unavailable locally.

Other forms of intervention include state intervention which occurs in a variety of ways. One is a bail out, in the form of financial handouts to selected companies who are considered viable and capable of contributing to the economy. It is not uncommon to see the government stepping in to provide bail out funds for companies whose collapse will pose a systemic risk to the economy. For example, during the global financial crisis, the US government bailed out institutions like Fannie Mae and Freddie Mac to prevent a systemic crisis.[60] Another common form of government intervention is the establishment of state-owned or state-funded asset management companies to take Non Performing Loans (NPLs) off the balance sheet of banks and resolve them. Korea, Thailand, Malaysia and Japan are examples of countries that created assets management companies for this purpose.[61]

59 *Ibid* 12.Also, some have stated that because they aim to exploit the going concern surplus in the distressed company, DDIs have the right motivation and incentive to maximise value. Paul M. Goldschmid, 'More Phoenix Than Vulture: The Case for Distressed Investor Presence in The Bankruptcy Reorganization Process' (2005) Colum. *Bus. L. Rev.* 264 They cite empirical research that suggest that supports the idea that aggregating claims in the hands of distressed debt investors maximise firm value.

60 Bail outs are not always restricted to domestic governments. International financial institutions have also been known to provide financial aid to struggling economies. During the Eurozone crisis which erupted in late 2009, highly indebted countries like Greece, Portugal and Ireland required urgent intervention from the IMF, the European Commission and the European Central Bank (the Troika) to ameliorate the adverse economic effects of the crisis on the whole of the European Union. The Troika exerted extensive structural reforms as part of the conditions for the bailouts. Philip Lane, 'The European sovereign debt crisis' (2012) 26 (3) The Journal of Economic Perspectives 49-67.

61 Malaysian DANAHARTA, Korean KAMCO and Industrial Revitalisation Corporation of Japan (IRCJ) were all quasi-governmental AMCs established for that purpose and were regarded as successful examples.

We now turn to discuss corporate rescue frameworks and strategies in the context of Nigeria.

Economic Crisis and Corporate Rescue in Nigeria

In the past decade, many African countries have experienced tremendous growth, characterised by an influx of foreign investment, huge infrastructural projects and cross-border financial transactions.[62] Converting this growth into sustained economic development has been challenging because of factors like weak institutions and economic volatility. Nigeria has been recognised as one of the fastest growing countries in Africa, earning her a place on lists of emerging and frontier economies like MINT, The Next Eleven, FTSE frontier markets list.[63] Like other countries on the continent, Nigeria has struggled to maintain her economic growth and translate it into sustained development for her citizens. In early 2016, following the oil price drop, the resumption of militancy in the oil-producing Niger Delta and a forex crisis, the country plunged into her first economic recession in twenty-five years.[64]

One vital lesson that economic crisis has taught the world is that no country or business is too big to fail. Rather, the bigger they are, the harder they tend to fall. Nigeria's recent economic crisis had an adverse effect on businesses of different sizes. Large

62 Regional Economic Outlook: Restarting the Growth Engine (IMF Regional Economic Outlook May 2017) <https://www.imf.org/en/Publications/REO/SSA/Issues/2017/05/03/sreo0517>

63 BBC News, The MINT Countries, Next Economic Giants? 6 January 2014 <http://www.bbc.co.uk/news/magazine-25548060>; Goldman Sachs, 'Beyond The Brics: A Look At The Next 11' April 2007

64 The Nigerian naira lost over a third of its value to the US Dollar in 2016; Yomi Kazeem, 'Nigeria has now been in recession for five quarters—but there may be light at the end of the tunnel' Quartz Africa (23 May 2017) <https://qz.com/989805/nigerias-economy-has-been-in-recession-for-the-fifth-consecutive-quarter/>; The country has now been declared as out of recession. Oladehinde Olawoyin,Nigerian economy out of recession — Statistics Bureau' Premium Times (5 September 2017) <https://www.premiumtimesng.com/news/headlines/242391-breaking-nigerian-economy-recession-statistics-bureau.html>

companies in the oil and gas, manufacturing, telecommunications, hospitality and aviation sectors, including the country's national carrier, Arik Airlines were affected.[65] This had a ripple effect on smaller companies who built their businesses around servicing the larger companies. Corporate financial distress was also passed on to banks and other financial institutions, creating the risk of a systemic breakdown. The aviation industry particularly stands a lot to benefit from sound corporate rescue laws. The precarious position in which Arik finds itself is sadly reminiscent of the collapse of its predecessor as Nigeria's national carrier, Nigeria Airways, over a decade ago.[66] In other parts of the world, businesses of such significance have long been able to seek protection under corporate rescue frameworks.[67] The crisis and

65 Arik Air, West Africa's biggest carrier recently went into receivership. Although the airline has been struggling with operational and management challenges for some time, the recent forex crises and tough economic conditions have been identified as the most immediate cause of the airline's near collapse. While corporate insolvency law will not solve management issues, it also offers safeguards against mismanagement by providing a formal platform and legal devices with which the insolvency practitioners can reach back and hold the company's management responsible for negligent, fraudulent and reckless actions. See sections 212, 213, 214,238 & 238 of the UK Insolvency Act 1986 for example.Yomi Kazeem, 'Nigeria's government is taking over the country's largest airline to try and save its aviation industry' Quartz Africa (9th February 2017) <https://qz.com/906521/nigerias-government-is-taking-over-arik-air-to-save-it-from-collapse/>

66 Emmanuel Elebeke, 'Liquidation of Nigeria Airways is Illegal' Vanguard (15 August 2011); Pearl Ngwama, 'Nigerian Airways shouldn't have been liquidated- Olubadewo' *The Authority* (2nd May 2017) <www.authorityngr.com/2017/05/Nigerian-Airways-shouldn-t-have-been-liquidated--Olubadewo/> accessed 3 June 2017

67 American Airlines files for Chapter 11 bankruptcy protection, The Guardian (29 Nov 2011)https://www.theguardian.com/business/2011/nov/29/american-airlines-chapter-11-bankruptcy; Dynamic International Airways, 'Dynamic International Airways Files Chapter 11 Petition' Cision (19th July 2017)https://www.prnewswire.com/news-releases/dynamic-international-airways-files-chapter-11-petition-300491244.htmlTriad airline files for Chapter 11 bankruptcy <https://www.bizjournals.com/triad/news/2017/07/21/triad-airline-files-for-chapter-11-bankruptcy.html>;Thomas Seal, 'Monarch Files for Insolvency in U.K.'s Biggest Airline Failure' Bloomberg (02 October 2017)<https://www.bloomberg.com/news/articles/2017-10-02/u-k-s-monarch-air-placed-under-administration-flights-canceled>; Air Berlin files for administration<https://buyingbusinesstravel.com/news/1527570-air-berlin-files-administration>all accessed 18th Nov 2017

its effect on companies emphasise the importance of corporate rescue mechanisms for the efficient resolution of financial distress. The effect of economic crisis on a country that lacks corporate rescue mechanisms can be compared to the effect of a fire outbreak on a community that lacks fire services and other emergency responders. It makes it difficult to contain the damage and prolongs the recovery time, sometimes irreversibly.

Unfortunately, like in most developing countries, the formal insolvency regime in Nigeria has not fully developed to accommodate corporate rescue and other alternatives to liquidation. Corporate rescue is not a predominant aim of the insolvency law regime and this can result in viable companies getting pulled apart and wound up indiscriminately, particularly when the nation is hit by an economic crisis. To provide an important background for the rest of the chapter, the next section briefly outlines the existing framework for the resolution of corporate distress in Nigeria.

An Outline of the Nigerian Framework for Corporate Insolvency and Restructuring

Corporate insolvency in Nigeria has seen very little development since the enactment of Companies and Allied Matters Act 1990 (CAMA),[68] which contains most of the legal framework for corporate insolvency.[69] The provisions on corporate insolvency have however remained untouched. Corporate insolvency and restructuring mechanisms discussed here include winding up, receivership, arrangement and compromise (schemes) and informal work outs.

68 Cap C20 Laws of the Federation of Nigeria 2004 (particularly sections 387 to 580)

69 (particularly sections 387 to 580). CAMA was significantly informed by the UK Companies Act 1948.

Winding up

A Nigerian company can be wound up in three ways — voluntarily, by the court or under court supervision.[70] The most common form of winding up for a distressed company is a court winding up.[71] A liquidator may be appointed to coordinate the winding up and is vested with powers such as audit of the company, asset sales, payment of creditors, litigation, etc. The effect of a liquidation is that the company's life comes to an end and its assets are sold off to settle its debts obligations.

Receivership

For former colonies of Britain like Nigeria, Ghana, Kenya, Gambia and Sierra Leone, receivership is a commonly used procedure. Receivership allows a creditor or group of creditors with security over the company's assets to appoint a receiver or receiver manager (RM) to take over the management of the company or assets until the debt is realised.[72] When a RM is appointed, the law makes it clear that he is not a mere debt collector or debt recovery agent. Both receivers and RMs are deemed to be agents of their appointor(s) but a RM owes the company a special duty to, 'act at all times in what he believes to be the best interests of the company as a whole so as to preserve its assets, further its business, and promote the purposes for which it was formed.'[73] Thus, even though not clearly spelt out as with the administrator

70 Section 401 CAMA; Section 7 of the Federal High Court Act, Cap F12, Laws of the Federation of Nigeria, 2004 and section 251(1)(e) of the Constitution of the Federal Republic of Nigeria, 1999 (as amended) confers jurisdiction to hear insolvency proceedings on the Federal High Court.

71 One of the grounds for a court winding up is the inability of a company to pay its debts. S486 CAMA. Alternatively, a distressed company may be wound up without the involvement of the court. This is called a creditors' voluntary winding up; Section 117, Companies Winding up Rules.

72 S 390 CAMA

73 Section 390 (a) CAMA,the law describes the 'best interest of the company' to include the interests of the employees, members of the company and finally, special (but not exclusive) consideration to a special class of members or creditors that he is appointed to represent.

under the English Insolvency Act, there is a scope for a RM to aim to rescue a distressed company. Consequently, a receivership is not intended to mean a death sentence for the company like the winding up procedure. In reality however, receiverships have been used mainly as a debt collection device and can have the same deleterious effect on a viable company as the winding up procedure.

Scheme of Arrangement

Nigerian law provides for an equivalent of the English scheme of arrangement, the arrangement and compromise (referred to here as a scheme). Section 537 of CAMA provides for an arrangement which is defined to mean 'any change in the rights or liabilities of members, debenture holders or creditors of a company or any class of them…other than a change effected under any other provision of the Act or by the unanimous agreement of all parties affected.' The 'change' referred to in section 537 may be in the form of a compromise by the creditors (and or members).[74] Creditors may agree to have their rights to receive interests deferred, let go of their priority rights, or agree for all or part of their debt to be converted to shares in the company. Like in English law, the courts have an important role to play in the process, to sanction the meetings and proposals after being convinced of their fairness.

The scheme of arrangement has always been an important restructuring tool for corporate restructuring in Nigeria. It proved particularly useful in the banking industry after the CBN's regulation that commercial banks divest from all non-commercial banking businesses.[75] Given their non-contentious nature and the fact that they were not distress related, these schemes were executed within as little time as possible.[76] However, this is not the case with distressed debt restructuring. Although the scheme

74 Section 539.

75 Many Nigerian banks executed compliance with this regulation by effecting restructuring via schemes of arrangement. Scope of Banking Activities & Ancillary Matters, No. 3, became effective on November 15, 2010, prohibiting banks from undertaking non-banking activities.

76 The timeline for the UBA (one of the banks) scheme of arrangement for instance was three months- between November 2012 and January 2013.

provides a framework within which the debts of a distressed company can be restructured, there are certain factors that reduce its utility for resolving financial distress. With the tension and pressures that come with economic crises, the seemingly straightforward provisions of the Act on schemes can quickly become contentious. Another factor is the lack of a moratorium, as the law makes no provision for a stay of actions by creditors of a distressed company while a scheme is being arranged. This is because schemes are not exclusively designed as a rescue or insolvency tool. Thus, the kind of philosophy that seeks to restrict the 'race to collect' by imposing a stay of proceedings is absent. Also, the voting mechanism in a scheme allows for a three-quarter majority within each voting class. This creates a challenge with cramming down a whole class of creditors who can essentially hold the whole scheme to ransom irrespective of the nature of their stake. Finally, the provisions for a scheme are highly court-driven.[77] The time and efforts required to obtain the different court and regulatory sanctions is counterproductive when the company is on the brink of insolvency and the economy is in decline.

Coping Strategies

It is clear that the frameworks described above are not rescue-oriented. As stated earlier, economic crises have the effect of inspiring retrospection within actors in the economy. One of the issues that come out very clearly is the need to explore other alternatives to liquidation, especially where companies that provide significant services are involved. While the country awaits legislative reform in this area, distressed companies, commercial banks, insolvency practitioners and regulatory bodies are now beginning to realise the importance of corporate rescue and they increasingly make efforts to achieve it. Some of the rescue-oriented strategies and the challenges that confront them, are discussed below.

77 Under Nigerian law, more complexity is introduced by the involvement of the SEC. Section 539, CAMA

Private work outs

Private restructuring is not an uncommon practice among commercial lenders in Nigeria. Increasingly, creditors would almost always explore the option of an informal restructuring before taking stricter measures against the company.[78] Virtually every commercial bank in Nigeria has a department dedicated to the informal work out of distressed loans.[79] However, certain factors militate against the effective use of private work outs in Nigeria. One factor is the rarity of multi creditor work outs. Except in few cases of syndicated loans or large companies, restructuring is typically carried out only between a bank and the distressed company. Even where the distressed company has several bank creditors, restructuring is still done unilaterally because there is no device for coordinating all the creditors and no agreed guidelines to which stakeholders can turn for direction. It is important to be able to synchronise the efforts of creditors, else, the restructuring efforts of one creditor would fail if the other creditors decide to enforce their rights.

While stakeholders have a heightened interest in corporate debt restructuring, what is visibly lacking are well-thought-out restructuring guidelines designed with the specific needs of the Nigerian society in mind. For example, factors such as the means and responsibility for supervising the restructuring process, information gathering, adhering strictly to timelines and monitoring of the distressed business must be determined. The benefits of informal restructuring are lost where there is no clarity on how to navigate these issues during and in the aftermath of financial distress. Finally, because private workouts often suffer enforcement challenges, they can only play a supportive role in the resolution of financial distress and their use does not eliminate the need for other corporate rescue mechanisms.

78 Information gathered from empirical study conducted by writer as part of doctoral research project in June 2017.

79 *Ibid.*

Rescue-oriented receivership

It has been noted that contrary to the spirit behind it, receivership has come to acquire the reputation of a death sentence. This is because of the focus on the realisation of the assets of the appointing creditor(s), often to the detriment of the company's health. In more recent times however, there is a gradual shift towards achieving corporate rescue within the receivership framework. This is especially so when large companies that provide essential services considered vital to the health of the economy, are involved. For example, Nigeria's Arik Air was recently put into receivership as a result of a massive financial debt burden and operational challenges.[80] The airline was adjudged the biggest carrier by numbers in the whole of West Africa. Owing millions of US dollars to a wide range of creditors, a receiver manager was appointed in February 2017 and it was clearly intended that the receivership would be used as a turnaround procedure.[81] As part of this, a new team of experts was appointed into management under the supervision of the receiver manager. This is in sharp contrast to the attitude surrounding the liquidation of Nigeria Airways, which suffered similar challenges as Arik.[82]

While the purposeful use of the receivership procedure for rescue is a step in the right direction, the impact of this approach can be limited.[83] Firstly, the reputation of the procedure as a

80 Chinedu Eze, 'FG Takes Over Troubled Arik, AMCON Says Airline Too Important To Fail' The Vanguard (10 February 2017) ; <www.vanguardngr.com/2017/04/arik-air-owes-creditors-n375bn-amcon/> accessed 8th June 2017

81 Nwafor Polycarp, 'We are not Interested in Liquidating Arik Air- AMCON' Vanguard News (27th February 2017)<www.google.com.ng/amp/www.vanguardngr.com/2017/02/not-interested-liquidating-arik-air-amcon/amp/>

82 Thirteen years after its liquidation and with the benefit of hindsight, stakeholders have expressed regret over the avoidable death of the airline. Abdullateef Aliyu, '12 Years After Liquidation, Nigeria Airways Echoes' *Daily Trust* (31st May 2016) ; Emmanuel Elebeke, 'Liquidation of Nigeria Airways is Illegal' Vanguard (15th August 2011) ; Pearl Ngwama, 'Nigerian Airways Shouldn't have been liquidated- Olubadewo' *The Authority* (2nd May 2017) <www.authorityngr.com/2017/05/Nigerian-Airways-shouldn-t-have-been-liquidated--Olubadewo/> accessed 3rd June 2017

83 Receiverships: The Safe Way Under'<www.standardmedia.co.ke/business/article/2000006175/receiverships-the-safe-way-under>

death sentence and resulting hostility towards the procedure is still dominant. Consequently, using the procedure as a rescue mechanism remains an uphill task. Secondly, efforts towards making the most of the receivership procedure have been directed mainly towards large companies whose collapse pose some systemic risks. It must however not be forgotten that these large companies represent only a small percentage of the total number of businesses that cumulatively contribute to the economy.

GOVERNMENT INTERVENTION

The Nigerian government has shown more concern for and intervened in the survival of companies in distress, notably by bailing out some ailing companies and establishing the Asset Management Company of Nigeria (AMCON).[84] AMCON was set up in 2010 to relieve financial institutions of their toxic assets and NPLs. In adverse economic conditions, AMCON has contributed to restructuring (particularly multi-creditor restructuring of large companies) in Nigeria.[85] AMCON can also be considered to have cleared the path for a distressed debt market in Nigeria.[86] The characteristics that make it possible for the corporation to facilitate restructuring include government backing, special powers to obtain expedited judgments from the courts and funds to finance a restructuring.

Government intervention has also come in form of a strong influence on the decisions of creditors regarding the fate of a distressed company. The involvement of the government in the

84 It is again noted that this intervention is often directed at large companies and financial institutions to prevent systemic risks and adverse effects on the economy.

85 Editorial, 'AMCON Records 57 Percent Recovery of Banks' Bad Debts'Daily Independent (Lagos, 26 May 2015) <http://allafrica.com/stories/201505270863.html.>accessed 1 September 2015.

86 By 2015, AMCON's portfolio revealed that it purchased more than 12,000 loans from industries including aviation, gasoline marketing and manufacturing for about N1.8 trillion, Editorial, 'AMCON Records 57 Percent Recovery of Banks' Bad Debts'<http://angelstartups.com.ng/industry-news/amcon-recovers-57-of-bad-debts-bought-in-crisis >accessed 1 September 2017.

restructuring process can provide some comfort to the company's foreign and local creditors and enable them show more commitment to the rescue of the company.[87] Recently, Etisalat Nigeria, one of Nigeria's top four telecommunications companies narrowly escaped insolvency proceedings because of government intervention.[88] Like the other strategies discussed, government intervention in the form it currently takes has its limitations. The use of asset management companies to clean up the balance sheet of financial institutions can be an appropriate emergence response to prevent systemic risk. As far as the root cause of distress is concerned however, it is ultimately a case of treating the symptom.[89] It has been advised that, 'Governments should forge stronger links between corporate and bank restructuring. It is impossible to address the problems of banks successfully without addressing the underlying problem of bad corporate loans. By contrast, rehabilitating banks with the intention of addressing corporate weaknesses at a later time is a recipe for failure...'[90] AMCON is now at a stage where it is caving under the weight of the distressed debts it has acquired and can no longer take on NPLs.[91] With AMCON no longer able to acquire more NPLs, and with the precarious state of the economy, banks may be back right where they started, desperately looking for means to deal with distressed companies.

---

87 Damilola Odetola, 'Corporate Insolvency Reform In Emerging Africa: The Need, Challenges And Prospects' (2017) ICCLR.

88 Three Banks Take Over Etisalat <www.proshareng.com/news/Debtors%20&%20Recovery/Three-banks-take-over-Etisalat---8206-Nigeria/33969>; CBN, NCC Launch Rescue Mission for Etisalat The Nation Online (May 2017) <thenationonlineng.net/ncc-cbc-join-forces-save-etisalat-creditors/amp/>

89 Thomas Laryea, Approaches to Corporate Debt Restructuring in the Wake of Financial Crises (IMF Staff Position Note, 26 Jan 2010) 17.

90 Micheal Promerleano, 'Introduction: Toward Better Practices in Systemic Corporate Restructuring' in Corporate Restructuring, Lessons from Experience (World Bank Institute 2005) 36

91 Beatrice Ada-Waku, 'AMCON Forecloses Fresh Acquisition of Bad Loans' Today (9 December 2015) <www.today.ng/news/nigeria/49388/amcon-forecloses-fresh-acquisition-of-bad-loans>accessed 17 April 2015

Similarly, bailouts are not a long-term solution even though they are not uncommon in many jurisdictions.[92] The overuse of bailouts as panacea for corporate failure can also pose its own risks to the health of the economy and may be an unrealistic solution during an economic crisis where there is immense pressure on resources. The arguments against a bailout culture range from the moral hazard that can result from the dependence of businesses on state rescue, an erosion of the nation's resources to a lack of transparency in the selection and implementation of the bailout process.[93] For African countries that are battling deep seated problems of corruption, economic instability, lack of corporate governance and lack of accountability, these risks become even more palpable.[94] It would be a better use of government resources to provide an enabling environment by putting up solid rehabilitative frameworks for viable businesses.

Distressed debt investment

As stated earlier in this chapter, one of the most important elements of corporate rescue is rescue finance. Additional funds that must be freed up or pumped into distressed companies to ensure that whatever restructuring plan is made, is effectively executed. Without this finance, a proposal for the rescue of distressed

92 The Ugandan government recently earmarked $300 million for the bailout of distressed companies who are considered viable and capable of contributing to the economy. Before it has gained traction, some factors already cast doubt on the efficacy of the program. For one, it has been criticised as politically motivated. Uganda Company Bailouts Politically Motivated <www.dw.com/en/uganda-company-bailouts-politically-motivated-critics-allege/a-19432023> accessed 8 June 2017; In Ghana, about 100 distressed companies have been identified as viable candidates for government bailout. Government to Bail Viable But Struggling Companies <http://ghananewsonline.com.gh/government-bail-viable-struggling-companies/>accessed 8 June 2017.

93 Block, Cheryl D. "Measuring the True Cost of Government Bailout." Wash. UL Rev. 88 (2010): 149; Rosas, Guillermo. Curbing bailouts: Bank crises and democratic accountability in comparative perspective. University of Michigan Press, 2009.

94 <www.iol.co.za/business-report/companies/abils-bailout-is-political---experts-1733837;<www.dw.com/en/uganda-company-bailouts-politically-motivated-critics-allege/a-19432023>

businesses, no matter how brilliant, can quickly become useless. Unfortunately, commercial banks (which provide the bulk of the funds for businesses in Nigeria) are often constrained to provide rescue finance, even under good economic conditions. This is because of regulatory capital and disclosure requirements, high cost of funds, their duty to depositors and shareholders, etc. The situation, as expected, gets even more constrained in dire economic conditions where the level of borrowing, even to healthy companies, plummets. At such times, the term 'never throw good money after bad money' becomes the mantra of commercial banks. This, more than any other factor, makes corporate rescue very problematic during an economic crisis. This is where distressed debt investment can come to the rescue. It is interesting to note that DDI is gradually becoming part of the conversation around corporate debt resolution in Africa. Investors, notably private equity funds from the US, the Middle East and China have shown some interest in exploring distressed debt opportunities on the continent.[95] In Nigeria, Duet Private Equity recently partnered with AMCON to establish an investment fund in the Fast-Moving Consumer Goods sector. The fund is fully focused on turnaround and distressed situations in the sector, with the aim of funding additional capital and debt restructuring needs of selected companies.[96] Stakeholders would naturally have some reservations about the use of DDIs in Nigeria, especially regarding fairness, valuation and transparency. There may also be concerns like those highlighted earlier in the chapter about the potential of DDI for true rescue after they have purchased a distressed company. Do DDIs truly revive the company to

95 South African Restructuring Outlook Survey Results (Deloitte 2016) <www2.deloitte.com/content/dam/Deloitte/za/Documents/finance/ZA_RestructuringOutlookSurveyResults_2015.pdf&ved=0ahUKEwjXoZ_lx7XUAhVRLFAKHZB0AsAQFgglMAQ&usg=AFQjCNHzXIJs_XzefoR4DFIreUdaO2_1gg> ; Irene Madongo, 'South Africa, Private Equity Eyes Distressed Debt Assets' Beyondbrics (April 19 2013) <http://blogs.ft.com/beyond-brics/2013/04/19/south-africa-private-equity-eyes-distressed-assets/ >

96 http://moderninvestor.com/news/duet-group-expands-african-pe-portfolio-with-new-acquisition/a867532

continue contributing to the economy or do they just bide their time waiting for the perfect moment to dismantle the company and sell its assets at a profit? These are indeed valid concerns that demand some thought as DDI activities gradually increase in African countries. It is likely that these reservations will matter less during an economic crisis where rescue finance is desperately needed and banks are eager to offload distressed companies to a willing buyer. Also, with AMCON no longer purchasing NPLs from them, banks would seek creative ways to deal with distressed assets on their hands. In a bid to fill the vacuum left by AMCON's cessation of NPL purchases, the CBN in April 2017 announced the licencing of Private Asset Management Companies in Nigeria.[97] For distressed debt investment to flourish however, it is important that clear frameworks and supporting institutions for resolving corporate distress are in place. This allows investors to estimate the risks they are undertaking and exit strategies with more certainty.

Towards Rescue-Oriented Insolvency Frameworks

While the downward turn in the performance of Africa's largest economy has ushered in a lot of pain and setbacks, it also presents an opportunity for deeper growth. In the face of economic instability, economies can confront deep seated issues and gaps in the economy. Policymakers will see the need to take firm steps towards diversifying the economy by developing and exploiting other potential revenue sources.[98] Also, stakeholders can clearly appreciate the need to set up efficient insolvency and restructuring laws. In the face of economic hardship, the importance of establishing rehabilitative frameworks to support

97 See<https://www.cbn.gov.ng/Out/2017/FPRD/EXPOSURE%20DRAFT%20OF%20THE%20FRAMEWORK%20FOR%20LICENSING,%20REGULATION%20AND%20SUPERVISION%20OF%20PRIVATE%20ASSET%20MANAGEMENT%20COMPANIES%20IN%20NIGERIA.pdf>

98 NEPAD, 'Economic Diversification in Africa: A Review of Selected Countries' <www.nepad.org/resource/economic-diversification-africa-review-selected'countries> accessed 9 June 2017

viable companies that have been pushed into distress because of volatile economic conditions, becomes clearer. From the foregoing discussion, Nigerian insolvency frameworks are not designed for corporate rescue. It is also evident that the coping strategies that have been emerged in response to financial distress have their limitations, one of which is the absence of legal frameworks, guidelines and supporting institutions for corporate rescue. It is therefore important that these debt resolution strategies are supported by restructuring frameworks and guidelines that are specifically designed to achieve corporate rescue. This chapter is of great practical relevance at a time when many African economies, including Nigeria, are undergoing bankruptcy reforms after many decades.[99] Unsurprisingly, a common feature of this reform agenda is an intention to introduce corporate rescue frameworks.

While this is a laudable move, it must be noted that introducing modern, sophisticated laws and frameworks is not sufficient in itself. The mere adoption of formal institutions and so-called best practices is not enough to yield the potential benefits that economies stand to gain from rehabilitative, rescue-oriented systems. It is therefore emphasised that reform in this area must take the development needs of the country into account. Corporate rescue frameworks should not merely be transplanted from the United Kingdom or the USA, but be adapted to fit the institutional capacity and development needs of the country. For instance, to circumvent overdependence on courts, the level of court involvement in corporate rescue frameworks should not be excessive. Reform in Nigeria also stands to gain a great deal from horizontal learning. This means learning from the reform experience in other developing or emerging jurisdictions.

Judicial independence from political influence, speedy, transparent and effective resolution of cases, experienced turnaround professionals, strong formal institutions and sufficient

99 Damilola Odetola, 'Corporate Insolvency Reform In Emerging Africa: The Need, Challenges And Prospects' (2017) ICCLR.

skills and resources to deal with the large volume of distressed loans are other important factors for effective corporate rescue. This paper acknowledges some of these challenges as part of bigger systemic issues that would takes years of effort and learning to change. It however concludes that the education of stakeholders, horizontal learning in the enactment and implementation of rescue laws are essential first steps towards consolidating the ongoing rescue efforts of stakeholders.

CONCLUSION

This chapter takes a critical look at economic crises and their effects on businesses. It examines how corporate financial distress impacts the economy and the role of corporate rescue in those times of hardship. It discusses rescue frameworks in jurisdictions like the UK, the USA and East Asia. In the Nigerian context, corporate rescue is at its infancy. The chapter analyses the current state of the law and the efforts of stakeholders outside the insolvency framework. It concludes by offering some thoughts on rescue-oriented reforms, emphasising that close attention must be paid to local peculiarities and institutional capacity, if proposed reform would offer the needed protection to business during economic crises and in their aftermath.

INDEX

About the Editor

Professor of Criminal Law and Criminology, Member of the Governing Council, University of Ibadan, Ibadan, Director of the Women's Law Clinic and Immediate Past Dean at the Faculty of Law, University of Ibadan, Nigeria with a continuous teaching academic career since 1987.

She is a member of the Governing Council of Chrisland University, Abeokuta, Ogun State.

She is an Associate Lecturer, Department of Geography, University of Ibadan, Nigeria (Legal Issues in Disaster Risk Management since 2010) and also an Associate Lecturer Department Of Psychology, University of Ibadan, Ibadan Nigeria. (Legal, Criminology and Security Psychology) (2004 – 2014).

She studied law at the University of Lagos Akoka and obtained her Master's degree from the same University.

She specialises in women, children and human right issues and has over 80 (Eighty) publications made up of articles, books and technical reports in English language and translation into other languages in both national and international outlets. Professor Bamgbose was a visiting Professor at Ogun State University, Ago-Iwoye, Ogun State 2000 (now Olabisi Onabanjo University); Marquette University School of Law, Milwaukee, Wisconsin USA 2001; Saint Louis University School of Law, Saint Louis, Missouri, USA in the 2008 and the University of Lagos, Lagos 2014.

Professor Bamgbose is a 2013 Teslim Elias Gold Medal Award of Excellence Winner: For Outstanding Merit and Significant Contribution to the Profession of Teaching and Law as a Law Teacher, a Fellow, Salzburg Seminar Austria (1996) and a Fellow, Chartered Institute of Arbitrators.

She was Chairperson, The Study Group on the Death Penalty for the Government of the Federal Republic of Nigeria; Member Nigeria's National Committee on the Reform of Discriminatory Laws against Women; Chairperson of the Constitutional Review Committee for Oyo State Nigeria; Member of the Technical Working Committee of the NHRC/ Ford Foundation Project "Fair Trial and the Right to Effective Participation in Governance: Independent Review of Evidence of Gross Violations of the Rights to Participate in Government and Fair Trial through the Election Petition Process in Nigeria.

She has delivered papers as guest lecturer at many international and national public lectures including the celebration of 100 Years of Women in Law, at Saint Louis University, Saint Louis, Missouri, U.S.A. on 'Privatisation of the Prison System: A Panacea for Nigeria'; Grace Foundation Day, Grace Leadership Foundation Ibadan on 'Child Trafficking: Creating a hopeless Future for Future Leaders'; 3rd Professor Jadesola Akande Memorial Lecture, the Nigerian Institute of Advanced Legal Studies, Lagos on 'Re-Evaluating the Juvenile/Child Justice System in Nigeria.'

Professor Bamgbose is a co-author of the book Criminal Law in Nigeria by Evans Publishers and Editor of the major publication in Nigeria to present legal materials from diverse fields of law in a single volume on disaster issues, *The Law and Practice of Disaster Issues.*

She is member: The Nigerian Bar Association; The Nigerian Association of Law Teachers; West African Research Association; West African Research and Innovation Management Association and the International Federation of Women Lawyers.